THE OXFORD COMPANION TO AUSTRALIAN

JAZZ

THE OXFORD COMPANION TO AUSTRALIAN

JAZZ

BRUCE JOHNSON

MELBOURNE
OXFORD UNIVERSITY PRESS
OXFORD AUCKLAND NEW YORK

Oxford New York Toronto Delhi Bombay
Calcutta Madras Karachi Petaling Jaya
Singapore Hong Kong Tokyo Nairobi
Dar es Salaam Cape Town Melbourne Auckland
and associated companies in
Beirut Berlin Ibadan Nicosia

Oxford is a trade mark of Oxford University Press

National Library of Australia
Cataloguing-in-Publication data:

Johnson, Bruce, 1943–   .
The Oxford companion to Australian jazz.
Bibliography.
Includes index.
ISBN 0 19 554791 8.

1. Jazz music—Australia—History—Dictionaries.
2. Jazz musicians—Australia—Dictionaries.   II. Title.

785.42′0994

Edited by Carla Taines
Designed by Guy Mirabella
Cover illustration by Regina Newey
Typeset by Asco Trade Typesetting Ltd, Hong Kong
Printed by Nordica Printing Co., Hong Kong
Published by Oxford University Press,
253 Normanby Road, South Melbourne, Australia

Contents

For Sam and Shannon

Preface

As a twentieth-century music which has evolved at a bewildering rate, jazz presents conceptual difficulties not encountered in the case of more venerable art forms, nor even in the case of others of similar vintage. There is little argument, for example, about where something ceases to be 'film' and becomes something else. The definition of 'jazz', however, is unceasingly fluid. The kind of music to which the word has referred at different times has altered so much that it is unlikely that its pioneers in the early years of this century would recognize what now passes under the name. In Australia the problem has been exacerbated by the discrete access we have had to the music, so that the continuity of its development has been obscured. As a consequence, at any given moment in this country, there has been radical disagreement as to what kind of music the word referred to. This conceptual problem is itself the subject of the first part of this volume.

The difficulty is compounded by the nature of the jazz industry. The majority of jazz activity continues even today to be 'underground', operating beneath the world of written contracts. There is little of the documentary apparatus that surrounds, for example, the publishing industry. Most jazz work is poorly paid and much of it is *ad hoc*. One consequence is that even as a bandleader, one can never be sure that the sidemen one has booked and advertised have not in the meantime accepted a more lucrative wedding, private party, club or session engagement. It is not uncommon for a bandleader to arrive at his own gig and find himself playing with musicians he has never seen before. Personnel and residencies come and go faster than the professional magazines and daily press can register, and one can rarely take at face value printed announcements as a way of situating musicians or reconstructing bands. The dedication which most jazz musicians manifest towards their instrumental development is frequently at the expense of much concern for their public visibility. Indeed, there is often a benign suspicion of the documentary spirit. Some musicians keep scrapbooks, but since these consist largely of news-clippings (often undated), they simply bring together the dispersed errors of the sources.

Such sources are numerous, scattered, and various in kind. They share, however, to a greater or lesser extent, the characteristic of unreliability. The material in this volume has drawn upon interviews, correspondence, copies of radio and television documentaries, films both commercially released (features, documentaries, newsreels) and private, sound recordings both commercial and private, photographs, scrapbooks and other private documents (letters, receipts, telegrams, contracts, diaries, appointment books) and a wide range of printed material. This last category includes articles, press releases, record covers, programmes, advertisements, professional and jazz club journals, reviews, and the handful of books on Australian jazz. Very little of the foregoing was prepared by its compilers with any sense of history or scholarly rigour. Much of it, like press material, thrives on hyperbole and other forms of distortion in the interests of newsworthiness.

The state of jazz historiography also makes the writing of a volume such as this a perilous exercise. Because so much primary material consists of ephemera and private records, it is safe to assume that there is a huge store of data scattered about the country as yet undiscovered but which would make some difference to points of detail in the overall picture. There are dedicated historians of the music, but little more than a handful. In spite of the quantity and the internationally recognized quality of the music, Australian jazz remains a neglected phenomenon academically. A book of this kind is usually constructed upon a relatively extensive and stable base of established scholarship. There is no such foundation for Australian jazz. Compared with more academically accommodated subjects, Australian jazz scholarship is negligible. This is not to demean what there is, but to note that there is so little. There is comparatively little mediating the raw primary data and this volume. This companion therefore occupies a slightly uneasy position between an extended inventory, and finished cultural analysis, of its subject.

For all these reasons, it is important to make some observations on what this book consists of and how it may most usefully be approached. There are many categories of information which another writer might have considered worth an entry or an essay: visiting musicians, Australian jazz musicians abroad, jazz on film or on television, original Australian jazz composition. That I have not set aside space for these and other subjects must not be taken to imply that they do not signify. But with limited space, and in the absence of any prominent historical review of such subjects signalling that they have extraordinary importance, they have had to be neglected in favour of what seem to be matters of greater priority. Considerations of space also limited the kind of information provided in individuals' entries. In general I have left out information regarding a musician's major influences and her/his first exposure to jazz, although it can be noted that Australian musicians have had a considerable influence on their peers and successors, and that radio and recordings have been the most frequent first point of contact with jazz. I have also not made a practice of indicating a musician's 'day gig'—his non-musical career—but would observe that the diversity of such commitments could hardly be greater. I had originally planned to indicate broadly the style of jazz favoured by each musician, but decided that, since general indications are provided by the individual's musical associations as listed in the entry, this was unnecessary. Except in special cases I have not given exhaustive or even representative coverage of recording activity; to have done so would pointlessly have duplicated in part the comprehensive Australian jazz discography prepared by Jack Mitchell and currently awaiting publication.

I have also refrained from providing any analyses or transcriptions of specific passages of music. This has been partly to maintain the book's accessibility, but also because Australian musicians have, in the global context, introduced no revolutionary innovations from the point of view of harmonic or rhythmic theory. Australian jazz is distinguished more by its quantity and durability. Although it is difficult to document such an assertion, I believe that there is more jazz played per capita in Australia than in any country in the world. There are other distinctive features of Australian jazz, but they are not amenable to illustration through the analysis of transcriptions. Above all, there is the belief, which I share, that at a particular point in its history Australian jazz developed a distinctive flavour, but as a 'flavour' rather than an implemented doctrine, it remains difficult to define, and the proposition itself generates contention. I have alluded to the matter in the course of this book, as I have also on occasions to the impact of Australian jazz musicians in other parts of the world, which has been more considerable than our often obsequiously parochial sensibility has allowed us to acknowledge.

The special subjects which I have sought to deal with in essays or extended entries, in some instances, surfaced as the research proceeded, and in most cases I have tried to indicate in prefatory remarks why it seemed necessary to devote particular attention to them. I mentioned above that most musicians' first contact with jazz in this country was through radio and recordings. This fact in itself seemed to demand some treatment of both those subjects. Although the two media are now taken for granted, they were crucial innovations in the early years of jazz, and assumed an importance in the dissemination of the music that distinguishes the origins of Australian jazz radically from its beginnings in New Orleans, where contact between audiences and the music was direct, where jazz in performance permeated all aspects of life. The implications of the fact that most Australian jazz audiences first came to jazz as a technologically reproduced, rather than as a visibly performed music, remain to be investigated.

The essays on radio and specialist jazz record labels are only two of a number of sections of this book which have been contributed either in full or in part by other writers. I wish to repeat the thanks to them which I have recorded in the course of acknowledging their contributions where they occur. I also wish to emphasize that, unless otherwise indicated, the opinions which they express are their own: I requested their assistance as well-informed commentators in those areas, and felt it would be to deny those credentials if I attempted to intrude editorially unless by agreement or invitation. Let me therefore stress that I do not always agree with their assessment of matters, but believe that to attempt to conceal these differences of opinion would be in the worst interests of fruitful and continuing debate. Such disagreements are testimony to the open-endedness of jazz historiography, and should be regarded as an invitation to further creative hypothesis formation.

Most of the shorter entries deal with particular musicians, and it seems likely to me that in the short term the most spirited controversies aroused by this volume will concern inclusions and omissions in this category. It is therefore useful to make some observations about the criteria applied. First, the entries are not an attempt to include all the most instrumentally accomplished or the most famous Australian jazz performers. Technical virtuosity is neither a necessary nor sufficient qualification for determining the effect which a musician might have had upon jazz in Australia. Many massively gifted jazz musicians have in fact rarely been seen in jazz settings, and have therefore had little or no impact on the development of jazz as such. 'Impact' is important, but it is complicated by being a relative matter. A less accomplished musician is likely to have greater impact in a provincial area than in, say, Sydney, where he would probably go unnoticed. This, however, then raises the question of how much national impact that region has had. Impact also operates in different directions, one measured in radius, the other in depth. A musician with a national reputation gained partly by a gift for publicity has frequently not become such a local institutional focus for jazz activity as someone who has remained deeply rooted in the one place, though virtually unknown elsewhere. Dedication has something to do with impact, and that too is not necessarily connected with instrumental virtuosity: an unskilled amateur who has struggled for years to keep local jazz interest going is in certain obvious ways a more important representative of the spirit of jazz in Australia than a brilliant studio musician who makes the very occasional highly paid guest appearance at an up-market venue for socialites and tourists. In some instances a musician is given an entry not simply for what he or she has achieved, but also for what he or she represents: a certain category of musician which has become important in maintaining jazz as a thread in the Australian culture.

None of these criteria alone accounts wholly for the inclusion of a given entry. Their variousness, however, is noted here in order to forestall any simplistic assumption that these entries comprehend in some mysterious sense the 'best' Australian jazz musicians. They constitute, rather, a broad selection of individuals who, because of infinite permutations of skill, durability, dedication, typicality, and other intangibles, represent the kinds of musicians who have had significant impact on Australian jazz.

This will become clearer if I indicate the opening procedure in the research for this book. This consisted of discussion and correspondence with interested persons throughout Australia, with a view to assembling a list of individuals, venues, bands, clubs, which should be investigated in any attempt to build up a comprehensive picture of the history of Australian jazz. Not all of those approached in this first round of consultation offered assistance, but the several score who did provided a basis from which to proceed. No name submitted to me was excluded from consideration (apart from my own, on the principle that one becomes a more objective judge of others' eligibility if one is not also a candidate). It should also be added, in the light of surprising omissions, that a number of individuals specifically expressed a desire not to be represented by an entry of their own. Leaving these examples aside, while the responsibility for final inclusions and omissions is mine, in attempting to avoid the limitations of an individual viewpoint, I began by conflating scores of perspectives, and then balanced these against the information in all the written, oral, pictorial, musical sources referred to earlier. I wish to record my thanks to those who helped to provide that foundation, and their names are provided in the first part of the Acknowledgements.

The individual entries attempt to incorporate as much information as possible, and certain forms of shorthand are employed. Where the name (or names) by which the musician is most often known is *not* the first Christian name, the name is given in brackets. It should not always be assumed, by the way, that a subject has only one Christian name where only one is provided: sometimes the subject wished to withhold the full name. Where only one instrument is listed, it should be remembered that most musicians have at least some basic facility on more than one. The data to be treated most carefully are the dates pertaining to a musician's career, although these have been cross-referenced as carefully as the scope of the project has allowed. Apart from the fallibility of memory and other sources, however, a musician's movements from band to band are often so gradual that no one is really sure when he or she ceased deputizing and became a member. Similarly, a band might evolve as a practice group over a year, with fluctuating personnel, with its name materializing at an unspecified point during that period.

The most useful piece of scholarly apparatus in this book is the index. The length of an entry is not necessarily a guide to the subject's importance or the eventfulness of her/his career, since it is often the case that experience in a particular band is expanded in the entry for that band or bandleader. If a reader wishes to assemble the fullest possible picture of an individual's career it is essential to use the index as a means of cross-referencing. Indeed, by this means the reader can put together a fairly detailed biographical sketch of scores of musicians with no entry of their own.

The most important thing missing from this is also the most vivid thing about the music: the sound and the sweat of people playing instruments, the palpable human presence. Ideally, to complement a volume like this, we need an Australian equivalent of *Heah Me Talkin' To Ya*—something that conveys the anecdotal richness of jazz, the smell of the music. We can trust words too much. Finally, the only way to absorb what is essential and compelling about jazz is to hear, or better, to play it.

In addition to the people mentioned in the Acknowledgements I wish to record my great indebtedness to Norman Linehan (Sydney), Jack Mitchell (Lithgow, NSW), Mike Sutcliffe (Sydney), and John Whiteoak (Healesville, Vic.), all of whom are engaged upon substantial research projects pertaining to Australian jazz, and who, with unstinting and unconditional generosity, made their private archives and their knowledge continuously available during the preparation of this encyclopaedia. Thanks also to Graeme Bell, who made available the manuscript of his autobiography currently in progress. Others who provided substantial material from their collections were Tich Bray (Brisbane), Nigel Buesst (Melbourne), Mal Eustice (Adelaide), Maurie Le Doeuff (Adelaide), and Roger Beilby gave open access to the Australian Jazz Archives in Melbourne. Research assistance, particularly in the form of information gathering, was provided by Des Camm, Shirley Horsnell, Alison Johnson, Paula Langlands, and Ron Morey. Other institutional repositories used include the various state libraries, the Mitchell Library (Sydney), the National Library (Canberra), the Adelaide Performance Museum, and the National Film and Sound Archive (Canberra). Without generous funding from the Australian Research Grants Scheme and the University of New South Wales Special Research Grants, particularly for the immense travel undertaken, this project simply could not have proceeded. My thanks also to Bruce Matthews and his staff in the Audio-Visual Unit of the UNSW Faculty of Arts for sound and film transcriptions, to the UNSW Audio-Visual Section for printing copies of photographs, to Carolyn Creevey for transcribing hours of taped interviews, to Susan Keen who typed the manuscript with such intimidating efficiency, and to Ruth Kiely, secretary of the School of English, UNSW, not only for typing voluminous correspondence, but also for her valuable recollections of Sydney's wartime ambience. Finally, I wish to acknowledge the assistance of Betty Briggs, who prepared the index, and to express my appreciation to Ev Beissbarth and Carla Taines of Oxford University Press for their wise and patient counsel.

Bruce Johnson
School of English
University of New South Wales
February 1987

Acknowledgements

From those invited to do so, the following responded with lists of names as an initial basis for consideration in the preparation of this volume (see Introduction): Tony Ashby, Roger Bell, Peter Brendlé, Sid Bromley, Errol Buddle, Brian Brown, Kate Dunbar, Malcolm Eustice, Duke Farrell, Daniel Fine, Alex Frame, Bruce Gray, Ron Gray, Clare Hansson, Harry Harman, Shirley Horsnell, Roger Hudson, Dick Hughes, Adrian Jackson, Herb Jennings, Norman Linehan, John McCarthy, Jack McLaughlin, William (Bill) H. Miller, Jack Mitchell, Ron Morey, Eric Myers, Ross Nicholson, Niels Nielsen, Ted Nettelbeck, Barry Pascoe, Ian Pearce, Tom Pickering, Rick Price, Alf Properjohn, John Roberts, Merv Rowston, John Sharpe, Neil Steeper, Mike Sutcliffe, Lloyd Swanton, Bruce Viles, Chris Welch, Ray Wooster, Bob Wright.

Almost every living individual given an entry in this volume provided information, in correspondence and personal interview, and I wish to acknowledge in general this category of assistance. There have also been scores of individuals (some with entries and some without), in addition to those acknowledged in the Preface, who were kind enough to provide information over and above basic curriculum vitae material, including photographs, documents, tapes, and research assistance in the form of interviews. My sincerest apologies to anyone whose name has inadvertently been omitted from the following list: Phil Batty, Betty Briggs, Sid Bromley, Eric Brown, Tony Buckley, Jim Burke, Rod Byrne, Ken Carter, Bruce Clarke, Percy Cramb, Warner Dakin, Tom Davidson, Kate Dunbar, Rick Farbach, Jim Fillmore, Marie Fisher, Win Frame, Percy Garner, Alan Geddes, Eric Gibbons, Ron Gowans, Ron Gray, Ernst Grossman, Bill Haesler, John 'Jazza' Hall, Harry Harman, Terry Hickey, Shirley House, Alex Innocenti, Arthur James, Moreton Johnson, Ita Kennedy, Ted Kennedy, Juliet Kitchen, Phil Langford, Doug Lloyd, Neville Maddison, Peter Magee, Clive Moorhead, John Morris, Peggy Morris, Ken Morrow, Mick Mulcahey, Frank Mulders, Ken Murdoch, Phil O'Rourke, Keith Pattie, Alf Properjohn, Peter Rechniewski, Leah Sharp, Tony Standish, Jim Smith, Neil Steeper, Harry Stein, Joe Tippet, Roy Theoharris, Mary Traynor, Ted Vining, Alan Whiting, Ron Wills, Alan Woodhouse, Bruce Wroth; from England, information was provided by John R.T. Davies, Humphrey Lyttelton, and Derrick Stewart-Baxter.

Abbreviations and Terms

Many of the stylistic labels used in this book have passed into standard international jazz terminology, and readers who are unfamiliar with such terminology are advised to read any or several of the standard histories of jazz, such as Joachim E. Berendt's *The Jazz Book—From New Orleans to Jazz Rock and Beyond*, trans. H. & B. Bredigkeit with Dan Morgenstern (Granada, London, new and revised edn, 1983).

ABC	Australian Broadcasting Commission/Corporation
AB & ON	*Australian Band and Orchestra News*
ADBN	*Australian Dance Band News*
AIF	Australian Infantry Forces
AJC	Australian Jazz Convention
AJQ	Australian Jazz Quartet/Quintet
AJS	Adelaide Jazz Society
a.k.a.	also known as
alt.	alto saxophone
AMM & DBN	*Australian Music Maker and Dance Band News*
APRA	Australian Performing Rights Association
arr	arranger, arrangement, etc.
ASO	Adelaide Symphony Orchestra
bar.	baritone saxophone
BBC	British Broadcasting Corporation
bjo	banjo
blow	as in 'to have a blow': to play one's instrument, usually used in connection with an invitation to join in with a band informally either for part of a night's gig, or at a private jam session
bs	bass; in general this term may be taken to refer to any or all of acoustic, amplified acoustic, and electric string bass, unless otherwise specified
c.	circa
CAE	College of Advanced Education
CAS	Contemporary Art Society
clt	clarinet
cnt	cornet
comp.	composer, composition, etc.
dms	drums; generally referring to the standard drumkit

edn	edition
el.	electric
fl.	flute
flug.	flugelhorn
gig	musical engagement or performance
gtr	guitar
hca	harmonica
incl.	including
jam jamming jam session	a reference to informal, often private performance, with the emphasis on unrehearsed and improvised material
JAS	Jazz Action Society
JB	Jazz Band
ldr	leader
LSC	Life Saving Club
MBE	Member of the Order of the British Empire
MBS	Music Broadcasting Society
MD	Musical Director
MJQ	Modern Jazz Quartet
modern	a very broad term referring in general to post-traditional jazz styles. The word has acquired this broader (and not always literally appropriate) sense than the term 'contemporary', which refers to the most recently developed styles.
mouldy fygge	a lover of earlier styles of traditional jazz to the exclusion of all else. See further, the third part of The Evolution of the Concept of Jazz in Australia. Also known either just as 'mouldy' or 'fig'.
NO	New Orleans
NSW	New South Wales
NT	Northern Territory
NZ	New Zealand
or	generally means 'alternating with'
p.	piano; can include other keyboard instruments such as organ and synthesizer
perc.	percussion instruments other than those usually used in the standard drumkit
pick up	a band having no continuous existence, but assembled for a particular occasion
PJJB	Port Jackson Jazz Band
progressive	when used without inverted commas this refers to an interest in recent developments at whatever period is under discussion
Qld	Queensland
reeds	the family of single reed instruments; generally, clarinet and various saxophones
righteous	a term used by members of the mouldy fygge fraternity to describe their preferred music and musicians

SA	South Australia
sax	saxophone; used without qualification, it can be taken to refer to any or all of the saxophone family
scene	the overall musical landscape, incorporating musicians, followers, work opportunities, venues
SE	South East
sideman	any member of a band other than the leader
SIMA	Sydney Improvised Music Association
SJC	Sydney Jazz Club
sop.	soprano saxophone
SSO	Sydney Symphony Orchestra
synth.	synthesizer
Tas.	Tasmania
tba	tuba; referring to any of the varieties of brass bass
tbn.	trombone; generally slide, unless otherwise specified
ten.	tenor saxophone
TH	Town Hall
tpt	trumpet; including any or all of trumpet, cornet, flugelhorn
'trad'	a version of traditional jazz which originated in England and became the foundation of the 'trad boom'
uke	ukelele
UK	United Kingdom
USA	United States of America
vcl	vocal (i.e., singer)
vibes	vibraphone
Vic.	Victoria
vln	violin
WA	Western Australia
wbd	washboard
WEA	Workers' Educational Association

The Evolution of the
Concept of Jazz in Australia

The First Wave

The history of jazz in Australia cannot be understood simply through a review of events, since it is equally a succession of attitudes. This point is more crucial in jazz scholarship than in the other arts. A person is either a painter, sculptor, a film-maker, or not. But there is deep disagreement concerning the jazz musician because, for reasons which this essay seeks to clarify, there is and always has been deep disagreement about what constitutes jazz. The fringes of the definition are broader than in most art forms, not only because of conceptual problems raised at the experimental leading edge, but because many have wished to legislate certain jazz idioms out of existence in order to purify their particular preference. Not only does this controversy exist at any given moment in the history of music, but the question 'What is jazz?' has received a bewildering range of answers throughout the twentieth century. To assume that everything presented under the name jazz is an example of the music would be to include at various times variety acts like juggling or even wrestling demonstrations and fashion parades. This essay is an attempt to trace a broad history of the perceptions of jazz held by musicians, enthusiasts, and the lay public in Australia, and the way in which those perceptions were conditioned by certain important moments in the development of Australian popular music.

The first major period in which Australians formed their attitudes to jazz can be taken as being from 1917, the first year in which jazz was widely spoken of in this country, to 1929, ushering in the economic Depression which was a major factor in the decline of this first phase. Visiting American entertainers brought news of the new musical fashion in 1917, and by mid-1918 entrepreneur Ben Fuller presented Belle Sylvia and her Jazz Band, billed as 'Australia's First Jazz Band' to Sydney audiences. It is difficult to overestimate the importance of the impression which these performances made. Here was Australia's first exposure to a style of music which, in more fully developed form, would do more than any other to establish the ambience of popular music in this century. As an international phenomenon, the spread of jazz was probably the first major post-Renaissance infiltration of the European sensibility— liberal, humanist, rationalist, putatively Christian—by the manifestation of a non-scientific consciousness.

We can never know now what Belle Sylvia's group sounded like, nor what the musicians actually thought they were doing in terms of harmonic or rhythmic convention. But we have some contemporary accounts of the music and the effect it had on audiences.

'Jazz' is a Negro expression for noise, peculiar to music. A 'Jazz' band is an orchestra composed of the following instruments: piano, clarinet, saxophone, cornet, trombone, violin and drums, and the drummer must be 'some class'. . . There is no music used. They all fake and make up their own combinations. . . the 'Jass' band consists of a pianist who can jump up and down, or slide from one side to the other while he is playing, a 'Saxie' player who can stand on his ear, a drummer whose right hand never knows what his left hand is doing, a banjo (ka) plunker, an Eb clarinet player, or a fiddler who can dance the bearcat.

review by B. [William] Dean
in *Australian Variety and Show World*, July 1918

many laughable effects are attained. There is a comic drum interlude which arouses shouts of laughter, and a farmyard selection, in which are heard the farmyard sounds from the cheep, cheep of a young chicken—by the violin—to the trombone lowing of cattle. . . . each instrumentalist is not only a skilled performer, but also something of a dancer and a humorist.

Table Talk, Aug. 1918

It requires a determined imaginative leap to see past the period quaintness of these impressions to the threatening anarchy, the menacing irrationality which this overtly comic spectacle must have presented as, during the coming years, it advanced beyond the protective framework of a stage and became the dance music of a generation. One recollection conveys the destructive mania of a concert by a group like The Who. But he is describing Belle Sylvia's Jazz Band in 1918. The percussion instruments

included a frying pan and saucepan. An operatic fantasy, played to a realistic accompaniment of thunder and lightning, with a few revolver shots thrown in, just to add to the general din. The act used to conclude with "Oh! You Drummer", a rag in which the band apparently went mad, playing with heads, arms, and bodies swaying to the music. The drummer concluded by throwing all his instruments up in the air. The pianist stood on his chair playing the frenzied melody, while the trombone and saxophone players fell over totally exhausted, gasping for breath.

AMM & DBN, Dec. 1936

If the audience laughed, they were also disturbed—it is, when visualized even at this distance, a disturbing spectacle. The knowledge that this was Negro music prepared the spectators for an ambiguous *frisson*, a glimpse into a fascinating but disquieting darkness. If the music made many people happy, it was an abandoned happiness, an ecstasy which ominously bypassed the rational intellect, a visceral, physical, and pentecostal indecorum. It undermined morality. In 1919 a locally made film called *Does the Jazz Lead to Destruction?* was screened as part of a 'jazz week' at the Globe Theatre in Sydney. The publicity for the event created a character called McWowse whose attitude before the event was quoted as follows: 'Oh, it's terrible! It leads straight to destruction. Somebody told me it did. And I can assure you that neither Mrs McWowse nor myself will be seen at these terrible doings at the Jazz Week at the Globe Theatre.' By the end of the week, however, McWowse had succumbed: 'I shall no longer Wowse, but will Jazz my way to destruction if I want to. And I don't give a dash what the congregation says.'

Contemporary advertisement for the film 'Does the Jazz Lead to Destruction?', 1919

Significant local exposure to live American jazz bands can be dated from 1923, with the arrival of Frank Ellis and his Californians. Before this, according to Frank Coughlan, Australian jazz was very elementary, with special emphasis on loud drumming (*AMM & DBN*, Dec. 1933). A writer in *Everyones*, 12 Dec. 1928 recalled that before the arrival of Joe Aronson in 1922, local musicians were unable 'to produce little more than blaring noise and uneven rhythm'. The effect of Ellis and subsequent imports was ambiguous. While it gave local musicians and audiences a taste of the more authentic sound, it also caused some alarm. *Everyones*, 10 Sept. 1924, reported concern in the Musicians' Union of Australia that visiting musicians were 'having a bad effect upon the welfare of the Australian musicians'. Something other than simple physical welfare was at stake, however, when in response to the suggestion that attempts were being made to import coloured musicians, Senator Pearce, Minister for Home and Territories, was quoted in reply to the union delegation, 'There is no need to worry about the coloured people. We will not admit them.' Although in the short term the touring Americans sometimes deprived the locals of employment, in the long term they helped to increase the public awareness of and appetite for jazz. By 1926 jazz was the dominant popular music and receiving commensurate publicity. The South Street Brass Band contests in Ballarat instituted a jazz section because of the 'trend of public taste'. Bernard Heinze defended the music against its more strident opponents, and other symphonic musicians and composers were seeking to accommodate aspects of jazz out of either sympathy or opportunism. It was a word that made money. A London violinist reported being asked to form a jazz band for £200 to £300 per week. Novelty firms advertised 'jazz caps' (like clown caps), and the Melbourne retailer Buckley & Nunn had even presented what it called the 'jazz corset' for the fashion-conscious woman. We have available abundant evidence of what the professional and the layperson considered jazz to be, and it is clear that, if the imported musicians had extended and refined the concept, they had not radically altered either its instrumental, harmonic, or moral complexion.

The evidence gains force by accumulation. First, instrumentation. The mainstays of earlier small dance orchestras, the strings, were still in evidence, but with louder instruments being increasingly used, the demise of their centrality was approaching. The professional journals of the day gave much attention to jazz band instrumentation, in feature articles, answers to enquiries, and through photographs of current orchestras. Just as the music represented the last word, so its instrumentation favoured the latest novelties. Saxophones, banjos, and extended drumkits were essential. Trumpets and trombones increased their jazz credentials by the use of mutes, not to reduce volume but to produce novel effects. Almost anything capable of producing a curious sound was seen as a jazz instrument; advertisements for kazoos refer to them thus, describing their 'weird' effects as 'Good for Fun'. An article on the subject in *AB & ON*, 26 July 1926, mentions the use of a bicycle pump in a jazz band. Other new instruments mentioned in connection with jazz were the portable melo-piano, the flexatone, and the frisco whistle (*AB & ON*, 26 Jan. 1926).

Advertisement for the film 'Should a Girl Propose?', In Everyones, *21 April, 1926*

Cover of Rube Bloom's 100 Jazz Breaks for Piano, *1926*

Percussion effects were extended, from sirens, bells and the firing of hand guns, to various accessories which have since become standard, such as the 'synco wire jazz stick' (presumably brushes).

The atmosphere of a jazz performance was loud and hectic, a high-energy exercise. The cover of a booklet called *100 Jazz Breaks For Piano* (one of a series for different instruments) depicts flames rising from the keyboard, on which sits the music of the song 'Nervous Charlie Stomp'. The combination of agitation, irreverent familiarity, and heavy footwork is apt. In Charles Chauvel's film *Greenhide* (1926) there is a sequence showing an afternoon garden party at which a jazz band plays. Although silent, the extravagant and often grotesque movements of the musicians convey the kind of outrageous energy which many writers of the period refer to. *AB & ON*, 26 July 1926, published the findings of two scientists in Switzerland who had measured the expenditure of energy by musicians in performance. The smallest was 49.1 units for someone singing a song by Brahms, to 290.2 units for the drummer in a jazz band. It was meaningfully noted accordingly that 'sometimes the rendering of music can have a dangerous effect on the organs of breathing and circulation'. The music was characteristically frenetic, loud, and highly physical. The first public dance halls had opened in Australia in 1913—the Crystal Palace in Sydney and the St Kilda Palais in Melbourne—not long before the advent of jazz. This coincidence no doubt strengthened the perceived association between jazz and dancing, an association so intimate that the term 'jazz' was often taken to mean not the music but the dancing which it accompanied. *AB & ON*, 26 Jan. 1926, reports that jazz is currently 'the most popular dance', and the word frequently did duty as a verb: an advertisement for the film *Should a Girl Propose?* in

Everyones, 21 April 1926, informed the public that 'The modern Girl jazzes, smokes, indulges in athletes [*sic*], enters law and politics.' This was not the elegant dancing of Edwardian ballrooms, but a frenzied, often immodest and uninhibited activity which in many ways emphasized the body's grotesquerie rather than its grace, the *trompe-l'oeil* effect of hands crossing from knee to knee in the Charleston being a well-known case in point. This aggressive physical non-sense was a parallel to the nonsense of the musical effects and, frequently, the lyrics of what were perceived as jazz songs like 'Yes, We have no bananas'. The fact that this was regarded as a jazz piece eloquently emphasizes what was and was not conveyed by the term; few, if any, today would agree that this inoffensive piece of idiocy had any merit as a jazz vehicle. Both as a music and as a dance, a certain kind of idiocy was a vital part of the spirit. Contortions, tricks, novelties, extravagance, insolence: these characteristics share a quality which was central to the first jazz phase in Australia, and that is defiance of received norms and standards. It threatened the status quo. *AB & ON*, 26 Feb. 1926, carried a report of two coal-miners in England being heavily fined for intimidating someone by performing on jazz instruments.

There is a moral dimension to all this. Jazz was dubious. Not only did it lead to degradation, it destroyed the capacity for remorse: 'Dash the congregation', as McWowse said. Much was made of the sound of 'sin' in 'syncopation'. *AB & ON* in 26 May 1927 reported sympathetically the news from England that the Lord's Day Observance Society protested the BBC's broadcasting of jazz music and music-hall fare on Sundays as profane and somehow 'continental'. Jazz was the sound of rebellion and iconoclasm, or of depravity, depending on where you stood. *AB & ON*, 26 July 1926 quoted the views of Fritz Hart that jazz is 'an almost entirely evil thing, something even actively fiendish in character', and a writer in the *Argus* that 'the noises known as jazz represent an imported vogue of sheer barbarism'. It was the musical expression of everything modern, of emancipation from the restraints of the nineteenth century, of youth and energy, of fun and irresponsibility, of disrespect for authority and conservatism and therefore, as far as authority and conservatism were generally concerned, it was evil, something from the heart of darkness. Its 'negroid' origins were often taken as a guarantee of its barbarity. Thus, from its first appearance in Australia, jazz invited attention from other inconoclastic attitudes developing in the wake of the First World War. The image of the flapper embodied a convergence of jazz with the 1920s version of feminism, and jazz provided one of the anthems of liberated womanhood throughout the immediate post-war period. The publicity of the film *Should a Girl Propose?* cited above makes the point. The sheet music for 'Flappers In The Sky' presents the same image of aggressively liberated women defying the constraints of pre-war role models by drinking, smoking, flying aircraft, and playing jazz instruments (in fact it was not remarkable to see women playing in and leading jazz bands).

The question remains, what was the musical content of this early jazz? We have no record of jazz in Australia before 1926 or, arguably, 1925. The first local jazz performance was in 1918, but Australians didn't hear any American jazz until the release in this country of the Original Dixieland Jazz Band's recording of 'Lazy Daddy' in 1921. What had they been playing before that? How did they interpret the first American sounds they heard? We can form some assumptions from the very earliest Australian hot dance records, but we must remember that these present an infinitesimal fraction of the music actually being played throughout the country. A writer in *AB & ON*, 26 July 1926, estimated that there were 1600 jazz bands currently playing in Australia. What did a band called The Black Cat Jazz Band in Winton, Qld in 1926 or 1927, for example, actually play in order to persuade its audiences they they were hearing jazz? Percy Garner, now in Brisbane, recalled this group in an interview. Formed by a local chemist, Percy Smith, the band's uniform featured a black cat motif; but when asked why it was called a jazz band, Garner answered, 'I just don't know how that came in. But he named it anyhow.' The instrumentation was piano, drums, violin and xylophone. It is not unreasonable to assume that this group, and the majority of the small, lesser known, and especially rurally based 'jazz bands', simply appropriated the name because it was the height of fashion, but that their repertoire remained essentially that of an Edwardian dance band with perhaps the occasional novelty number as its 'jazz' component. To assume that every so-called jazz band in Australia was presenting to its audiences the sound of Red Nichols is dubious in the extreme, and depending on where they lived, the Australian public was probably coming to perceive the musical characteristics of jazz as anything from the (at that time) innovative sounds of Frank Ellis to honks, whistles, farmyard effects, and antics by musicians in funny costumes like clown suits. In 1939 (*AMM & DBN*, 1 Feb.), Donald Williams recalled that in the '20s, 'a jazz band was still a stage

'. . . antics by musicians in funny costumes like clown suits.' The Charleston Jazz Band, 1927, with, l. to r.: Joe Lewis, Les Clements, Frank Wilson, Bob Graham, Arthur Dewar, 'Tiny' Douglas

act with lots of clowning and mirthmaking', and that Boyd Senter 'was regarded as the epitome of great clarinet playing' (1 March); Senter's great specialty was novelty effects like slap-tonguing and elaborate physical contortions while playing. In Perth, Ken Murdoch recalled in interview that even as late as 1930, in Theo·Walters's Knickerbockers the trumpet player's contribution to the 'jazz' spirit of the band was to juggle his mutes. I would suggest that for the vast majority of the non-playing public, and even for a considerable majority of the thousands of musicians who wished to capitalize on the fashion, jazz was understood primarily to be weird novelties and strange, hectic instrumental sounds, with its primary musical component being a rhythmic one. At any moment in the history of a popular music, however, there would be a core of musicians who were professionally interested in the specifically musical challenges presented by the new vogue. While the public probably failed to appreciate the formal musical characteristics of jazz—they were still too new to be recognizable as such—at the professional level some attempts were being made to determine its musical conventions. This was aided of course by the growing amount of American jazz available on record and through the performances of visiting musicians. For all listeners, lay or professional, the most prominent distinguishing musical feature of jazz was its rhythm: aggressive, disordered. We must recall the rhythmic conventions which preceded this African-inflected fashion—the undisturbed regularity of the waltz, the pride of Erin, the gypsy tap. The extent to which we have now, sixty years later, habituated ourselves to time displacements in popular music, makes it difficult to retrieve sympathetically the disrupted impression created by syncopation. This was the most widely recognized musical innovation in jazz. The *100 Jazz Breaks for Piano* referred to earlier is an inventory of two-bar phrases whose common features are comparative harmonic orthodoxy and remorseless syncopated patterns which have since become mechanical clichés. The author urges that the player should give careful attention to the 'accents and phrasing'. Harmonically there seems to have been very little understanding of what was going on (and it is important to remember in this connection that during most of this early period, even the American jazz being heard on imported recordings leaned heavily towards the novelty style of white New York bands). The fact was that even professional musicians with an interest in theory were not sure what harmonic conventions informed the new music. Roy Delamare felt that there was almost no understanding of the harmonic basis of improvisation in Perth throughout the '20s and '30s, and Merv Rowston remembered that the 'blues' which often appeared in dance band programmes was just any slow foxtrot: that is, 'blues' signified a tempo, not a harmonic progression. Clearly, there was some violation of the received notions of melody and harmony, but it was some time before these could be perceived as anything other than nonmelodic cacophony. The idea of embellishment was important, in the form of grace notes or breaks, but there seems in 1926 to have been very little, if any, grasp of the idea of improvisation over chords. The consequences are complex. Because improvisation was not understood, it tended to be perceived as an arbitrary departure from melody. And because it was perceived as an arbitrary exercise, it is almost certain that for many musicians attempting to play jazz, it *was* conducted more or less arbitrarily. The charge that jazz was cacophonous was frequently, therefore, not just a deficiency in the listener's education, but an accurate perception of deficiencies in the performer's education. Much of the time, local performance probably was actually discordant, in the sense of having nothing to do with the chords, especially for instruments which did not play chords—that is, in particular, the horns—and even more especially for the instrument least frequently used to harmonize, which came increasingly to be the trumpet. In the 1920s, and particularly in the first half of that decade, improvisation, one of the most significant criteria of excellence in the music in later years, was scarcely even grasped as a component of jazz by many, and understood to a limited extent by only a few. The available models provided little assistance here. The pianist Rube Bloom, author of the 100 jazz breaks booklet already referred to, advises the reader to commit the breaks to memory and also warns against their excessive use since there is a danger of obscuring the written melody. He then says that his purpose as author will have been served if his book 'assists the player to PLAY HOT and acquire the art of IMPRO-VISATION'. With such advice from an American pianist associated in the public mind with jazz, it is hardly surprising that improvisation, with its foundation in harmonic theory, should continue to be a mystery to the majority of musicians well into the next decade, and that even some of the hotter swing bands of the late '30s were still presenting solos transcribed note for note from recordings and calling them 'improvised'.

It is certain that by 1926 the local understanding of jazz had made advances since 1918, particularly in the larger capital cities where, in addition to the increasing accessibility of sheet music and recordings, there were more frequent visits from American musicians. The evidence suggests that, notwithstanding these advances, to the general public jazz was dance music (or a dance) characterized by violent movement, abandoned delight, profanation of established decencies, the essence of vulgar, modern America. Even musicians with a serious interest in the technical challenges presented by jazz were likely to be fixated upon mastering the rudiments of syncopation or the physical problems of producing unusual sounds from their instruments. The 1926 Belle Vue Band contest adjudicators were much exercised when a trombonist used a 'jazz mute' to simulate a donkey's bray in one of the Australian National Band's items. One witness marvelled at this demonstration of 'fine technique', while others considered the introduction of 'jazz effects' to be deplorable (*AB & ON* 27 Dec. 1926). The harmonic challenge of improvisation was grasped by only very few, and certainly not mastered to anything like the degree evident on American recordings made in the same year (though not yet heard in Australia) such as the masterpieces directed by Jelly Roll Morton. The point is succinctly summarized if we consider the adjudicator's comments on the jazz band contest held as part of the South Street Competitions in Ballarat in this year, and reported in *AB & ON*, 26 Oct. 1926. Eight bands entered, and Bert Howell, then musical director of the Victory Theatre, St Kilda, and later involved closely with the development of the hot dance bands, was adjudicator. His comments indicate two important aspects of the Australian professional musician's perception of jazz at that time. One, that, whatever the public continued to think, musicians had made considerable advances since 1918 in discriminating certain formal characteristics of jazz; two, that they were scarcely aware yet that one of those characteristics included an improvisational component based upon clearly understood, even if relatively recent, harmonic conventions. Howell's comments deal with such orthodoxies as dynamics, pitch, tone, constancy of rhythm. There is also mention of specifically jazz effects: 'Trumpet out of time in stop chorus'; 'after beats not emphasised', 'trumpets slightly out with mutes'; 'originality in breaks'. The writer reporting the event observes that

before a man can become a jazz player he must have a sound musical education to draw upon. The days when jazz was played on a variety of kitchen utensils and when the player who made the most discordant sound was acclaimed a star, are gone forever.

Bert Howell expressed approval at the institution of the jazz band category in the contests, and predicted that 'before long the theatrical and dance palais managers would go to South Street looking for their star bands instead of looking abroad'. The implication is that, whatever advances the local musicians had made in jazz performance since 1918, the Americans were still seen as the pacesetters. But Howell's comments overall also suggest that in the view of this professional, keenly interested in the latest developments in the execution of popular music, the players at the South Street competition, who at no point are given credit for extended improvisational skill, are close to the equal of the Americans. In other words, formal improvisation is scarcely recognized as an element in jazz performance, except perhaps as effects with mutes or trombone glissandi, in which case it is their novelty rather than any harmonic foundation which marks them as jazz effects.

1926 was the peak year in this first Australian jazz era. In 1927 an import tariff on records diminished exposure to the source, leaving the rather superficial local product as the main model for jazz. In 1928 the talkies arrived in the major capitals, and their effect on the entertainment industry militated against the music business in general and the jazz fad in particular. Ironically, the first of those talkies was *The Jazz Singer*. Jazz activity diminished, first in the cities and moving like a ripple into the rural area as, over the next two years or so, the technology for talkies spread beyond the urban centres. Letters of enquiry to *AB & ON* in Oct. 1928 included one from a musician who had hitherto been engaged in jazz work, but now wished to know the standard routines for old-time dances. In 1929, studio jazz hands were being laid off. The US stock market crisis of 1929 heralded the Great Depression. Apart from generating an atmosphere in which the flippancy of jazz was wholly inappropriate, the associated labour problems helped to dry up the last trickle of American visitors as well as leading to severe unemployment among local musicians. By 1930, municipal councils' subsidies to local bands were being reduced or abolished, and increasing numbers of musicians were busking or, having sold their instruments, simply begging in the streets. The growing popularity of radio also had the kind of effect on recreation that television and later video were to have. One could entertain and even dance at home. Radio also hastened the decline of this first jazz era by glutting the air waves with the music when it was becoming more superficial and mannered in the absence of continued fertilization from American sources. *AB & ON*, Oct. 1932, noted the shrinking sales of jazz sheet music and opined that the music had been killed by over-exposure on radio. By the early '30s, jazz had suffered the worst fate of any fashion: it was passé. The word no longer described something up to the minute, but the faded vogue of a vanished ambience, as quaint and embarrassing as platform shoes or *très chic* flared trousers of the '60s had become in the late '70s. Popular music had moved to a sweet style that evoked all the domestic security threatened by the Depression. The titles make the point more powerfully than any commentary. In August 1932 the country was dancing to 'Home', 'Love Goes on Just the Same', 'Snuggled on Your Shoulder', 'Tell Me with a Love Song', 'There Never was a Love like Mother's Love'. No one 'jazzed'. They waltzed or fox-trotted. The strident novelties of jazz were replaced by the syrupy and more conventional harmonies reaching back to the old-time ballroom tradition. The professional musician had embraced the sweet style. Those who retained an interest in challenging instrumental and harmonic developments kept in touch with 'hot rhythm' releases. But this was no longer part of one's livelihood. And for the general public, jazz was dead.

The Advent of Swing

By the very early 1930s, jazz as publicly understood in Australia during the 1920s had receded as unequivocally into history as other period curiosities like the penny farthing bicycle. The cacophony of the small jazz group was old fashioned, and the professional musician as well as the public had passed on to the style of the day, which was for a sweeter sound that would be more consoling during the care-worn Depression. Australian dance band leader Al Hammett, writing from America in 1933, concentrated on the sweet bands of men like Guy Lombardo, and while he mentioned some of the black bands as being hot, dismissed them as too loud and coarse. In the same year Henry Hall, of the BBC Dance Orchestra, reported that the trend to sweet and sentimental music was world-wide. The public appetitite for songs redolent of domestic stability continued to be fed by items like 'My Blue Heaven', 'Goodnight Sweetheart', 'When Day is Done' and 'I'll See You in my Dreams'. In such straitened times it was more than ever necessary for the professional musician to keep his eye on the market, and the dance band industry was generally at some pains to distance itself from the jazz fashion of the '20s, now so inappropriately indecorous and so embarrassingly dated. In December 1932 Eric Sheldon of *ADBN* began a history of 'Modern Dance Music' with the observation that 'the apprenticeship period has passed'. Like so many spokesmen for dance music of the time, he is rather defensive on behalf of what was a relatively young musical phenomenon, and eager to stress the gains it had made in terms of current musical respectability. In particular he was anxious to underline how far it had progressed since the raucous irresponsibility of the jazz craze. Much of the admiration lavished upon Paul Whiteman, not only in Australia but in the US and Europe, sprang from his having attempted to give the music a symphonic respectability. A writer in *AMM & DBN*, Aug. 1933, dismissed as 'utter rot' a clergyman's charge that dance music was a menace to the younger generation, and shuffled a little guiltily at the suggestion that 'music of this nature was created by negros [*sic*]'. 'As a matter of fact,' he wrote, 'it springs from quite a different source, but if it were, it is no crime.'

In this climate, the situation of jazz was extremely confused. There was on the one hand a distinct odium attached to it. Sir Hamilton Harty, conductor of the London Symphony Orchestra, spoke for the Establishment when he described jazz as 'slovenly and immoral', and the saxophone as 'indecent' (*ABDN*, June 1934). There was, however, fundamental confusion over the connection between jazz (that is, the now dead fashion of the 1920s) and modern dance music as currently played. This confusion would persist in many quarters throughout the '30s and in some cases well into the '40s, and for this reason it is this period which has generated the most active controversies over the history of jazz in Australia. The problem arises out of a number of circumstances already glimpsed: the bad odour of the word 'jazz' in the early '30s among large sections of the public and the profession; the fact that by the time of the trailing off of that first phase of jazz, few musicians, including most of those who appropriated the term 'jazz' as a description of their music, actually understood what improvisation involved; the association of jazz with certain effects which survived, even if in diluted form, into the dance music of the '30s—effects relating to rhythm, instrumentation and instrumental timbre; and the frequent equation made from the earliest days between jazz and modern dance music. Thus, although the dance music of the '30s had passed well beyond what was regarded as the barbaric cacophony of the '20s, it still used, in modified form, other innovations associated in the public mind with jazz—syncopation, saxophones, brass with mutes, a range of percussion effects—and it was music for modern dancing. The only major jazz component missing was extended improvisation, either solo or collective. But most of the public and the profession had not yet grasped that this was a formal characteristic of jazz—indeed, as they had most frequently experienced the local equivalent of improvisation, it was precisely that 'cacophony' which had been left behind as an anachronistic barbarism. It is not surprising then that for many whose interest in jazz was at that relatively ill-informed level occupied by most witnesses and even exponents of a fashion merely because it is fashionable, there was widespread confusion as to the connection between the jazz of the '20s and the dance music of the '30s. Views ranged from the belief that the latter was jazz cleaned up and made a lady, to the strident denial that it was to be confused with jazz, though it was a legitimate development incorporating some of the innovations of the previous decade. In the '20s, there were those who loved and those who hated jazz, but they were generally talking about the same thing even if they did not understand its musical conventions. Following that first phase there was still the same aggressive partisanship, the same general incomprehension, but the situation was imponderably complicated by the fact that different people meant very different things by the word, and many of its most knowledgeable followers discovered that it was better to avoid the word altogether. We thus have the tangled situation of many people using the word but not agreeing as to what it meant, while a minority had a clear grasp of what it meant, but insisted on using another name, such as 'hot rhythm'.

Let us turn to those *cognoscenti*. We have been talking at the level of fashion, not because the fashionable attitude is necessarily a reliable one—quite the contrary—but because it has the strongest influence on the broad perception of a phenomenon. It is necessary to notice, even at this early stage, an underground stream. A minority of both the public and musicians had developed an attachment to jazz, not as a fashion, but as a musical form. This group pursued its interest with dedication and often, in the case of musicians, talent. Against the increasingly bland background of current dance music, this minority became more visible, especially as an emerging younger generation sought the revitalization of its musical entertainment. In *ADBN*, March 1933 we have one of the first references to a revivalist spirit which is more frequently seen as dating from the end of the decade. Frank Coughlan and Benny Featherstone were heard playing at a 'session' in which they played 'The Sheik of Araby' in 'the style of eight years ago'—that is, a recreation of the first 'Australian jazz age' at or about its peak. While the public demand for jazz had died away to almost nothing, we are seeing the first evidence of one of the new strands in the history of Australian jazz. Broadly speaking, the '20s phenomenon was at least as, if not more, important as the expression of a fashionable attitude than as a serious musical exploration. The reference to Coughlan and Featherstone is an early glimpse of an interest in jazz not because it was fashionable—far from it—but for its inherent musical possibilities, as a musical genre rather than as a period mannerism.

This double strand—of a general public uninterested in jazz, and of a small but growing minority of laypersons and musicians becoming knowledgeable regarding the music—continued to characterize the situation until 1936 when, as we shall see, a new musical vogue complicated the issue even further. Reviewing ten years of record releases in 1935, Ron Wills traced the decline of jazz from the pre-eminence of Red Nichols in 1926 to the dead years of 1929 to 1931, but observed an improvement in the jazz content of record catalogues and a revival of 'hot music' interest from 1934. The use of the term 'hot music' should alert us again to the confused state into which the word 'jazz' had fallen, and how it now stood in relation to modern popular music. One of the first 'Modern Music' clubs was the Westralian MMC which held its first meeting in Perth on 24 June 1934, and its organizers included Ken Murdoch, Sam Sharp, Merv Rowston, Wally Hadley, Viv Nylander, all of whom demonstrated during their careers an interest in jazz of one kind or another. Indeed, on that first night, Murdoch presented records by Ellington, Cab Calloway, the Boswell Sisters, Eddie Lang and Lonnie Johnson and a number of other 'rhythmic releases'. But on the same night were also presented a tap dancer, zither player, Hawaiian guitarist, and George Watson performing 'eccentric dancing'. Musicians themselves continued to exhibit confusion about the situation and character of jazz. George Dobson estimated that less than 10 per cent of currently working trumpet players could swing as well as Lew Stone (so how many fewer could swing as well as members of, for example, Fletcher Henderson's band in 1934!). Lynn Miller, with Dick Freeman at Sydney's Ginger Jar, wrote that most dance band musicians did not currently understand chords, without which 'rhythmic style' playing was impossible. Benny Featherstone who, with Frank Coughlan, was probably the most perceptive musician writing and thinking about jazz during the period, lamented that local 'hot music' was of a very low standard, having been enervated by the shortage of records (as compared with English catalogues), the lack of visiting American bands, and the general fashion for sweet dance music.

Witnesses like Featherstone and Miller, and record collectors like Murdoch, reflect the ambiguity of the situation of jazz in a number of ways. Their comments indicate just how slightly jazz appreciation had apparently penetrated the profession and the public, the former trying to earn its living by pandering to the sentimental preferences of the latter. But in their own activities, Featherstone et al. reflected a small but dedicated hot music following; they also gave notice of the shape of things to come, as did other events of 1935. For example, Sonny Brooks was imported from America as the epitome of the sweet style to lead the orchestra at the Sydney Palais Royal. He didn't even last the year before returning to the US.

A change of taste was in the air. 1935 was later recalled as the year that the word 'jitterbug' first appeared in Australia, and in November the 2UW Sydney Swing Music Club held its first meeting, with a reported attendance of 70, increasing to 110 the following week. If that sounds like a trickle, there was no doubt about the torrent of 1936. The haven of sweet dance music which the jazz fans of the 1920s had retreated to with the onset of the Great Depression in 1929 was too confining to a new audience of seven years later. 1936 was the year swing overwhelmed popular music in Australia.

Although we have noted evidence of growing interest in 1935, as a public fashion the flood-tide of swing was in 1936. In Sydney, it was given impetus by the opening amid much publicity of the Trocadero. The coincidence was good for both, and Frank Coughlan's appointment as the musical director of the new ballroom encouraged the public to regard him as the leading exponent of the latest in hot dance music. This explosion of publicity also did something to enhance Sydney's more progressive reputation, and in fact the sweet music of the early 1930s was not overwhelmed in Melbourne as fully nor as quickly as in Sydney. Nonetheless swing clubs blossomed in such rapid profusion that demand for the music outstripped supply. Jack Spooner's band in the Ginger Jar

Frank Coughlan and the Trocadero Orchestra, c. 1936 or '37, with, l. to r.: Don Baker, Dick Freeman, Reg Robinson, Reg Lewis, Frank Scott, Jack Baines, Bunny Austin, Frank Coughlan, Ted McMinn, Frank Ellery, Colin Bergerson, Jack Crotty, Dave Price, Stan Holland, Billy Miller

was probably the first in Sydney to bill itself as a swing group, but others were hot on its heels. If we now tend to associate the word with big bands like those of Goodman and Dorsey, in 1936 most of the groups that took up the title were the usual five or six piece groups of the cabarets and restaurants. Record reviews suddenly sprouted swing supplements, Coughlan addressed the Sydney Swing Music Club on the nature of swing, a topic taken up in Melbourne some months later by Harold Moschetti in a lecture delivered to the Victorian Society of Dancing, an organization of dancing teachers anxious to supply the huge demand for instruction.

While it was clear that 'swing was the thing', there was very little agreement as to *what* thing it was, and in particular how it related to jazz, the last major innovation in modern dance music. Bands and musicians thrown together in the category of swing included the Hot Club of France, Coleman Hawkins, Mills Blue Rhythm, Fletcher Henderson, Joe Venuti, Nat Gonella, Ivor Mairants, Freddy Gardner, Arthur Young and Jack Miranda. Some of the early patrons of the Sydney Swing Music Club stopped attending because they were not hearing the Harry Roy and Victor Sylvester records they had hoped for. Yet at the same time, the club devoted one of its early sessions to the rhumba, and 2UW's programme *Red Hot Rhythm* featured such artists as Peter Dawson and the London Piano Accordion Band. The word 'swing' had spread so rapidly among the general public that it had not had time to achieve a clear definition. At its loosest and broadest conception, it became, as jazz had in the '20s, virtually any modern dance music. As such, it embraced that residue of jazz effects that had survived: prominent syncopation, modern instruments like saxophones, strong brass, mutes, excursions away from the melody and peppiness, bright tempos. The word 'jazz' itself in some quarters enjoyed some degree of rehabilitation, sometimes being simply equated with swing. Others preferred still to distance themselves from the crudeness of jazz and used terms like 'hot' to distinguish the new fashion from the rather saccharine style it had displaced. The range of relationships postulated as existing between swing and jazz could not have been greater: Coughlan's band in the Trocadero was called a jazz band, but so was the band of Howard Jacobs. Jacobs's orchestra presented what was billed as the 'First Australian All Dance Band Concert' at the Sydney Town Hall, a concert repeated in other eastern states. One reviewer in a professional magazine praised it as 'the first programme of all-jazz music heard in this country' (*ABDN*, July 1936). Items on this 'all-jazz' programme included Fritz Kreisler's 'Caprice Viennois' arranged for six violins by Harry Bennett (*not* Ferde Grofé, as the reviewer believed), 'Ay, Ay, Ay', 'Black Eyes', 'Schon Rosmarin', 'On with the Motley' and 'Liebestraume'. Gladys Moncrieff was a featured vocalist.

Finally, of course, a word means what the majority agrees it shall mean. The difficulty here, however, is not that a majority in the late 1930s agreed that 'jazz' and 'swing' meant something other than what the words mean to us now, but that there was very little agreement at all, and that the range of opinions clearly extended to music that is far more firmly situated in such categories as the light operatic tradition and the music-hall repertoire. We cannot accept the labels at face value as they were being bandied about at the height of the swing phenomenon. One writer draws a clear (but unexplained) distinction between swing and hot music; another equates swing with jazz; bandleader Al Hammett writes that 'jazz' is just the old name for dance music; Ron Moyle, the most important dance band leader in Perth, names rhythm and extemporization, the latter *arranged*, as being the essence of swing. Howard Jacobs, defending his concert programme, maintains that he was trying to demonstrate that swing is not new, and indeed has strong connections with 'the beauties of the classics'.

In all the confusion, however, there is evidence again that those interested in the phenomenon as a music rather than as a fashion were coming closer to its distinctive features. In all the conceptual agitation, certain important notions were drifting to the surface. A new word, 'jamming', was coming into use in direct connection with swing. An enquiring reader of *AMM & DBN* (Oct. 1936) was informed that a 'jam band' involves musicians improvising their parts for the 'sheer joy' of it at a 'jam session', and that only 'top-notch' players are capable of this, the 'true source of swing music'. (This is not to say that many musicians yet understood the principles involved, and there is considerable evidence that much 'jamming' was still illiterate exhibitionism. Furthermore most jam sessions—attempts at more or less pure jazz performance—were still private.) Swing arrived so suddenly that, initially, the big names in the professional dance band scene continued to be leaders from the sweet era trying to accommodate and associate themselves in the public eye with this new, highly commercial idiom—Al Hammett, Harold Moschetti, Charles Rainsford, Bill O'Flynn, Bert Howell, Cecil Bois, Hal Lloyd. Being in the spotlight, such men were often required to make public utterances on swing, and being caught somewhat by surprise, naturally they often committed to print curious misunderstandings which tend to make the *whole* profession look like a stronghold of conservatism and ignorance. But, as at any moment in the development of an art form, there was a younger generation of professional musicians, not yet as established or publicly visible, eager to succeed in the business by keeping abreast of developments before they even became widely fashionable, and the more able to do so for not being obstructed by old and entrenched musical habits. Because they had not yet become established, we tend to hear less of them at this time. Although Coughlan's first Trocadero orchestra consisted, naturally, of experienced musicians, they were young turks as compared with the better-known bandleaders mentioned above—their average age was around twenty-four—and their preferences in music showed a clear bias towards jazz musicians: Fats Waller, Frank Trumbauer, Benny Goodman, Louis Armstrong, Duke Ellington, Red Nichols, Jimmie Lunceford, Teddy Wilson. However generally misunderstood, swing did lead to a renewal of interest in jazz, if not always by that name.

It was evident to many that elements of this new, aggressive, 'peppy' music had much more in common with the jazz of the 1920s than with the recent fashion. As a consequence, jazz itself became the subject of new interest. If 1936 is the year swing hit Australia, it was also the year in which, a little later, black bands and white dixieland bands began to receive new or increased notice. *AMM & DBN*, Nov., carried in its swing record section, a review of a group named as the Windy City Five, led by Bud Freeman.

In the years leading up to the Second World War, swing consolidated itself, and indeed came perilously close to ossification in Australia. The ABC acknowledged the music with its own programme on 2BL, Sydney; there were calls for a national federation of swing clubs, and a concert played by Coughlan's band at the Sydney Town Hall in 1937 drew an estimated 3000 people. The phenomenon quickly became national. Swing clubs proliferated in every capital and many rural centres, many in the wake of Jim Davidson's national tour in 1938. The fashion did not spread evenly, nor always without resistance. Musicians committed to the older dancing styles claimed that swing was completely undanceable. This resistance was less evident in Sydney than in Melbourne, where the entertainment business was in general more conservative. Bearing in mind post-war developments, it is worth stressing this impression of greater conservatism in Melbourne than in Sydney. It meant that there was less of the latest swing available in the former city. Nevertheless, there was a very real demand for a more virile, high-energy music, especially among the perennially irrepressible young. It is reasonable to assume that the greater frustration encountered by this small but significant audience in Melbourne should lead to an even more fervent support of the comparatively few venues that supplied what they wanted, and that more energetic and devoted loyalties would be inspired at the grateful discovery that there were, after all, a few places that generated more excitement—thus, the extraordinarily rapid growth of the 3AW Swing Club, the frenzied popularity of the Fawkner Park Kiosk and certain of the coffee lounges.

The increasing popularity of swing did not always lead to increasing consensus as to what it constituted, especially in relation to jazz, and it remains essential when dealing with this period to remember that the labels 'jazz' and 'swing' did not necessarily convey what is now generally understood by them. One pattern remains clear, however: the spread of swing continued to reawaken interest in the term 'jazz'. Writing in *AMM & DBN*, Jan. 1938, Steward Edwards identified swing as 'just Jazz—Jazz in its purest and most essential form'. For him, the essence of jazz was the 'hot' chorus. In the light of these comments it is also interesting to note his assertion that no band currently heard on Australian radio exemplified swing, that 'local musicians have religiously copied the notes they have heard played by American bands'. A writer in the same issue included Jelly Roll Morton, King Oliver, and Red Nichols among those most responsible for the development of swing, and in June, Nichols was referred to as one of the few survivors of 'the golden era of swing'. Beneath these disparate views however, there were certain areas of agreement, the most prominent one of which is a direct echo of the 1920s: that is, that an essential feature of this new development in popular music was a matter of rhythm. To define what it was 'to swing' has always been nigh on impossible, but if it was still rare among Australian musicians of the 1930s, it was instantly identifiable to the educated ear. In Hobart in 1937 Ron Richards's orchestra presented itself visually in the manner of Benny Goodman, but only one member of the band, Alan Brinkman, was noted as actually having a sense of swing. Likewise in the following year in the same city, it was remarked that newcomer to the dance band scene, Tom Pickering, made the band he was in swing with his 'hot expertise'. The prominence of rhythm provided dance bands wishing to capitalize on the lucrative fashion with one of several focal points. When one wishes to copy an effect without really understanding its principles, then one will tend towards mannerism, and such was the rush to participate in the swing vogue that the period of transition from innovation to stale and overworked mannerism, surface tricks, was unusually brief. As embodiments of the rhythmic heart of the music, drummers in particular quickly tended to flashiness. George Watson noted that 'jazz' had created an interest in drum 'tricks', and there was during 1938 and 1939 an efflorescence of drum forums, correspondence columns, workshops. But just as drum techniques quickly descended into mannerism, so did the general repertoire of swing effects, so that as early as June 1938 the music was spoken of as over-exposed, cliché-ridden, 'showmanship jazz'. In January 1939 Jim Davidson perceptively observed that swing had perfected itself and had no further to go. Whatever was meant by swing, it seemed to many by 1939 that it was going to be a very brief episode in modern dance music in Australia. And but for the outbreak of the Second World War they might very well have been correct.

Before reviewing the impact of the war on the performance and perception of popular music, it is important to reaffirm that, however misunderstood the term 'swing' might have been, it led indirectly to the spread of an intelligent interest in jazz. I mean by this a more discriminating understanding of its various manifestations to that time, and an increased awareness of it as music rather than as a passé novelty. In February 1937, *AMM & DBN* had noted and analysed a 'Chicago' style of jazz. In October, a reader's letter asked how to organize jam sessions, how harmony functioned in improvisation, and how to co-ordinate a rhythm section. Small band jazz, in particular of the style known loosely as 'dixieland', was being brought back into prominence in association with swing. As far as the general public was concerned, there was still some suspicion of the word 'jazz' with its associations of cacophony, ratbaggery, and an era that didn't really know what hardship could be. The further one moved from any centre of jazz interest, naturally the more evidence there was of confusion over what jazz, hot music, swing were. And the swing clubs themselves continued to present a range of material that would shock the purists of the coming decade—variety acts, freak bands, comedy presentations.

Less publicly, the number of interested enthusiasts for jazz was increasing. They were still very much an underground minority. Jack Spooner's band was congratulated for playing to the public rather than to a minority of 'jazz hounds'. 2UE was taken to task for broadcasting early 'negroid' jazz. The reviewer 'Guv'nor' (Terry Pierson), noted in February 1939 that while a lot of musicians liked jazz, the public wanted corn, and musicians at the Ambassadors in Perth would play 'hot records' in the band-room between sets to let off steam. When Charlie Lees started holding swing socials in mid-1939, musicians would queue up for a blow. In addition to some musicians, there was also an increasing audience who found something lacking in the standard commercial offerings. Because the general level of swing energy in Sydney music was higher than in Melbourne, the audience for that music was not so visible as a distinct entity. In Melbourne the channels for gratifying a desire for 'hot' music were fewer and narrower, and therefore the current running through them was more turbulent. The Fawkner Park Kiosk became a venue for audiences and musicians who sought the thrill of individual solo improvisation, the excitement of a form of music which represented a current form of jazz, or free blowing, small band swing. A clear distinction was drawn between the 'commercial stuff' played in Melbourne and the unrestrained musical self-expression to be heard at Fawkner Park, referred to in April 1939 as that 'notorious swingery'. The decision taken by some later purists that the music at Fawkner Park was not jazz because it drew so heavily on swing devices like riffs and close harmonies rather than collective improvisation is, in the larger conceptual and historical perspective, indefensible, or at least a perverse use of the word jazz. In the late '30s, beneath the public confusion or indifference about the matter, there was a clear, if contained, revival of interest in jazz, and an increasingly perceptive grasp of various of its stylistic forms. New Zealand jazz broadcaster Arthur Pearce arrived for an extended visit in Sydney in 1938, and in October published the first of his articles on jazz for *AMM & DBN*. His analysis of the lineage of swing, his knowledge of the early black jazz bands, and of the relationship between Bob Crosby's Bobcats and the older New Orleans jazz, proclaim him, along with Ron Wills, as a major early scholar of the music in Australia, as important as a source of knowledge in Sydney as Bill Miller was to become later in Melbourne. His writing and that of Wills both stimulated and reflected a growing and intelligent curiosity about different forms of modern American popular music and jazz in particular.

The War

Although the new interest in jazz in the late thirties was stimulated by the swing phenomenon, from the early years of the war the two forms of music developed increasingly divergent followings. Enervated by over-exposure and emasculated by the surface mannerisms indulged in by inferior groups, swing came to be absorbed into the larger dance music scene. With the exception of special events like Coughlan's Swing Jamboree at the Melbourne Trocadero in 1940, swing became a special, but circumscribed, item in the overall programme of an evening's public dancing. No dance band in the normal course of events ever presented a full programme of swing or jazz. Even Bob Gibson's band, regarded as a first-rate progressive orchestra, included items by, for example, Delibes. These bands were not there to give swing concerts, however much a few might have liked to, but to provide dance music for the general public. At the same time, perhaps because of wartime anxiety, there was initially a reversion on the part of that public to a taste for the sweeter style of the early 1930s. The swing clubs continued to be the main live forum for the music, primarily through small group swing-style jam sessions participated in by that cadre of modern dance musicians who continued to be interested in what was still a major new challenge: swinging and improvising. But as professionals, even their interest often lay in the fact of the challenge itself rather than in the inherent properties of the music, and, as the swing club programmes indicate, many were equally interested in other musical innovations for the same reason—the pride of the professional in keeping abreast of the state of the art. The point here is that as we analyse swing in Australia asking how far it was a manifestation of an enthusiasm for jazz for its own sake, we have to discount much of the activity that went on in its name.

In the meantime, interest in jazz continued to develop along a path that for many led away from swing and other commercial dance music, partly because the latter was becoming less satisfactory to those with a taste for 'hot rhythm'. While the public and in many cases even the professional attitude to the genres remained vague, a minority of enthusiasts was already in the late 1930s using the term 'righteous' jazz, an interesting phrase in foreshadowing the evangelical stridency that would become increasingly common in the next two decades. Arthur Pearce approved the growing following for dixieland which he perceived from the beginning of 1940. While many professionals such as Wally Norman insisted that swing could be equated with jazz, there was a vocal movement under way to separate the two, affirming that jazz was first embodied in the early New Orleans bands and was currently alive in the white

Chicago school of Condon, Freeman, and their confrères. In the light of later developments in the '40s and '50s, with their altered emphases, it is very important to stress two points. One is that we are witnessing here the beginning of major schism within a fraternity which regarded itself as committed to jazz. The differences of opinion between people like Wally Norman and Arthur Pearce were not a simplistic debate between an anti-jazz and a pro-jazz faction. They were both committed and, in their own way, well-informed enthusiasts for what they called jazz. The disagreement was over what *could* legitimately be called jazz. The second point is that at this stage, the proselytizers for 'righteous' jazz (i.e., something other than swing) did not at the very first generally regard themselves as being antiquarian for its own sake. As far as they were concerned, they were dealing with a current, living music form. Theirs was not at the outset a self-consciously revivalist movement in reaction against the whole modern music scene, but more essentially a reaction against a vitiated swing in favour of a more authentic and muscular jazz currently being played primarily by white Americans. As with so many schismatic movements, however, it would very shortly breed more.

For the moment, the new feature in the evolving conception of jazz was a dedicated minority wishing to distinguish its interests from the commercial component of modern popular music. The inauguration in Melbourne of *Jazz Notes* in January 1941 was a sign of things to come. In March when Graeme Bell's group began at Leonards, St Kilda, it was reported as 'the only band playing real jazz' (*Music Maker*, March 1941). In the face of the claim sometimes made that this was the first unadulterated jazz residency in Australia, it should be pointed out that the jazz was in fact leavened with Olive Lester, *inter alia*, presenting interspersions of 'sweet vocal renderings'. Similar activity was appearing in other cities, in particular Hobart, and Adelaide where, in addition, the Adelaide Jazz Lovers' Society was formed in October. This new interest was in part sustained by a new kind of audience and musician. Earlier committed jazz players like Frank Coughlan had been primarily interested, as professionals, in the most recent accessible examples of the music. Coughlan knew jazz from the Red Nichols material onwards. He was *not* an antiquarian, not a rigid purist, not what became known very soon as a 'mouldy fygge'; likewise, most of that minority of the listening public who were attracted to the hot music at swing clubs and on some radio programmes. These people liked the energy, the excitement of the music. But with very few exceptions they had little interest in researching background. What began to emerge as an identifiable and vocal movement around 1940 was a corps of almost vocational jazz enthusiasts, deeply interested in its past, its development, what they felt to be its most authentic manifestations. These were not people for whom jazz was simply another of several stimulating forms of entertainment. It was a uniquely obsessional object of study and affection. By December of 1940 reference was being made to two kinds of enthusiast for the various forms of jazz or jazz-inflected popular music. On the one hand, the jive-talking hipster whose essence was to be at the leading edge of a given fashion; the latest with the latest. But there was now also the hot jazz fanatic, a walking discography of a music fastidiously sealed off from all commercial contamination. I suggest that this species first appeared in its most articulate form in Melbourne because that commercial scene as embodied in, for example, the dance bands was more bland than in Sydney, but that at the same time Melbourne was sufficiently larger than the remaining other urban centres to generate enough interest in the matter to become visible. But it was developing elsewhere. In Newcastle, for example, a small group of 'self-styled John Hammonds' criticized the Newcastle Hot Jazz Club for not distinguishing carefully enough between swing and jazz, and thus betraying the club's name (*Music Maker*, March 1941). The intimations in all this of principle in addition to taste, the evangelical tone, the growing insistence upon artistic authenticity and the diatribes against commercialism, the sense of a lone voice calling for integrity in a commercial wasteland, inevitably alerted kindred minds in other arts. On 28 October 1941 the Contemporary Art Society presented a jazz concert featuring Graeme Bell's band, Don Banks, Ron Howell, and a jam session with guest musicians. The printed programme included a fervent unsigned manifesto. Among the points it made were: that the essence of jazz was improvisation; that 'crooning' and 'swing' were commercial, bastardized, by-products of jazz; that jazz was also to be distinguished by its emotional sincerity. The Contemporary Art Society was itself fiercely anti-Establishment and something of a rallying point for dissident intellectuals. The association between jazz, in particular of a traditional and self-consciously non-commercial style, and other radical movements, has had a strong effect upon the way the music is perceived, not only by the general

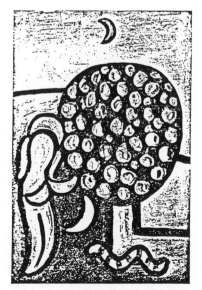

MUSIC

PAINTING

THE
CONTEMPORARY
ART SOCIETY

FIRST CONCERT
OF

HOT JAZZ

1941

HOTEL AUSTRALIA, MELBOURNE
BY COURTESY OF THE DIRECTORS
OCTOBER 28
3

Cover of the programme for the CAS 'Hot Jazz' concert, 28/10/'41

public, but often by its supporters. The fortunes of jazz and of radical thought in Australia in the subsequent years constitute a revealing *pas de deux*, but which is unfortunately beyond the scope of this essay. For the moment it is sufficient to record this new variation in the image of jazz. (I stress the word 'variation'. There is a clear conceptual connection with the anarchical dimension perceived in the music in the '20s, though the intellectual foundation is different.) This radical tinge has had important effects, often subliminal, on the ambience of the music; witness for example the persistence of chthonian imagery in connection with jazz venues—The Basement, the Cellar, the Catacombs, the Snake Pit. Sometimes of course the places are, literally, basements, but the question still remains of why they should be. I suggest that an important contributing factor was this burgeoning movement that consciously sought to isolate jazz as an underground art form, something peripheral to the commercial cultural mainstream.

In the meantime the war indirectly influenced jazz through its effect on the popular music industry. The public appetite continued to be for nostalgia, an atmosphere of domestic security and patriotic sentiment. In response, the established bands became more bland and genteel. This tendency was accelerated by the depletion of the ranks of professional musicians because of enlistment (which of course was also likely to remove the more adventurous temperaments). This meant both that band personnel became unstable, and that the replacements were inevitably somewhat less than the best. They were often veterans called out of retirement, or youngsters still developing their technique. The war also brought import restrictions. In 1940 the government determined that England would be the sole source of imported instruments, a decision that particularly hit dance bands because of their heavy reliance on American products. At the same time the demand for entertainment increased, with fund-raising charity functions and the recreational requirements of service personnel. In other words, the available musical resources, already diluted, had also to be spread more thinly. All of these disparate factors conspired to the same end, which was to take the edge off dance music. Complex, up-to-the-minute arrangements were not easy for semi-retired musicians habituated to earlier harmonic and rhythmic conventions, and while the apprentices might have had the spirit, they lacked the execution. The mixture of musicians was also less cohesive. The general standard of professional dance bands inevitably suffered, and in the insipid musical climate, the free-wheeling sound of jazz—increasingly being identified with what was most commonly called dixieland—picked up an increasing following. Its audience ranged from the high-spirited young whose response was pre-eminently visceral to the more cerebral, even ideological, mouldy fygges. The latter tended to work with more narrowly circumscribed definitions of the music, especially as they became more aware of the early 'classic' black jazz of the 1920s, now becoming available on the important Columbia reissue series. Although the general public was likely to think that 'jazz' was anything from a discordant fad in the '20s to any modern music, there was a growing audience, mostly young and in a few, but important, cases highly intellectual, who saw jazz as a distinct strand in contemporary music. As small group improvisation, it was being performed at private jam sessions involving professional musicians, through dixieland features in the public dance programmes, and growing out of an amateur scene around groups like the Bells and the clubs. The clubs themselves varied from 'modern music', where jazz would be just one element in a programme that might include concert pieces, variety acts, and Latin American, to hot jazz societies with their much more specialized interest.

These effects of the war were both extended and complicated as the American presence in Australia became more prominent. Those musicians who entered the services benefited from contact not only with their colleagues from other states, but particularly with visiting American musicians. Although Artie Shaw's navy band was the most publicized, it had less impact than it might have had because although a few locals managed to gatecrash concerts, audiences were officially restricted to American service personnel. There were, however, other American service bands which mingled professionally with Australian musicians and these, as well as American servicemen on leave, did much to change the face of local entertainment, not least in stimulating a resurgence of swing. The style of the military entertainment bands is exemplified on acetates from 1942 of seven members of the 32nd Division Special Services Band, known as The Red Arrows, at a jam session in Adelaide. To put it as briefly as possible, the strongest influence is the Count Basie small groups, and in turn they transmitted this influence to the locals, strikingly indicated in acetates of Adelaide musicians before and after exposure to the Americans. The 'jamming' scene sustained by local professionals was richly fertilized by the visitors. The latest harmonic developments (in which there is evidence that local front-line players were generally seriously deficient), dynamics, and the importance of extroversion, were absorbed. Chord symbols were increasingly accepted and understood as foundations for improvisation, freeing soloists further from mere melodic embellishment. The musicians from the two countries sat in together, a form of education which is incalculably yet indefinably superior to anything that can be learned from records or sheet music. This was the first extended personal contact between Australian and American jazz musicians for nearly twenty years.

For the general public, it was the American servicemen on leave that had the biggest effect on entertainment. These men brought with them a demand for a particularly energetic form of swing. It is almost certain that much of their enthusiasm for this extroverted American music was exaggerated, rather in the way so many Australians become instant and highly vocal devotees of Fosters as soon as they set foot in a foreign country. To jitterbug wildly to loud, up-tempo swing or jive bands was to emphasize one's American-ness, often a high-ly attractive quality to the local sweater girls and bobby-soxers. Ironically then, it is probable that the NCOs and enlisted American servicemen, as a group more given to public excesses in their leisure than officers, created a demand for an even more fevered version of swing and jazz than they left at home. This would have been even more striking in the light of the comparative gentility into which many of the older bands had settled. But the Americans were a lucrative market, and it did not take long before the bands that were capable of it were happy to cater to this demand. Such bands developed increased competence and energy, not to mention in many cases a degree of vulgarity that offended the more conservative. They also extended the range of music being presented; swing, jive, dixieland, were all words appearing on the same bill, though whatever their name, the leaning was towards swing rather than in the direction of 'righteous'. Wally Norman was leading a 'jive' band for allied NCOs in 1944. At about the same time a jitterbug championship was held in Leichhardt Town Hall. The estimated 4000 in attendance saw the orchestra discard its music and just jam through the contest. In Brisbane a writer lamented in 1944 that the presence of the Americans and their demand for music for jitterbugging had led dance music away from the English style to an extroverted barbaric style conducive to immorality. Such comments underline the powerful effect of seeing and hearing current dance and its music in the flesh rather than through records and films. Maurie Le Doeuff recalled in interview his most memorable musical experience as hearing the extreme dynamic range from triple f. down to triple p. on The Red Arrows' introduction to 'String of Pearls'. It is likely that the wartime resurgence of swing and jitterbugging was of even greater magnitude in its effect on the post-war dance scene than the first wave of 1936.

To the more conservative, and that was still likely to be a majority, this American-inspired craze was as much of a threat to civilization as jazz had been in the '20s. Indeed, the public showed little inclination to distinguish the two. It was generally and correctly felt that one was a development of the other, but more acute observers could see in the loudness of the music, the exhibitionism of it, the same inclination to stock mannerism that had characterized its first wave. Its followers dressed grotesquely, behaved indecorously, indulged in argot, and danced with an abandon that threatened life, limb and morality. It was just the kind of fashion to generate in reaction a purist clique. Ironically, again a swing craze helped to nurture a divergent jazz movement partly out of contempt for the former's excesses. The mindless hysteria of a bobby-soxer who thought that everything played loud and fast was the acme of jazz inevitably aroused irritation in a discriminating clique with, it must be admitted, its own intellectual preciousness. And as professional dance musicians, responding to the American taste, attempted to provide them with a full range of modern hot music from swing to dixieland, and irrespective of whether they could actually play it, it was natural that the more serious devotees of small band jazz would turn their back on an increasingly large proportion of modern popular music as crass and inauthentic. It led to its own excesses, especially among the non-playing fans. Between the musicians themselves there was a certain amount of cross-over, a recognition of jazz talent wherever it was to be found. Stan Chisholm, bass player with the Bells at Leonards, was a professional dance band musician working at the St Kilda Palais de Danse. But the non-musicians, perhaps subconsciously resenting the sight of a skill they did not possess being squandered for commercial reasons, tended, to varying extents, to condemn those associated with the dance band industry. It was they more than any musician who fostered schism.

As the war drew to a close the jazz movement differentiated more clearly into separate strands. The jazz versus swing issue developed considerable heat, especially in the pages of *Jazz Notes*. But if this publication became for some of its contributors a platform for narrow-minded denunciation, for others it provided a medium for discriminating analysis and for national, even international, propagation of the faith. It helped develop a national perspective and solidarity in the followers of 'righteous' jazz. The war itself also contributed to this by throwing together enthusiasts from different parts of the country. Interestingly, a certain parochialism also grew out of the process by causing people to notice not only that there was a range of jazz interest in various cities, but also the way in which the interest exhibited different emphases from place to place. The main contrast was felt to lie between Melbourne and Sydney, and it is a contrast which persists as a perception, if not also as a fact, to the present. Ade Monsbourgh reported in 1945 that Sydney leaned to swing while Melbourne had a stronger 'righteous' element, thus tending to confirm a pattern which had been implicit since the mid-1930s. That is, that the Sydney professional scene was more advanced, more conscious of the technical innovations introduced by developments in jazz and swing, and that there was in Melbourne more in the dance band industry for jazz followers to react against. They would have a clearer idea, by contrast, of what distinguished the music they preferred from the usual commercial fare. Even as the 'righteous' fraternity indignantly distanced themselves from swing, they began to show signs of internal sectarianism as well, as increasingly holy relics were discovered and showed up the comparative inauthenticity of imitations or bastardizations. If much of the debate now seems petty and ungenerously pedantic, it was nonetheless achieving something of profound significance in the evolution of the concept of jazz in Australia. The pages of *Jazz Notes* brought together the kind of discussions that had appeared discretely in professional magazines like *Music Maker*. *Jazz Notes* gave jazz a distinctive status and definition, concentrating attention upon it as a separate and significant muscial form in the larger context of a modern music landscape which was frequently banal and characterless. At the same time, the musicians who were playing the righteous jazz were disseminating a sound that, to intelligent ears, was distinguishable from the small band jive style that most of the jazz-literate professionals indulged in in jam sessions. In the latter, the general approach taken would not have been radically different from the large swing orchestras—solos, riffs, close harmonies, with the same basic rhythm section. In other words, these sessions would encourage the view that there was no great difference between swing dance music and jazz. The righteous fraternity were supporting a style of jazz (and calling it jazz with a purist exclusivism) that sounded more distinctive. In the manuscript of his autobiography, Graeme Bell distinguished between the 'Jump, Rhythm and Blues stuff or small Swing Combo' and what his group was after. The former would operate with two horns and a 'loosely swinging rhythm section'. Bell's idea was the three piece front line in the older New Orleans tradition, playing in improvised counterpoint that was often tense and quasi-discordant, unlike the more seamless unison work of small band swing. The foundation was a solid, 'uncomplicated' beat. It was for the opportunity to play this without any concessions to current dance band fashions that, in Graeme's absence, Roger Bell took the band into the Heidelberg Town Hall in July 1943. It was this kind of music which led Bill Miller to record in *Jazz Notes*, May 1944, that genuine jazz activity had at last begun in Melbourne. In spite of the success of the Heidelberg Town Hall residency however, the music still remained a source of puzzlement to the majority, including often members of the dance band profession who regarded themselves as capable jazz players. Hobart musician Roy Sutcliffe returned from a visit to Melbourne bewildered at the popularity of what he called Bell's 'Dixieland' outfit. The fact that he so termed the band is a clue to why he should be puzzled. The majority of professional dance musicians had not made the same obsessive study of the subject as the righteous coterie, and were often unable to see any significant difference between their own jam session activities, the dixieland features increasingly popular at dances, and what musicians like the Bells were playing. Indeed, often the most apparent difference was that the righteous players were inferior, in terms of formal musicianship, to the professionals who could reel off their version of dixieland higher, faster, more loudly, and without cracking notes. This lack of understanding was not total: it is always essential to remember that there *were* musicians able to move freely between the two camps. But at the same time it was equally true that those camps were being pitched farther apart as the 1940s drew on. The great *cause célèbre* which brought the issue to a focus was the release in 1945 of three recordings—six tracks—of a Sydney band called George Trevare's Jazz Group. Nothing more succinctly summarizes the distance between the jazz stream embodied in the professional swing musician and the one being charted by the mouldy fygges than the different discussions of these records in *Tempo*, *Music Maker*, and *Jazz Notes*. The musicians in Trevare's band were professionals with what would now be considered impeccable jazz credentials, including Wally Norman, Don Burrows, Rolph Pommer. Launched in *Music Maker*, July 1945, as the 'first hot jazz discs' to be made by Columbia in Australia, they were condemned energetically by Ray Marginson in *Jazz Notes*. These records, with their swing riffs, their lithe solos, their tightly parallel ensemble work, are very important as rare remaining examples of what many in the profession took to be hot jazz, and what those in the righteous fraternity positively repudiated. I suggest, in fact, that these records, launched with such fanfare, contributed significantly in the long term to the way in which the Melbourne jazz movement came to regard jazz in Sydney. The furore certainly overshadowed the very important activity at this time of Sydney's Port Jackson Jazz Band, working as successfully in Sydney as the Bell band and in a style so musically and ideologically kindred, that I believe the standard version of the history of the traditional movement as being a Melbourne phenomenon requires revision. But the Trevare material and the publicity surrounding them eclipsed all else in the eyes of the Melbourne jazz writers. Furthermore, the most vocal chroniclers and spokesmen of that traditional movement were in Melbourne. How differently might things now look if Bill Miller had lived in Sydney. To a large extent the issue of the Trevare sides became a matter of Melbourne versus Sydney and amateur versus professional, as much as traditional jazz versus swing.

Part of an all-female jazz band of the second world war period

The situation for Australian jazz at war's end was pregnant with possibility and tension. The simple fact that it *was* ending was significant: musicians who had travelled far and mingled much would return. Material shortages (including that of shellac) would ease, thus stimulating the local recording industry which in turn would spread an awareness of what was happening in Australian music. Imported records would become more easily available, bringing the latest sounds from the US. This was all potentiality rather than actuality. For the immediate moment, the Yanks, their tastes, and their money, were disappearing. Their patronage had changed the character and reputation of many places of entertainment, especially in Brisbane where they had been such a prominent presence. Would the locals return from whatever alternatives they had discovered? And jazz itself seemed, in the scarcely caring view of the public, to be poised on its pre-war base. A jazz player was still the exception rather than the rule in the dance band profession. Rhythm clubs, insofar as they had survived, were more modern music venues than jazz platforms. 'Jazz' itself was still a word barely understood and having the vaguest and most generalized associations. But it *was* 'poised'. The growing traditional-ranging-to-dixieland movement was something that was scarcely visible before the war. And something else was intriguingly hinted as well. In mid-1945 it was reported that sax player Les McGrath gathered some musicians together at a private party and played some strange experiments with unorthodox chords and harmonies on 'Cherokee', producing the impression of constantly moving away from a fixed key.

Post-war to the Advent of Rock 'n' Roll

During the 1920s the public music fashion and jazz were virtually the same thing. With the onset of swing in 1936, and the resurgence of interest in jazz, it was true again that hot music, even if somewhat tepid to an embryonic purist jazz fraternity, was also being played, if only featured, by the more progressive dance bands catering to popular taste. The underground jazz movement which gained strength during the war years gradually came, in the post-war period, to place increasing distance between itself and the established professional music landscape, especially as represented in the dance bands. It is important to remember that Australia was by no means a 'youth' culture. Popular music was still dominated by the tastes of adults settled into stable domesticity, with a fondness for a comparatively sweet style together with novelty material. Glenn Miller arrangements continued to be popular at public dances for the rest of the '40s, and the equivalent through that period of the 'hit parade' was dominated by musically very unchallenging material: cute, sentimental lyrics which made no attempt to question the values preserved at such cost by the war, in harmonic structures of the most naive kind, sometimes little removed from the standard of nursery rhymes. The general level was much lower than that of the popular material of the previous decade and more. No one with the wit, urbanity, and compositional sophistication of a Porter, a Gershwin, a Kern, was able to achieve the same kind of public popularity which they had enjoyed. The period 1946 to 1956 was probably the most vapid decade of popular music of this century. Almost anything by Cole Porter could be resurrected today as a vehicle for the most advanced approaches to music. But nothing from the following list could be revived except as barely credible kitsch: 'I Didn't Know the Gun Was Loaded', 'Manana' (1949), '"A" You're Adorable', 'If I Knew You Were Coming I'da Baked a Cake', 'I've Got a Loverly Bunch of Cocoanuts', 'Be My Little Baby Bumble Bee' (1950), 'Molasses Molasses', 'Abadaba Honeymoon', 'Bibbidi-Bobbidi-Boo' (1951), 'I Tort I Taw a Puddy-Tat' (1952), 'How Much Is That Doggie in the Window' (its double entendre almost universally unnoticed) (1953), 'Never Smile at A Crocodile', 'I Want a Hippopotamus for Christmas', 'Sippin' Soda through a Straw' (1954). Songs like these were the signposts in popular music of a profound public conservatism, and since the professional dance musician is not only himself a member of the public, but entirely reliant on that public for his livelihood, it is not surprising that he too will generally exhibit the same characteristics. Although a pool of potential creativity had accumulated during the war, it was dammed as far as the dance band industry was concerned by the older style groups that had for so long found such success with tested formulae which they showed little inclination to alter. At the same time, as Merv Acheson noted in *Syncopation*, Dec. 1946, the return of ex-servicemen musicians was leading to fierce competition and frequent undercutting, which was being felt especially by some of the lesser musicians who had gained a niche left vacant by enlistment. These dug more firmly into the style that had given them employment, and in many cases, as the only way of holding on to employment, became bandleaders and, as such, dictators of a style which was itself a subservient reflection of public taste. The conservative ambience extended from matters of taste to matters of social values. In early 1946 the secretary of the NSW branch of the Musicians' Union moved a resolution that Communist Party members should declare their affiliation before seeking office. Those who had developed a certain amount of security in the profession during the war now experienced a general anxiety for their position, and this was not eased by the new developments being hinted at in the music industry. Popular music could now be divided into the following more or less distinct categories: commercial material, as represented by the songs listed above; swing, ranging from commercial pop that had absorbed some of the rhythmic feel of the pre-war and wartime swing bands, to the aggressive, stridently brazen and percussive sounds of the new 'progressive' big bands like those of Stan Kenton and Woody Herman. The latter were not popular with the majority of dancers, especially following the departure of the Americans, and most established commercial bandleaders reverted to sweet and 50/50 (old time and modern) during the post-war years, with feature spots for the next category, which was an attempt, and often the poorest pastiche, to present a form of jazz most frequently known as dixieland. It had an air of comparative respectability as long as it was kept in its place, which for the general public was the occasional raffish, high-spirited feature. It had a long pedigree, being recognized as having its roots in the 1920s, and enjoyed an increasing vogue during the war. Throughout the whole country, promoters were finding that in the post-war period, this continued to attract growing audiences and they therefore showed more interest in advertising it as music for public dancing. Dixieland shaded off into the righteous or the mouldy movement, the dedicated retrieval of the most authentic examples of jazz. Its members ranged from positions of discriminating open-mindedness (which generally characterized the musicians) to the most narrow exclusivism. Although this activity was evident in most capital cities, its intellectual centre was in Melbourne. Finally, and embryonically, there was bebop, often associated with the 'progressive' school, and the least publicly accepted style of popular music. In all this, there were tensions as various musical idioms pulled in different directions. Clearly, for one thing, only the more complacent musicians could be satisfied with the current commercial diet. Those committed to the idea of challenge, and especially those who had been exposed to the more stimulating aspects of wartime music, inevitably looked elsewhere. So too did the younger audiences with energy to burn in a settled peacetime environment dominated by conservative adult tastes. While those adults frequently retained rather hazy notions about the relationship between jazz and dance music, to adventurous musicians and audiences, the two bifurcated. As commercial music descended deeper into banality, jazz moved in a different direction. We are here on the brink of one of the most important developments in the history of the concept of jazz in Australia, when it comes to be perceived as an entity distinct from all other modern music. At the same time, a new kind of musician is appearing: the jazz musicians, the men who are seen by the public as specifically concerned with jazz performance. Names like Graeme Bell, Ade Monsbourgh, Jack Brokensha, Tom Pickering began to attract national attention after the war. Simultaneously within the dance music profession itself, some established musicians were being thrown into relief as having important talents as 'take-off' men or jazz improvisers—Alan Brinkman, Rolph Pommer, Charlie Munro, Alf Holyoak, Maurie Le Doueff. Also, among the new generation entering the professional scene of night-clubs, cabarets, dances, there are some who have a specific reputation for their jazz abilities, the successors of Bennie Featherstone, Bob Tough, Frank Coughlan; men like Bob Limb, Errol Buddle, Ken Brentnall, Ron Falson, Clare Bail, Don Banks, Ron Loughhead, Doug Beck. In 1946 we witness the beginning of a significant increase in the definition of jazz by virtue of the widening gulf between popular dance music and jazz performances which are distinct in instrumentation, format, presentation, venue. The term is less frequently misunderstood, and its musicians are becoming more distinct, especially the traditionalists, who are less frequently than the boppers, full-time members of the dance band profession.

It was the traditional style which quickly came to dominate the public perception of jazz. It is a curious and often anomalous strand in the development of jazz awareness in Australia. It was a movement of profound importance to the music, and yet its primary emphasis was not on the advancement of instrumental sophistication. Especially among its non-playing disciples, the major concern was with increasingly discriminated definitions and concepts, with what actually constituted jazz. It was through the public analysis of this matter, in *Jazz Notes*, then *Australian Jazz Quarterly* and other periodicals, that the notion of jazz as a distinct music, and essentially traditional in style, filtered into the broader Australian consciousness. The effect of this movement on the history of jazz in Australia was so profound that it is important to reiterate and summarize much of what has already been dealt with. 'Traditional jazz' is a broad term embracing many sub-categories, including 'dixieland', 'trad', 'New Orleans'. This is not the place to become involved in a debate which is as widely geographically dispersed as jazz itself. By 'traditional' here, I mean, more or less, loosely structured, collectively and individually improvised small group jazz, usually with up to three or at most four front-line instruments, up to three or four rhythm instrumentalists; sparing use of riffs and unison work, and the main format being alternation between free ensemble and solo work. Traditional jazz, as thus loosely defined, arrived in Australia as we have seen towards the end of the First World War. Most dedicated professionals registered the fashion, without necessarily understanding its harmonic principles.

'Hot' music died around 1929, but staged its comeback in big and small band swing. Modelled on groups like Goodman's small bands, the latter involved anything from trios to sextets, but most frequently three-piece rhythm and two-piece front line using a reed and a trumpet. The music thus produced differed from the jazz of the 1920s in featuring tighter heads (that is, co-ordinated parallel lines in the ensembles; in fact very little collective improvisation, thus foreshadowing a major characteristic of bop). Its improvised solo work showed greater facility than in the '20s, and a diminution if not disappearance of novelty effects. To the Australian musicians, this represented a new challenge, in particular the harmonic one. This kind of music would be featured as a component in a dance programme in the late '30s, but again, jamming was more a feature of parties or swing club functions.

The other main reincarnation of 'hot' music was dixieland, which was often regarded as the legitimate heir of the jazz of the '20s, truer jazz, by those discriminating enough to make the distinction. There was also the black jazz of the '20s. This began being heard towards the end of the '30s for the first time in Australia, picking up especially with the record reissue programmes during the war. It was perceived in very different ways. The public, if exposed to it, found it barbaric. But it was not a public music; it was underground, tending towards cultism. The professional musicians presented a wide range of responses. To one end of the spectrum it represented the worst features of '20s jazz: discordant, coarse, incipiently immoral. Professional dance musicians had learned some interesting techniques from it, but had passed beyond its adolescent excesses to a stage of superior refinement and respectability. Towards the other end of the spectrum, there were professionals who showed interest, but always with a legitimately trained professional ear forming judgements, and holding the thing at arm's length in accordance with their responsibilities to supply a popular market. These musicians could hear the superior instrumental sophistication of, say, Goodman as compared with Cecil Scott. Their professionalism often overrode their sensitivity to the raw spirit, the folk appeal of this music. With hindsight, it seems therefore inevitable that the earlier black jazz (ten to fifteen years behind latest developments by the time Australians were first hearing it), would appeal most strongly to non-professional musicians, ears which didn't have as a professional necessity such a highly developed susceptibility to matters of technique and which therefore had nothing to obstruct their response to the spirit rather than the letter. Even within the Bell group, it was Roger, the relatively untrained musician, who finally converted an initially unreceptive, legitimately trained Graeme to the qualities in the music.

Thus, within the broader dixieland movement there evolved a group with what amounted to an ideological commitment to certain manifestations of traditional jazz, especially the classic corpus of the '20s, centred on Chicago. These enthusiasts brought something to the appreciation of the music which most professionals had not: an intellectual consciousness, a historical sense, an attitude about the place of jazz in the larger artistic and cultural picture. In particular they sought to distance themselves from the commercial music being peddled under the name jazz. While representatives of the profession often berated those 'juvenile prodigies' who were dragging the music back twenty years, praising 'feeling' at the expense of musicianship (*Music Maker*, March 1947), the defenders of the music pointed out with some justification that instrumental sophistication did not guarantee an authentic jazz sound and that much so-called dixieland being presented by dance bands was a superficial caricature relying on exhibitionistic effects which, if instrumentally facile, were aesthetically clumsy. At the outset this movement was open-minded and musically broad based. The Bell coterie drew its inspiration from a wide range of traditional jazz: Muggsy Spanier's so-called 'ragtime' sessions from 1939 were extremely influential notwithstanding the musicians' colour, the comparatively recent date of recording, and the unashamed dixieland flavour (dixieland was not a very acceptable term to the righteous following). Equally, there was musical mingling with members of the dance band profession who demonstrated an intelligent interest in jazz, as well as with that group that was becoming the nucleus of the bop movement. Not surprisingly, with their anti-commercial posture, the traditionalists included in their image the somewhat radical complexion which had earlier allied them with the Contemporary Art Society. The souvenir programme of the first Australian Jazz Convention, discussed below, was issued as an *Angry Penguins Broadsheet*. *Angry Penguins*, edited by Max Harris and Harry Roskolenko, was as trenchant a voice for modernism in literature as the Contemporary Art Society was in painting. The broadsheet included essays analysing the history of various aspects of jazz; they show an impressive conceptual grasp of what was, after all, still a relatively undocumented musical phenomenon, and the general tone is compatible with Harris's anti-philistinism. The radical intellectualism of members of this group was also advertised in its association with the communist Eureka Youth League which, through the Eureka Hot Jazz Society, sponsored jazz activity, including suggesting and subsidising the Bell band's tour to the World Youth Festival in Prague in 1947.

As seems to be almost inevitable with intellectually self-aware movements, the traditional fraternity gradually attracted increasingly purist and exclusivist attitudes. The term 'mouldy fygge' referred to those who narrowed their range of interest in their search for the most authentic jazz. Like the proliferating sects in the wake of the Reformation, the movement gradually turned in upon itself. In subsequent years Graeme Bell was attacked by Frank Johnson for having ceased to swing, and then, extraordinarily, Johnson in turn was condemned as a bopper. Increasingly in the late 1940s and 1950s any divergence from the strict patterns laid down by the black bands of the classic period was a heresy. The main forum for this ideology was the *Australian Jazz Quarterly*. This is not to say that the *Quarterly* itself was thus aligned editorially, but that it gave space to a range of opinions which included the mouldy position. Its spokesmen were active and important in flying the flag for major jazz masterworks of the '20s at a time when the general public still identified jazz with 'The Golden Wedding'. At the same time, however, they unwittingly did some damage to their own music in the long term. Their refusal to countenance any jazz beyond certain conceptual and historical boundaries effectively alienated many musicians interested in expanding their vocabulary, and I believe there is a serious danger that this 'wall' which some influential proponents of the mouldy fygge sensibility built around themselves could become an influence in limiting the duration of the traditional jazz movement. It is also likely that the suspicion in which many professional musicians hold any attempt to deal intellectually with jazz was exacerbated by this righteous movement. The first sustained attempts at intellectual analysis of the music were conducted by traditionalists, many of whom alienated professional and modern musicians by their strident narrowness. The distrust of intellectualism thus engendered has made the task of documenting Australian jazz history unnecessarily difficult, in some cases quite impossible. (It should be added that this distrust has been compounded by what is almost an opposite problem. Intellectual analysis expresses itself most overtly in print and talk, not music. Most of the print and talk about jazz has been conducted by non-musicians, and in particular journalists who often have little knowledge of jazz and no understanding at all of why a person plays music. They have inspired considerable contempt among musicians. What musicians have most commonly seen in print, then, is either pedantically narrow or hopelessly ill-informed.)

While the traditional style was enjoying increased attention both from the general public who absorbed it as something hazily called dixieland, as well as from a more discriminating following, another specifically jazz phenomenon came into clear focus immediately after the war. Within any profession there are those who wish to keep abreast of the latest developments, even if those have not yet become publicly marketable. The rising vogue for dixieland provided some stimulation but it was not precisely a new development and was never seen as such in the post-war period. The latest challenge was 'progressive' music (as represented by Stan Kenton) and, even more, 'bebop' or 'rebop'. Here was another movement which presented distinctive features, not only in terms of its inherent musical characteristics, but for the way it related to the larger musical context. The young musicians who entered the profession with the bebop style were playing in a way that was never really assimilated into the doomed public dance scene, and that was another reason why these players often had a more distinctive jazz reputation than earlier dance band musicians who could also play with a strong jazz tinge. Someone like Reg Lewis could play in a jazz style during the '30s, but it overlapped with swing. The bebop which developed in the '40s was felt to be undanceable, and therefore those who played it have since been seen more specifically as jazz musicians rather than as dance band musicians who could play hot.

One of the early champions of the new music was Wally Norman, who in August 1946 defined its characteristics as: fast, with the rhythm maintained on the drummer's high-hat, the bass holding a consistent four to the bar (that is, instead of two to the bar which dance music and dixieland often revert to), the piano playing broken accompaniment, with displaced front-line phrasing and unusual chord substitutions (*Music Maker*, Aug. 1946). It is also interesting that he received the music as though it were a fully articulated form with premeditated rules, rather than as an evolving form. For this reason, he and other Australians then and subsequently tended to overestimate its revolutionary character. In spite of the American presence during the war, Australia was isolated from the unfolding developments taking place at the experimental level in the USA. There *was* continuity between the 'jive' music of the late 1930s and bebop. We are able to see that more clearly as the latter has been conceptually accommodated, but when the first records of bop began to arrive in Australia, our preceding period of musical quarantine left us comparatively unprepared. Accordingly, bop's distinctiveness from other modern music, including dixieland, was accentuated, so much so that it was frequently denied the status of jazz (a situation which has never altered in the minds of many followers and even players of traditional). The practitioners of bop can be distinguished from other current and previous jazz interested musicians in another way, however. They represented the first concentrated group to focus on the very latest innovations in the music simply for the sake of their aesthetic interest, their inherent harmonic and rhythmic character, notwithstanding the almost total lack of commercial appeal. The first jazz phase of the '20s and the swing fashion had a commercial incentive at least as strong as the musical challenge. Even the jamming scene of the late '30s and the war had a following, and in any case involved more the increasingly competent implementation of established concepts rather than the working out of what were felt to be radical new ones. The traditionalists never saw what they were playing as musically revolutionary; their radicalism was more intellectual, a matter of the way they perceived a commercially enslaved music industry, and their solution was to reach back to a more authentic tradition. The boppers were seeking to push musically forward in defiance of, not in pursuit of, a popular fashion. This too helped to underline the distinctiveness of their music. Its separateness from dance music (and therefore from dixieland which, for the public, was something featured at public dances) was further emphasized by the fact that it was nurtured mostly in the coffee lounges and cabarets of the period. The former, particularly, underlined the status of bop as something to be listened rather than danced to, and in fact helped to establish the ambience which has continued to surround the music. In Adelaide Jack Brokensha, with Clare Bail, Ron Lucas and John Foster, opened his own coffee lounge in October 1946 specifically to create a venue for 'progressive' music. Brokensha's most important activity in Melbourne in 1947 as a focus for the new music was likewise at the Plaza coffee lounge. The connection between this music and this style of venue was so clearly established in the public mind that when Brokensha's band played a Swing Show concert in Adelaide in late 1947 or early 1948, his contribution was reviewed as 'a coffee lounge' segment (*Music Maker*, Feb. 1948). Although short on commercial appeal, bop was penetrating the larger community consciousness and in the late '40s, lay journalism carried the occasional article, naturally playing up the newsworthy eccentricity of the music, but as always, having a strong influence on community perceptions. The magazine *Glamour* in September 1948 noted that the new American music ('bebop'—the term 'rebop' was now square, unhip) was revolutionary, discordant, unsuitable for dancing, and not only non-melodic and formless, but any hint of recognizable melody was frowned upon. It was also, the writer noted, confined to jam sessions, and was chiefly to be found in Sydney. From the other side of jazz awareness, Bill Miller, speaking for traditional jazz followers, would probably have agreed with most of that analysis; certainly he divorced jazz from 'rebop' (writing in *Australiam Jazz Quarterly*, October 1948), the latter being derived from mere 'swing riffs', with empty displays of technique and melodic incoherence.

There is one further point which must be made in connection with the early experiments with bop. Some of the most independent minded of these were conducted by a group that included Splinter Reeves, Don Banks, Charlie Blott, Lin Challen, Ken Brentnall, Doug Beck. Among them are some of the major pioneers of bop in Australia. Yet in spite of the chasm which so many wished to dig between the traditional and the modern camps, all of these musicians had been to varying extents actively involved in bands led by Graeme Bell during the war. There is no question that for most of the public, for many jazz followers, and indeed for a certain proportion of musicians, the chasm existed. But it is equally important to note that the musicians held to be at the centre of each camp frequently felt no constitutional alienation from each other. Perhaps, unlike others, they were too secure in the knowledge of the honesty of what they were doing to feel threatened by the pluralism of the music. In any event, this circumstance further strengthens the proposition that much of the profound difference perceived between bop and the other jazz and jazz-inflected music of the time lay in the extrinsic or incidental features discussed above rather than in intrinsic musical properties. Many 'bop' solos of the time would scarcely sound out of place in a swing or even dixieland group.

1946 was a big year for both these developing jazz streams. For traditional jazz, much was brought to a focus at the 1st Australian Jazz Convention in December, held in Melbourne. Apart from creating publicity for the music, it inspired those participating and also created an *esprit* among the musicians which transcended statehood while at the same time bringing regional stylistic differences into view. Following this event, jazz musicians in most cities knew they were part of a national movement. 1946 was also a boom year for reissues of recordings and for local jazz recording, the latter on the new Ampersand label which now made its first releases. Bop saw a similar efflorescence, gaining in magnitude throughout 1947. By 1948, the notion of jazz had gone a long way to detaching itself from the established dance band scene. The central antithesis in the music was in fact not between jazz and popular, but between two styles of jazz: traditional and bebop. Jazz had gained an unprecedented visibility and definition. More people, especially in Melbourne and Sydney, knew what they meant by the term, albeit somewhat narrowly, than ever before. In the past (and still, to varying extents, outside the major centres), there had been a tendency to think of all modern dance music as jazz, and many professional dance musicians in the '30s thought of themselves *ipso facto* as jazz musicians. The full range of modern musical entertainment could be comprehended under the heading of jazz. By the late '40s there was a substantial fraternity for whom the matter was the other way round, and many stridently dismissed huge tracts of music as having nothing to do with jazz. There was still public confusion of course; it still persists. But the increasingly educated jazz following, although often repellantly narrow-minded, were giving a new distinctiveness to the concept. As we shall see in connection with the imminent jazz concert scene, the notion of what jazz was still ranged from any high energy music to a revered, tightly circumscribed folk art form. But a discriminated understanding was spreading. Dixieland or traditional, combining a nostalgic appeal, a comparative musicial simplicity, a danceable rhythm, had the larger lay following. The most usual inclination in the public mind was to think of jazz in terms of this style, partly because the first bands to advertise themselves as jazz groups, and the first unalloyed jazz performance programmes were thus. Bebop was still not thought of as jazz in many quarters, not only because of its unfamiliar source, but also because it happened to have this other name. The main bop spokesmen were within the professional ranks, though the majority of professionals had not proceeded beyond swing, with special reference to Glenn Miller. They came to prominence on that style, and receded from view as swing began to give way to dixieland and bop or progressive. Towards the end of the '40s, various small signs cumulatively pointed to the passing of a major phase in popular music. The big dance bands began to be reduced. In 1947 Clarrie Gange left the Melbourne Trocadero after five years when asked to take a cut in salary. Strings virtually disappeared from the dance bands. The brass band, before the war a cornerstone of local music, lost its popularity and *Music Maker* dropped its brass band section. During an evolution of Jazz concert in Sydney in 1948, a Goodman-style item received no audience response. In the same year The Royale band in Hobart was rent by hostility between the 'jive artists' and the older musicians who just liked to read the written parts. A new era was beginning, and a new generation of musicians was coming into visibility with it—amateurs and professionals attracted to traditional, and others, mainly professionals, with an interest in bop, the current musical challenge. The latter required a high degree of theoretical and instrumental sophistication, and was very much musicians' music (and only the most adventurous musicians). Although successors to an earlier generation of cabaret and dance musicians, these young men would become identified more specifically with bop (or jazz, as bop gradually became recognized as such). There were now, in the late '40s, tensions, lines of demarcation which gave clearer definition to jazz. First, some tension between older professionals—accomplished musicians, but with a rather dated sense of hot playing—and young turks who wanted to jive it in a way that bewildered and irritated their seniors (as, in some cases, it was intended to do). There was tension also between bop and dixieland at certain levels: although many musicians were not fussed by the issue, some were, and many followers especially assumed antagonistic postures. This confrontational image is more powerful in the post-war period than earlier, and is not the same as the much older debate between conservative and modern dance music. It is between two forms of modern music, one outrageously new (as it seemed in Australia), the other, the legitimate contemporary manifestation of jazz. The opposition was often fostered on non-musical grounds, such as for entrepreneurial reasons, as in the 'band battles' looming. But in any case, 'jazz' was central and distinct. This was also because it flourished in new venues. The old theatres, cabarets, and dance halls generally continued as before, presenting swing and sweet and attracting the same clientele. The new vogue found new venues. There was some exposure in the clubs and cabarets, but the audiences were not hip. Furthermore, the hip audiences were neither wealthy, nor were they very interested in alcohol-centred night life: they were youthful, in a middle-age dominated society. It was out of these circumstances, among others, that the jazz concerts grew.

The 'jazz concert' phenomenon was, as far as the public and many musicians were concerned, the main national jazz platform for a decade after the war, reaching its peak in the early 1950s. In 1947 there were concerts in Melbourne and Sydney which were causally unrelated, though they all featured jazz. Graeme Bell performed a concert in Sydney on 9 April, playing, in the words of a reviewer, 'pure unadulterated Dixie' to a 'youthful' audience (*Music Maker*, April 1947). On 23 May a fund-raising jazz concert was presented at the Brunswick Town Hall, Melbourne, to raise funds for an Australian delegation to the World Youth Festival in Prague. Although it was billed as 'The Story of Jazz', the posters listed only traditional bands. In August a parallel but stylistically separate event took place at the New Theatre in Melbourne. Two groups of progressive musicians who had been holding jam sessions in their homes had acquired such a following of listeners that they hired the threatre and gave a concert. The musicians included Don Banks, Bob Limb, Jack Brokensha, Charlie Blott, Ken Brentnall, John Foster, Doug Beck, Splinter Reeves. In March 1948 the Port Jackson Jazz Band gave a concert at the NSW State conservatorium of Music with the blessing of Sir Eugene Goossens. Wherever or whenever a witness would wish to set the beginning of the concert era, the first Battle of the Bands promoted by Ellerston Jones in the Sydney Town Hall signalled that it was under way in early 1948. On that occasion, the town hall was filled to capacity.

Who were these people who suddenly were found to constitute such a large and enthusiastic market for what was billed as jazz? Again, we must remind ourselves of the alternative styles of live musical entertainment to locate the answer. The old established dance bands had, since the war, scarcely altered their approach. This was partly because the most dynamic innovations in popular music were considered undanceable by the settled middle-aged clientele of the public dances. But this audience was growing older and less committed to recreations like the big dance halls. The business entered a slump, the old ballrooms went into a decline that would lead ultimately, in most cases, to their demolition. The suburban dance circuits became the centres of middle-of-the-road dancing, using a pool of small orchestras. Although later in the '40s and '50s they included the occasional jazz band in their stables, having seen the drawing power of the word, there was very little in the public dance scene of this period to attract the young. The jazz concert filled this demand. As it developed, the whole spectrum of jazz styles was presented, so that a wide range of youthful audiences was attracted. The relatively uninitiated young could relate to most forms of jazz, and it was this audience, one which made no strident distinctions between 'true' jazz and everything else, that made the concerts financially possible. Here, the music denied them at dances could be brought together under the one banner of 'jazz' (or sometimes 'swing'—they didn't care too much). The concerts brought in an audience that wanted the excitement of jazz—young people bored by the middle-of-the-road dance bands, but stimulated by the swashbuckling character of dixieland, the defiance which progressive music showed of decorum; the romantic image of the soloist going for it, the gladiatorial character which characterized the publicity surrounding these 'battles of the bands'. And by placing their vote for the best band in a ballot box as they left the hall, they could vicariously participate in that competitive spirit.

The growth and spread of the 'jazz concert', under that or similar names, was extraordinary. Bill McColl promoted regular concerts at Sydney Town Hall, but there were other regular functions at different venues like the Assembly Hall, to the extent that at their peak there would be an average of at least one each week and often more. In Melbourne the New Theatre concerts continued until what was ostensibly overcrowding (though in fact other reasons were involved) forced a move to the larger Princess Theatre. The backbone of the Melbourne concerts became the Downbeat series promoted by Bob Clemens. In Adelaide Bill Holyoak promoted the Swing Shows, in Perth Harry Bluck promoted the Jazz Jamborees (although these were comparatively infrequent, they came to be supplemented by similar functions including concerts at the Playhouse and the Youth Australia League), and in Brisbane there were Jim Burke's Public Command Concerts. The nearest regular equivalent in Hobart was the long and successful Pickering residency at the 7HT Theatrette which began in 1949, though this was more a jazz dance than a concert. But the ones

Programme for a Bill McColl Jazz Concert, 27/3/'50

listed were only the most prominent functions in a national movement that became the primary form of musical entertainment for the younger generation, on a par with the later discotheque phenomenon. It was the most public forum for jazz. What is relevant here of course is the question: what sort of music was being presented under that billing? It varied from city to city, depending on the progressiveness of the local musical awareness. As a general principle it is fair to say that, outside of Sydney and Melbourne, the percentage of jazz diminished in favour of variety and novelty material. At the same time, even in the two largest capitals there was some leavening of the jazz content, and that even the jazz itself was often more in the nature of caricature: it very soon became apparent that this youthful audience would respond most enthusiastically to obvious effects, so, as reviewers lamented frequently, anything played loud and fast had an inbuilt advantage when it came to votes and applause. For all that, however, these concerts did present the fullest possible range of styles that the broadest sympathies might wish to call jazz, from swing in the tradition of the late '30s, to progressive small bands in both the advanced swing mould of John Kirby and the more explicitly bop style. And dixieland or traditional was a perennial feature, together with a style owing much to rhythm and blues and looking forward to the advent of the earliest semiacoustic rock 'n' roll. (I suggest that the performances of groups like those of Les Welch so habituated audiences to hearing this kind of music under a jazz banner that, initially, rock 'n' roll was rather easily assimilated into the jazz concert format.) Among these styles, dixieland tended overall to be the most popular, although there were certain feature numbers presented by other bands that could always be counted on, almost ritually, to bring the house down. These, like 'Boogie Blues', were accordingly played to death. The vote in the first Battle of the Bands in Sydney was more than twice as much for the Port Jackson Jazz Band as for its nearest competitor. Similarly, Frank Johnson's band won in Melbourne three times in 1949. The relevant point is that these jazz concerts made jazz (thus advertised) the major youth music from the late '40s to early '50s. Furthermore, once again traditional jazz was the most central style in that public image, but an image which also extended to include other forms on the same stage, as well as implicating the music in aspects of comedy, variety, quasi-vaudeville performance. Doubtless the public didn't assume that a man telling *risqué* jokes was a jazz experience, but it certainly encouraged them to assume a spirit common to the two. Jazz remained essentially *déclassé*, racy.

The effects of this phenomenon on the state of jazz in post-war Australia were profound. At the most obvious level, it extended that tendency to give the music public prominence which had started with swing in 1936. This was a process further encouraged by Rex Stewart's tour in 1949. The word 'jazz' on an advertising bill became a crowd-puller, even though within the informed ranks there was still bitter debate about what the word applied to. Jazz was pulling interest away from the commercial dance band industry. Frank Johnson presented the ingenious but unconvincing theory in *Australian Jazz Quarterly*, Feb. 1949, that bop was being pushed as jazz by commercial interests in order to bring the latter into such disrepute that it would cease to compete with commercial music. Other venues and media for music began to capitalize on its popularity. The Sydney Trocadero advertised Thursdays as jazz concerts in 1949 (though in fact the music was heavily influenced by Glenn Miller and was interspersed with cowboy novelties and Jerry Colonna imitations). The High Hat dance circuit signed the Riverside JB and the Port Jackson JB for guest spots at their suburban dance halls. The jazz concerts also created a context in which participating musicians could establish a public reputation as *jazz* players as opposed to dance or cabaret performers. The centre of jazz activity was shifted from private or after-hours jamming to this new scene in which particular musicians were identified with jazz, and this also further widened the gap in the public perception between jazz and other forms of modern music heard elsewhere. Many new bands and musicians rose to national prominence on this wave.

Inevitably this all led to a certain amount of opportunistic bandwagon-jumping. An attempt to promote a jazz festival in Brisbane in 1951 foundered under the weight of groups seeking to participate, but only able to deal with sweet stock orchestrations. This was not just an attempt to get aboard something that was doing well: it was also often a case of abandoning a sinking ship. The popular and dance music industry descended into a serious slump from 1949. The popular scene had fallen to a nadir of blandness, in response to which various fads were taken up in an attempt to revivify the business.

Hawaiian and Latin American music were among those which enjoyed brief vogues, and probably the most durable of these ephemera was square dancing. By 1953 this was enjoying a brief burst of popularity so widespread that callers were making small fortunes, compared with standard entertainers' rates of the time. It attracted an age group concentrated on mid-twenties to forties, and fulfilled some of the function of the sentimental fashion of the Depression years, even though economically this was something of a boom period. In the climate of the Cold War, however, square dancing asserted traditional values, its patterns created a feeling of belonging to a larger community with agreed standards, it combined elaborate order with conceptual simplicity, and at the same time was a novelty yet without threat. I spend a moment on this because it has a prophetic bearing upon the way modern popular music was moving, and upon the position of jazz in that movement. Square dancing could hardly have been farther from the aggressive, individualistic, abandoned energy of the jazz concerts and the jitterbuggers who, in time, took to dancing in the aisles, necessitating police intervention on at least one occasion. The gap between square dancing and the jazz concerts also divided along generational lines, foreshadowing a youth culture and a music industry developed so specifically for it that it bore no trace of the history of local popular music. It even adopted a new name: rock 'n' roll.

In the meantime, from the late 1940s the commercial music industry was less and less able to support those employed in it. *Music Maker* noted in February 1949 that the dance hall business was being killed by a 30 per cent entertainment tax—one either absorbed it, narrowing the profit margin, or passed it on to clientele and saw attendances therefore drop. Theatre bands were being laid off, to be replaced by solo organists. A prolonged coal strike affected all entertainment, to particularly damaging effect if it was a sector already weakened. One of Queensland's most venerable ballrooms, the Coolangatta Jazzland, folded up in 1950, and in the same year many city bands began seeking work travelling in the country. A Royal Commission into the liquor trade in New South Wales in 1952 caused destabilizing anxiety, and the situation for professional musicians was not helped by the introduction in the same year of import restrictions on instruments.

There was every reason for musicians to look at what was going on in the jazz concert phenomenon and to wonder how they could get a share of the cake. But, as we have seen, there had been increasing bifurcation between jazz and commercial music since the war, and the number of men employed in the latter area who could convincingly play the former was not great. In the smaller capitals many of the attempts to stage jazz concerts using the local dance band personnel produced unsatisfactory results. In this situation there was another prefiguration of the future; we are approaching another watershed dividing two eras. On the one hand, there was that era, slipping into the past, when an ability to play jazz was not necessary in order for a popular musician to rise to the top ranks of his profession. In the future lay that era in which this was reversed. Today, most top ranking professionals are able to deal with jazz—improvisation, chord substitution—frequently more competently than those who are identified as jazz musicians. This is a reminder that the most important developments in modern popular music have come from jazz, so that while one might not be working in a perceived jazz context, one is most employable if able to play jazz.

As this watershed approached, however, the attempt by many otherwise competent dance band musicians to jump on the jazz concert bandwagon led to a drop in musical quality. Naturally these lapses of authenticity did not go unnoted, especially by that fraternity of dedicated and knowledgeable jazz lovers that existed as a separate body within the larger youthful jazz audience. What they were seeing was an equivocal situation. On the one hand, the word that described their favourite music had achieved unprecedented publicity. On the other, much non-jazz activity was being indulged in under its name; musicians who could only produce a caricature of jazz were being applauded just as vigorously as others with more integrity and less flamboyance. As in the past, this was a more vocal issue in Melbourne than elsewhere. As we enter the 1950s, the most vociferous spokesmen for the mouldy fygges in Melbourne were Tony Standish, Bill Haesler and Norman Linehan. Although the overall editorial policy of *Australian Jazz Quarterly* was relatively liberal in stylistic terms, it also incorporated much of the writings of these spokesmen and therefore contained the most comprehensive evidence of their opinions. By now they had narrowed their focus to such an extent that very little, if any, music not played in, or by black natives of, New Orleans was to be countenanced under the label 'jazz'. In September 1951 Standish dismissed all white bands (including that of Graeme Bell and one of the most evangelically mouldy of the early Australian groups, the Southern Jazz Group from Adelaide), he denied that Ellington, Calloway, Lunceford, *inter alia*, led jazz bands, and proclaimed that only New Orleans produced jazz: Dodds, Bunk Johnson, early Louis Armstrong, Morton. The flame was now being carried in Australia by a new generation of musicians (even though they *were* white) among whom Standish included the Barnards, Max Collie, Wocka Dyer, Frank Traynor, Geoff Kitchen, and Peter Cleaver. They now represented the local manifestation of authentic New Orleans jazz, and the idea of 'Australian jazz', in terms of a distinctive style, was a myth.

Thus, a schismatic spirit persisted with particular abrasiveness, deep within the ranks of traditionalists. It was scarcely visible to the audiences supporting the jazz concerts and cheering the dixieland bands, and it continued to be sustained more by non-playing scholars than by musicians. The situation of which the mouldy movement was the centre is in fact far more complicated than the publicly visible tokens of its activity would indicate, and it is an area of Australian jazz history which requires a study of its own. The gradation of attitudes was not only of vernier-scale subtlety at any one moment, but constantly changing as well. There were those who simply ruled out of court on an absolute basis all music not consistent with their preferences. There were others who, for example, simply felt that different kinds of jazz should be segregated. On occasions these attitudes, although different, would find common cause, as when AJC committees censored or censured peformances felt to be inappropriately traditional in style. Some of this disapprobation was based on a dislike of modern jazz, some of it came from people who enjoyed it, but felt the AJC was not its platform. Both manifestations of this attitude, however, could look the same in reports published in journals insensitive to the finer points of the debate. Evidence of this tangled situation came out of the AJC of 1952, at which Bell's band was dismissed by Linehan as being 'the worst exhibition of taste', partly because it was comparatively large, and because it interpolated written arrangements between the solo sections (*Music Maker*, Feb. 1953). The suspicion of Bell, which congealed at that AJC, and the identification of a new generation of traditional jazz musicians more acceptable to the ideologues of the movement, define the end of its first phase. It has been noted that during the late '30s and early '40s, jazz activity was not so easily differentiated into disparate styles: simply having a talent for jazz was enough, and one would not expect to be interrogated over minutiae of stylistic preferences. In retrospect, however, it was, by the early '50s, possible to see that a distinct strand of jazz development had begun around 1940 characterized by a disposition in favour of traditional, and having a particular appeal to intellectuals. The controversies arising from the 7th AJC could be seen as the beginning of a new stage in its development.

In spite of sometimes internecine disputation, however, traditional jazz, or dixieland to use the term by which the public more casually referred to it, was still the main flag-waver for jazz. It continued to increase in popularity throughout the early '50s. By 1952 it was estimated that Frank Coughlan's band at the Sydney Trocadero was devoting a third of its programme to dixieland; in the same year it was reported that virtually every show presented in Perth was billed as jazz, by which was understood the traditional variety. The general tendency, both within the fraternity and on the part of the public at large, was to make this equation. Within some sections of the profession this style of jazz attracted some contempt for its comparative simplicity, but musicians like Charlie Blott, Splinter Reeves, Ken Flannery, Graeme Bell, Jim Somerville, represented the broader view which seems to characterize those who actually play jazz more often than those who only talk about it. It was a view which enabled them to draw upon jazz talent wherever they found it, and often alienated the fygges. In its most studied and pure forms, traditional jazz was still somewhat hidden from the public view, which tended to think more in terms of the style represented by Bob Crosby's Bobcats. At both levels (that is, public and initiated), however, it flourished, in the continuing jazz concert scene and with the foundation of the instantly successful Sydney Jazz Club in 1953.

Because of the range of music presented at the jazz concerts, other forms of the music also enjoyed increased exposure. The short-lived magazine *The Beat* and many of the releases on Bob Clemens's Jazzart label both reflected and contributed to the dissemination of bop. As always of course, there was public confusion as to what was implied in this new music, and various forms of histrionics were often more eagerly applauded than demonstrations of genuine cognizance and skill. It was reported in *Music Maker*, March 1950, that even in America the term 'bop' was so imperfectly understood that Muggsy Spanier, Armstrong, Jimmy Dorsey and even Sammy Kaye had been accused of indulging in it. The following month saw another instance of the frequent assertion that it was an 'aberration' in jazz. Misunderstood or not, the first phase of bop in Australia was comparatively brief, and, in retrospect, clearly bounded both in terms of periodization as well as in terms of the musicians involved. The controversies over the music died down in the early 1950s, which was a sign not only that many aspects of it had been accommodated into the musical consciousness, but also that its first highly visible and distinct vogue was fading. The departure of Don Banks for England in 1950 deprived the movement of one of its primary local sources of inspiration. Jim Somerville, a musician of very broad sympathies, observed that bop and the Kenton style were in retreat by August 1950, and certainly by 1953, the more overt bop mannerisms were scarcely in evidence as such on the concert programmes. Intimations of the next phase of progressive or modern jazz in Australia surfaced in the mid-'50s (by which time Wally Norman, a self-proclaimed pioneer of bop in Australia, had for some years been earning his living from swing era music) as names like Brubeck and Shearing began to engage local interest. But we are talking here of a style which, while drawing upon bop innovations, has qualities which differ significantly, including its greater accessibility to a lay public. The first phase of bop in Australia was so brief in terms of public visibility, its devices so discontinuous with what had preceded it in Australian jazz, that it had little opportunity to establish its credentials as jazz. This was exacerbated by the determination of many of its lay disciples to distance themselves from the rest of the music scene as being square. One consequence is that for many years, and even down to the present, many who regard themselves as well-informed jazz followers categorically dismiss bop as having nothing to do with the music.

In the larger historical context, the jazz concert industry peaked at about the same time as bop. In Melbourne, the decline began a little earlier than it did in Sydney; indeed, there was a distinct feeling at the time that the jazz centre of gravity was moving from the former to the latter. Writers were reporting a slump in Melbourne in December 1950, aggravated by the usual fall-off in support experienced over summer. In 1951 the number of town hall jazz concerts was down—and one in September included light classics from the tenor Albert Argenti. Geoff Kitchen dissolved his band for lack of support. Although the Victorian branch of the Musicians' Union decided to undertake sponsorship of jazz concerts in 1952, in retrospect it appears that their conservatism prevented them from catching the movement at its flood. Holding the 5th AJC in Sydney (the first time it left Melbourne) helped to shift the focus, and generated considerable local interest. By the early 1950s Sydney was being widely regarded as the centre of Australian jazz activity, and was certainly pre-eminent in terms of professional musical opportunities in general. Similarly, it was where the jazz concert scene lasted a little longer, though its content as such would become increasingly debatable. The dilution of jazz material was reflected indirectly in a reviewer's praise of a Sydney Town Hall concert in mid-1951 on the grounds that it had no vaudeville or night-club acts, that it was pleasingly reminiscent therefore of the old Ralph Mallen days. Material presented in concerts that year included comedy, hit-parade material like Frankie Laine's 'Jezebel', light classics such as de Falla's 'Ritual Fire Dance', and Latin American music from Ernesto Rittez (a.k.a. Ernest Ritte), which a reviewer felt was more appropriate to cabaret. By December a writer in *Music Maker* estimated that McColl's concerts were now 75 per cent variety and commercial pops.

Similar complaints were being aired in other cities, and frequently the promoters agreed that the 'jazz' billing was inaccurate, but claimed that while current audiences would be drawn to such a bill, they would not in the event want to hear a straight jazz programme. It is possible that, in addition to the sheen having worn off the idea, National Service was altering the composition of available audiences, and that middle-of-the-road programmes were aimed at bringing in older people. It is certainly difficult to imagine young bodgies and widgies jitterbugging to the spectacle of mannequin parades and burlesque, reported as part of the August City Hall concert in Brisbane in 1953. At the same time, the jazz itself began to grow stale. Towards the end of the bop phase, the music as presented at jazz concerts had frozen into mannerism. Big band material was becoming dated and repetitive—material presented by Bob Gibson's band at the Perth Jazz Jamboree in 1953 included 'The Golden Wedding' and 'Sing Sing Sing', both more than a decade old, and the dixieland offerings were often rather tired war-horses based on the fast/loud formula. The concerts were still making money, but their offerings were increasingly a combination of predictable jazz, commercial music, and non-musical components. It appears that many non-jazz performers were jumping on the band-wagon and presenting a pastiche of sentimental or histrionic mannerisms under the name of jazz.

The more discriminating enthusiasts were understandably offended at the contamination. The concert scene was approaching a crisis. The youthful jivers were still in attendance, but there was a tiredness about the whole thing. The attempt to revivify the programmes only led to more bombast in the 'jazz' component, and more non-jazz. There was a real danger that the advances made since the war in the public perception of jazz would be lost by the increasing connection between the 'jazz' billing and the vaudeville, variety, and cliché-ridden music actually presented. Promoters saw audiences falling off, and tried new approaches, including new billing altogether such as 'Parade of the Disc Jockeys' and 'Parade of the Singers'. In terms of jazz, the most significant of these new strategies was the imported concert package which McColl inaugurated to great acclaim in 1954 with Ella Fitzgerald, Gene Krupa, Buddy Rich—but also Jerry Colonna. In general however, although these shows would have a significant impact, by 1955 the concert hall had ceased to be the main public focus for jazz in Australia. New developments were in the air: changes to pub licensing laws, local jazz dances, a growing jazz club scene. And of course, 1955 saw the beginning of a new era in popular music, with 'Rock Around the Clock'.

From the 'Trad Boom' to the Present

The period from the end of the war to the arrival of rock 'n' roll was the most volatile in the evolution of Australian jazz and of our attitudes to it. Forces that had been building since 1936 suddenly surfaced and jostled in a free market of ideas until achieving a more or less stable relationship. By the late '50s, jazz was widely regarded as a distinct strand in the tapestry of modern music, and in turn, as itself many-stranded. Much of the credit for this refinement in discrimination goes to the group known as the mouldy fygges, with their determined preoccupation with conceptual categories, which gradually filtered into the wider social consciousness, already alerted by the highly visible jazz concert phenomenon.

Although these concerts continued into the late 1950s (the cut-off point, like the onset, becomes a question of definition), they had ceased to be a focus for jazz performance some time earlier. They had been progressively invaded by material having nothing to do with jazz, attracting contempt from reviewers: 'mediocre corn', 'a vaudeville show plus talent quest', 'utter vaudeville', were descriptions applied to Jim Burke's Public Command Jazz Concerts throughout 1955. The February edition had presented concert pianist Isidore Goodman as 'The Maestro of Jazz', another featured operatic tenor Donald Peers with Bob Gibson's band. Infiltrated by comedy, burlesque, juggling acts, fashion parades, the Brisbane concerts were condemned by Sid Bromley in 1957 as damaging to the image of jazz, 'the most degrading exhibition of junk'. If these Jim Burke concerts attracted especially virulent comment, they were only the most obvious embodiment of a tendency throughout the country, one which was exacerbated by the simultaneous descent of the jazz content into staleness. The various styles of bands had fallen very largely into formularized repertoire which had established a ritualized relationship with its audiences. From 1955 the source of whatever virility the concerts contained was not jazz, but the new phenomenon of popular music, rock 'n' roll. In spite of the advances made over the previous decade in the public discrimination of jazz, in its first phase, rock 'n' roll was scarcely separated from it. This should not be surprising, since the 'jazz' played at the concerts had much more in common with this new music than it had had with the commercial material of 1946–55: strong rhythm, frequent use of blues structures, aggressive horns (the saxophone had always been an important feature of the rhythm-and-blues tradition which helped nourish the rock 'n' roll movement), and the early rock bands were not as conspicuously electrified as was to be the case later. Bill Haley's Comets used an acoustic bass. It was well before the evolution of the

Motown beat and the rhythmic pattern was frequently indistinguishable to the uninitiated ear from that of the dixieland groups. Performers like Les Welch had long been presenting what amounted to rhythm and blues, and indeed no one seemed to think it out of character when he began recording material associated with the new music. Rock 'n' roll concerts regularly included jazz groups: Will McIntyre played a rock concert in Melbourne in 1955, another in Adelaide included Stan Kenton arrangements and much-applauded dixieland. The reverse was equally true: Johnny O'Keefe and the Blue Jays played a Brisbane jazz concert in April 1957 and in the same year Graeme Bell recorded a skiffle session, a style perceived as associated with youth-oriented rock. Indeed, the audiences towards the end of the concert era were those same bodgies and widgies who wrought such havoc in the cinemas showing *Rock Around the Clock*. Violence equally came to characterize many of the later jazz concerts. Adelaide's Swing Show of June 1956 was disrupted by bodgies and widgies jitterbugging in the aisles (and the Adelaide concerts maintained a strong authentic jazz component as compared with other smaller capitals). The association saw the renewal of old claims about drugs and delinquency, as civic leaders read of fights breaking out at the December 1955 Downbeat concert in Melbourne. The rock phenomenon certainly damaged the existing popular music industry, but its biggest initial effect was on the already weakened ballrooms and dance circuits. Indeed, rather than hurting the concert scene, it seems likely that rock 'n' roll briefly breathed some new life into it, and thus provided work for jazz musicians for a little longer in a format that had almost exhausted itself.

But the word 'briefly' is important. In 1957, Bill Haley and the Comets toured Australia, supported by Johnny O'Keefe, and Little Richard also toured. The extravagant visual impact of these relatively outlandish performers —the clothes, the choreographed movements, the grotesque contortions— began to provide a basis for distinguishing what they were doing from jazz, and accentuated the relatively slight aurally perceived distinctions. Jazz was old, this was new. Many professional musicians were gradually diverted from the former. The personnel of Adelaide's main rock group, the Penny Rockets, included men earlier associated with jazz, such as Slick Osborne who had served with the Southern Jazz Group; likewise, Darcy Wright and Graham Schrader with the Hi-Marks, Dave Owens and John Balken with the Blue Jays (indeed, the Blue Jays were largely made up of bop singer/drummer Joe Lane's group). Wally Norman billed his band as a rock group, and Bruce Clarke released recordings called 'Rock Crushers' and 'Rockin' Like Wow, Dad'. The attention of the general public was also diverted from jazz. The audience that had supported the jazz concerts at their peak was now significantly older. As they began consolidating family and career, their tastes in recreation turned elsewhere, a shift made easier by the arrival in the late 1950s of the long playing record and, more significantly, TV. The live performance of jazz dipped briefly in the late '50s with a hiatus between two generations of jazz enthusiast. The old concert audiences found themselves well served with recordings presenting reissues and Australian musicians. In 1957 local releases included recordings of Pat Caplice, Clare Bail, Roger and Graeme

Bell, Graeme Coyle, Ade Monsbourgh, Les Welch, the Australian Jazz Quintet, the Cootamundra Jazz Band. But the market for this music was essentially of the stay-at-home variety, and Pat Caplice, whose records of this period attracted such favourable comment, had to disband his trio for lack of work.

While we watch the concert scene dying as a jazz forum, out of the corner of our eye we can see something else beginning. Scarcely noticed at the time, with hindsight it can be said that the opening of Jazz Centre 44 in Melbourne, the instigation of a jazz policy at El Rocco in Sydney, and the formation of the Melbourne Jazz Club in emulation of Sydney's, in 1959, were foreshadowings of the last period to date when jazz was to enjoy the status of youth pop music. Interest was not spread evenly across the jazz spectrum, and the emphasis was so much in favour of traditional that the most usual description of this phenomenon is the phrase the 'trad boom'. But various forms of progressive jazz were also about to enjoy an upsurge in public favour. Although bop in its aggressive earliest manifestations had virtually disappeared, developments and (in some cases) dilutions of it were making their appearance. It is a curiosity associated with our geographical position as well as our national sensibility that many of the American jazz musicians who have exercised the strongest influence in Australian have been lesser lights in the history of the music's development: popularizers rather than innovators. These include Red Nichols, Muggsy Spanier, Artie Shaw, in their particular eras, and Dave Brubeck from the late '50s. Brubeck, and his style of jazz, achieved massive visibility here, and helped to create a public interest in our own progressive scene. As a member of his best-known group, Paul Desmond became one of the most influential saxophone stylists among the modernists who were gathering in venues such as Jazz Centre 44, El Rocco, and somewhat later in other cities: the Cellar (Adelaide), the Hole in the Wall (Perth), La Boheme (Brisbane). Hobart has never been able to sustain this minority style of a minority music, until perhaps the 1980s. Few of the first wave of modern jazz musicians—the first generation boppers—were visible during this new phase. Some, like Don Burrows, remained visible and influential, but for the most part the names are new: Brian Brown, Bob Gebert, Dave Levy, John Pochée, Keith Stirling and arrivals from outside Australia like Bob Gillette, Judy Bailey, Colin Bailey (no relation), were just some of the rather sudden influx of new blood from which Australian modern jazz would be sustained and developed for the next twenty years. Not until the late '70s has there been such a sudden and significant large-scale transfusion into the stream of our post-swing style jazz.

As usual in this country, however, it was the traditional form of the music that would enjoy the lion's share of attention in the imminent boom. While the jazz concerts were subsiding in the late 1950s, the more dedicated followers of traditional, the mouldy fygges, continued to sustain their campaign for authentic jazz. In a periodical as mainstream and non-specialist as the *Australasian Post*, February 1955, Frank Johnson lamented the local taste for corn in the name of jazz, and the lack of recognition given to top-ranking Australian musicians. The *Australian Jazz Quarterly* continued to air the fraternity's internal differences of opinion, from the assertion that even Frank Johnson's music had become heretical, to such open-minded discussions as one by Frank Haywood (in the same April 1956 issue) who noted that both traditional and modern jazz were sustained by similar attitudes. The spirit of these relatively discriminating devotees of traditional jazz was brought to a focus in jazz clubs which began to spring up around Australia. These were not replications of the swing clubs of the 1930s, in that they concentrated with greater precision upon jazz rather than on the modern music in general which had characterized the interests of their predecessors. They grew, rather, out of the post-war traditionalist movement, and their spiritual forebears were more the occasional 'hot jazz' societies of the '40s. But while this committed and informed following maintained a continuity of interest, as the larger and less intellectually self-conscious concert audience diminished, a new group was gradually replacing the latter, and with a marginal increase in its perceptual discrimination. In Melbourne this group became the clientele of what came to be called the casual dance scene. From about 1958 until 1964 the casual dances fulfilled the same kind of function in keeping jazz alive as the concerts had in the past and the pubs would do in the future. Some idea of the magnitude of this phenomenon can be gained from the listing in *Jazz Notes* July 1960 (that the magazine had been successfully revived with this issue following its demise in 1950 is another indication of the resurgence of the music).

The jazz dances were far less prominent in Sydney, indeed, almost negligible by comparison with Melbourne. The full range of reasons for this probably takes us beyond the subject of this volume, but one of the more obvious was the earlier liberalization of the licensing laws in New South Wales where 10 p.m. pub closing had been introduced on 1 February 1955. The fact also that the Sydney Jazz Club had been operating successfully on a large scale since 1953 meant that the traditional jazz following which was building up as the boom of the early '60s approached had ready-made venues for their music, and the pub scene in Sydney fulfilled much the same function, though on a smaller scale, as the casual jazz dances in Melbourne as far as the exposure of traditional jazz was concerned.

At the beginning of the 1960s, these tendencies exploded in the most widespread fashion for jazz *per se* since the 1920s. It became again (with rock 'n' roll) the major pop music genre. Activity was being observed as far afield as Rockhampton. Important factors relating to this boom were: the growing casual dance movement outlined above; the International Jazz Festival promoted by Lee Gordon, in many ways an extension of the old jazz concert concept; the ebbing of the novelty of rock 'n' roll (though, as indicated below, the audiences for the two genres were in important respects separate), the worldwide popularity of Dave Brubeck, who in addition toured Australia in 1960 as also did George Shearing, and the importation from England of 'trad', a distinctive form of traditional jazz which was essentially an English creation based on the collectively improvised jazz of Chicago in the '20s. The boom embraced the whole established spectrum of the music and also stimulated some experimentation. Generally its followers divided into modern and traditional camps with varying degrees of exclusivism.

One thing which united them however was their alienation from rock 'n' roll, and this also distinguishes their collective sensibility from that of the young audiences at the tail-end of the jazz concert era. This alienation represents a further refinement in the evolving conception of jazz and it can take the responsibility for the large blind spot which many jazz followers have in respect of the rhythm-and-blues tradition which fed both jazz and rock. One reason for this new discrimination lay in the inherent distinctiveness of the most publicized new sounds in progressive music. As *Music Maker* editorialized in June 1960, with the sounds of Brubeck and Shearing being aired under the jazz banner, the Australian public was becoming aware of the difference between it and rock. In the same year, issuing a permit for a dance at the Paddington Town Hall, the council approved 'jazz, new vogue, or old time variety', but specifically prohibited rock 'n' roll. In increasing its distance from rock, jazz was also achieving marginally greater respectability. Even John Rayment, the *Music Maker* correspondent in Adelaide who, for years, had virtually ignored the music, suddenly began providing heavy jazz coverage. He was also at pains to distinguish it from rock 'n' roll or the 'big beat' as he sometimes called it. Rayment's columns for January and February 1961 suggest something of a sudden forced diet. His rather determined enthusiasm is made quaint by his desire to appear *au fait* with a music he clearly knows little about, but which, one suspects, his editor has directed him to stop ignoring. Jazz was becoming acceptable, not simply as a commercial force, but also because of the general socio-economic complexion of its audience. The intellectualism of the early Melbourne followers now extended itself to larger tracts of the music. One of the images which jazz now evoked was of three-button suit sleekness associated with middle-class intellectualism. It was a music that appealed particularly to the young products of the 'baby boom' for whom the university system suddenly expanded through the 1950s. Jazz had a strong undergraduate image, a development of the situation which had led to the establishment by Ade Monsbourgh and others of the Melbourne University Rhythm Club back in 1937. Bands associated themselves with the undergraduate set in names like the Varsity Five, the Campus Six, the Melbourne University Jazz Band. Jazz concerts became commonplace at universities and colleges. Bryce Rohde organized a programme of them in 1960–61. The anomalous position of the music emerged at a performance he gave at the University of New England in 1961. The mayor of Armidale tried to ban the concert, then refused permission for the use of the piano. He was overruled by the vice-chancellor. Perhaps the mayor—elected representative of the city's civic and commercial interests—was responding to another aspect of the intellectual image of jazz. It was also closely associated with the major movement of youthful intellectual radicalism and dissent of the period: the beatniks and the 'angry young men' (the hero of *Look Back in Anger* was an amateur jazz trumpeter). This alliance was more with the progressive and experimental jazz styles of the time. The Perth correspondent for *Music Maker* reported (in September 1962) that there was a strong local jazz following, both traditional and, less strong, progressive, with the latter patronized by 'Paris Left Bank' types. When 'poetry and jazz' experiments were begun at Jazz Centre 44 in 1960, writers like Ginsberg and Ferling-etti were prominent. Spanning jazz styles, there were similar exercises at the AJC in the same year. This alliance with experimental and radical intelligentsia and the arts was a natural extension of that first movement in the late '30s and early '40s which asserted that jazz was an art form distinct from the undiffer-entiated mass of commercial, philistine Australian culture. Clive James's *Un-reliable Memoirs* refers to the patronage of jazz by intellectuals, writers, actors, in Sydney in the late '50s at the Royal George Hotel. In 1963 the Jazz Art Gallery in Melbourne combined modern art, pottery, and modern jazz. In March 1964 Adrian Rawlins, writing in *Music Maker*, recorded the association between the contemporary art movement and the jazz spirit. In 1964 John Allen and Dave Levy collaborated on the music for the film *Abstractions*, a documentary on Australian painters. In 1968 Michael Dransfield published a poem inspired by Errol Buddle. In trying to establish a connection between jazz and non-musical art forms, jazz followers have been asserting that it has more in common with legitimate artistic considerations that with the other streams of twentieth-century commercial music. In the early '60s in particular, the growing numbers of young people who regarded themselves as members of the intelligentsia wished to distance themselves from the naiveté of rock 'n' roll and its 'vulgar' audiences. Jazz had romantic and visceral appeal, but it also had a certain intellectual mystique; improvisation represented a valid form of creativity and an individual statement which differentiated the music from the endlessly reduplicated surfaces of despised urban materialism. Jazz consoli-dated further its association with an intellectual and artistic reaction against blandness or complacency (which is not, of course, to deny that it often gener-ated its own forms of intellectual complacency).

It is therefore not a little ironic that through the early 1960s jazz was the biggest thing in commercial popular music. The public developed the clearest notion to date of a music called jazz. The television series *Peter Gunn* was extensively publicized as the first of its kind to feature a jazz soundtrack, and others followed: *Johnny Staccatto* (whose hero was a jazz pianist), *M Squad*, as well as films like *Anatomy of a Murder* which used the music as well as the person of Duke Ellington, and *Odds Against Tomorrow*, with music by John Lewis. The foregoing featured the more progressive jazz styles, and associated them with a moody, internally driven but often outwardly cool personality, a rebellious individualism contained within a controlled composure. Traditional jazz also conveyed something of an iconoclastic image, though not so shadowed and tormented. The film *It's Trad Dad*, released in 1962, emphasized the generation gap as two young people set out to mount a concert primarily devoted to traditional jazz in the face of opposition from older people in author-ity. Even the title embodies an impudent gesture to fusty age. Jazz was again a youth phenomenon, and very much a youth seeking to define itself in reaction against the past, in rebellion which was either high spirited or sullen. And since this was the beginning of the discovery of an affluent youth market, that made jazz big business. By the end of 1960 *Music Maker* had virtually changed from being a broad-based popular music journal to a jazz magazine. It introduced a regular guide to jazz in Sydney, it featured jazz musicians on its covers with unprecedented frequency, it brought jazz items towards the front of the magazine, discussed the music editorially, devoted more review and advertising space to it, and now employed among its correspondents Alan Lee, Sterling Primmer, Ken Herron and David Bently, all of them jazz musicians. Its December issue (with the 3-Out trio on the cover) editorialized: 'During the last 12 months jazz has certainly 'arrived' in our fair country—and appears to have reached unprecedented success both artistically and commercially.'

Many of the old debates about what constituted jazz were now swamped under the new wave. Some of the intensity and aggressiveness disappeared from the more extreme mouldy fygge position, partly because the widespread confusion about what distinguished jazz from other modern music was less extreme than before, but also because, from being voices crying audibly in a rather silent wilderness, they were now being drowned out. Although inevitably generating stereotypes, jazz was now being recognized as some kind of distinctive twentieth-century art form. The kind of attention being given to the music was a mark of both its increased respectability and profitability. Graeme Bell appeared (albeit on the fringe) at the Adelaide Festival of Arts in 1962. A Sunday jazz concert in Newcastle in 1963 attracted such interest that the city council began investigating the idea of a similar function for its band and orchestra fund-raiser. Television became interested in jazz. Brisbane's Varsity Five was given a monthly programme, the Sydney Symphony Orchestra combined with jazz musicians for TV broadcast, and Graeme Bell was signed for his own show, all in 1962. In the following year the ABC began shooting material for *This is Jazz*, to feature local musicians. Jazz festivals sprouted overnight, including in country centres like Tamworth, Townsville and Glen Innes (though often these were jazz in name only; the point is the currency of the term), and the undergraduate connection was consolidated with university jazz clubs, university jazz bands and concerts, and the Intervarsity Jazz Convention, the first of which was held in Adelaide in 1963. The Pelican Jazz Club in Brisbane acknowledged the alliance by co-ordinating its operations with the university term.

If jazz thrived in this climate, it was at some cost. The public's formation of a clearer picture of the kinds of music understood by the term also tended to produce a freezing of styles to the point that further organic evolution became difficult. One of the great strengths of the traditional movement of the early '40s was that in most cases the musicians were relatively eclectic. They wanted to play traditional jazz, but beyond that fairly loose categorization they left their options open. One consequence was a supple flexibility in the music produced. It remains a contentious proposition, but I incline to the view that a distinctively Australian sound was developed at the time, and one reason was that musicians like Bell, Pickering, and their colleagues assimilated rather than pedantically copied their sources. The mouldy fygges gave jazz more definition as a concept, but in so doing they narrowed its possibilities. When this degree of clarity, or something approaching it, percolated more widely among the public in the '60s (and based on different models) it often led to stylistic ossification and caricature. This was particularly so where the 'trad' or English influence (primarily transmitted through the bands of Kenny Ball and Acker Bilk) was strongest. In general, traditional jazz enjoyed the greatest popularity, so much so that many musicians associated previously with more modern styles were working in traditional settings. Three strands could be distinguished: simple traditional, ranging loosely from dixieland to Chicago classic jazz, New Orleans, in which the drive for authenticity was assuming cult-like obsession, and 'trad'. This last was particularly strong in Perth, where Kenny Ball's visit in 1962 had great impact in a city not exposed to the same range of visiting stylists as the eastern states enjoyed. The rather brittle banjo-dominated sound of many traditional rhythm sections over the last twenty years is very much a legacy of the English 'trad' style. The unwavering succession of strict and basic chords made up of only four notes imposes very severe limitations upon front-line opportunities for rhythmic and harmonic adventurousness. For both conceptual and instrumental reasons, Australian traditional bands since the '60s have generally been fixed in stylistic amber.

The progressive scene enjoyed nothing like the same public reception accorded to traditional, but nonetheless experienced a modest boom of its own, particularly in Sydney. Most modernists had to supplement their incomes in non-jazz contexts like session and studio work, teaching, clubs, cabarets. But the concept of modern jazz was haloed by a certain grave intellectual kudos which had not previously been a significant element in its public perception. The style presented under its aegis was predominantly cool, with the Brubeck-Desmond influence most apparent. But frequent turbulent eddies disturbed this often self-consciously composed surface, indicative of shoals of experimentation going on in the more fervent centres. Ornette Coleman was an important influence, especially among the younger musicians. Avant-garde experimentation was active in such essays as free jazz, poetry and jazz, and in collaborations with symphonic musicians. This helped to project an image of intellectual seriousness and artistic integrity which traditional jazz forfeited during the 'trad' boom, partly because of its commercial success and partly because of the slightly mindless good-time atmosphere with which it frequently associated itself. I have stressed the undergraduate appeal of jazz at this time. Certainly, traditional jazz was felt to be a far more intellectual recreation than rock 'n' roll. But as compared with the cerebral experiments in the modern area, and the latter's more overt instrumental sophistication, traditional jazz with its high spirits and straw boater public image lost ground to the modernists in terms of artistic gravity. The aura of intellectual radicalism which surrounded the traditionalists through the '40s has, since the early '60s, almost wholly transferred itself to the modern styles of jazz, and this has had a deep influence on the bias towards the latter in government funding and in music education programmes.

The boom came to an abrupt end in 1965/66. As a commercial phenomenon, it succumbed to commercial forces. It was shaken slightly throughout 1963 and 1964 by the advent of the folk music fashion, which bred its own magazine and appealed to similar audiences in similar venues. But the two forms of music were generally compatible, and indeed many of the 'folk' singers of the period were so closely associated with traditional jazz that it was difficult to make the distinction. Judith Durham, Paul Marks, Pat Purchase and Judy Jacques all built their folk following through performing with jazz bands. Although the folk scene led to some jostling and rearrangement, it was not what killed the jazz boom. Put simply, jazz fell victim to a new development in the rock scene: the aggressive, fully electrified sound triumphantly embodied in the Beatles. From the beginning of 1965 the jazz industry collapsed. Jazz clubs folded up, venues closed or altered their policy. Melbourne's centre for modern jazz, the Fat Black Pussycat, closed after a brief flirtation with rock. The Criterion Hotel in Sydney discontinued its eight-year-old jazz policy. Ray Price finished at Adams Hotel in July 1966 after five years. One of the winning bands at the 1965 Glen Innes Jazz Festival was the Impacs: keyboard, drums, two electric guitars and bass guitar. Brubeck bombed in a Perth concert, likewise Thelonius Monk in Brisbane. The Sky Lounge closed its Sunday jazz dance in 1966 after ten years. It was reported in October of that year that Ray Price hadn't worked for a month, Bell was out of work, Burrows had only one concert in two months, Acheson was down to one night a week. And these were the biggest names. Musicians retreated to clubs, sessions, pit-bands, or joined rock groups.

The boom was finished. There was still jazz activity and in fact the 'low water mark' generally remained higher than it had been before 1960. But the market could no longer sustain the artificial peak of the early '60s. Jazz and jazz musicians, however, had achieved total public visibility, had become a distinct feature of Australian popular music. Never again would it be confined to something smuggled in after hours or between cabaret acts. Although this boom was a defined era in Australian jazz, in isolating the music so clearly in the context of modern music in general, it was developing and consolidating the tendencies begun in the immediate post-war period. It would not again be confused with 'pop', until jazz itself enouraged the confusion in the form of a deliberate merger, as in jazz-rock. From the time of the boom on to the present, there would always be certain venues and musicians specifically identified with jazz by the public. The new generation of jazz musicians brought into prominence by the boom made up the visible corps of jazz players for the next fifteen to twenty years, until the next major influx in the late '70s.

It is worth noting as well that the advent of electric pop completed the dismantling of another, longer musical tradition. The cabarets, night-clubs and ballrooms, which in their time had made their own contribution to jazz in Australia, now virtually disappeared. Already weakened by the perennial effect of ageing generations, by TV and by the jazz boom itself, they now began succumbing in such numbers that it scarcely remained even newsworthy. In 1964 in Sydney André's closed, Prince's was converted to a function centre, and Romano's was put up for sale, to close after more than twenty years in 1966 following a brief fling as a disco. The Adelaide Palais Royal, a forty-year-old institution, followed shortly, and in 1969 one of the most atmospheric links with a bygone era, the Sydney Trocadero, closed. The old entertainment centres had not had much effect on the public consciousness of jazz because they had rarely presented music under that name. But they nurtured the music, through the jazz that was played as part of the general diet of cabaret music; they provided a place for professional musicians to develop their jazz skills at times when there was little opportunity to perform their style of jazz billed as such, and they also gave informed followers including apprentice modernists a place to find inspiration.

By the mid-1960s electrified pop and Bob Dylan's style of folk (which could not be accommodated in jazz in the way the more spiritual/gospel tradition had been) overwhelmed the popular music industry. Jazz returned to its minority status as audiences who were committed to fashion *per se*, as opposed to dedicated followers of jazz, turned elsewhere. Jazz entered a period of quiescence, the longest since the Depression, though the level of activity was considerably higher during the later period. The gains the music made in terms of public awareness during the boom, however, were quietly consolidated. One of the reasons for the transition in popular music styles was the cyclical one—the succession of generations of taste. The supporters of the jazz movement were entering their careers and embarking upon domestic commitments by the mid-1960s. But bearing in mind the socio-economic composition of that group, there was an unseen promise in this development. In the short term, obviously, the jazz market contracted as far as live performance possibilities were concerned. In absolute terms this was especially true of traditional jazz because, in enjoying greater commercial success, it was hit harder by adverse commercial forces. But the core of followers that remained was stronger in its commitment than had been those who drifted away. Furthermore, they were of that section of society that would move into positions of influence. Jazz had made friends who would ultimately be in high places. This had happened before—Clement Semmler, for example, one of the founders of the Adelaide Jazz Lovers' Society, was later able to do much for jazz in his senior administrative capacity with the ABC. But what in the past had been isolated and fortuitous now assumed the proportions of a minor tendency. From the mid-1970s we shall start seeing the influence on jazz of some of those who first came to the music during the boom of the late '50s to early '60s.

Although the fuss died down through the late 1960s, an imprint remained and jazz activity persisted at a lower, but relatively stable level. Although rock had been the main culprit in terminating the massive vogue for jazz, the youth consciousness which expressed itself in rock during the subsequent decade made fashionable certain attitudes which fertilized jazz as well. The exploratory, decategorizing pop 'philosophy' of the period created a favourable climate for further experimentation in the jazz field. Especially in Sydney, the avant-garde became more prominent and was appreciated with increasing intelligence, in spite of its bewildering pluralism and frequent conceptual unfamiliarity. Although the avant-garde is often thought of as a young person's province, two of the musicians who spearheaded the movement in Australia were veterans of earlier eras, John Sangster and Charlie Munro. Younger contributors included Dave Levy, Bruce Cale, Phil Treloar, Roger Frampton, and in Melbourne, Brian Brown was pivotal. The movement was also fortunate in having John Clare as its chronicler. Clare has been a broad-minded spokes-man for jazz in general since the 1960s, but has shown particular sympathy in dealing with the difficult avant-garde. In 1969 he wrote a series of articles on 'the new jazz' in which he demonstrated a conceptual flexibility and boldness unprecedented in Australian jazz writing (with the possible exception of Adrian Rawlins), and which lifted his work so far above most other commentary on the music as to be, in effect, a new approach. The boppers in the late 1940s, when they had expressed themselves on the matter, based their thought on musical theory. The mouldy fygges began with historical premises. Clare brought to bear a larger aesthetic and cultural awareness which marks an important expansion of the range of jazz appreciation. It is likely, unfortunately, that his often poetic grasp of the subject is unique rather than the beginnings of a movement, since his remains an almost solitary voice. The experimental movement which he has done so much to document continues to be a small but very active element in the overall Australian picture, with Keys Music Association incubating a new generation of avant-gardists for the 1980s. While inherently there are large dissimilarities between the staple fare of the '60s jazz boom and experimental jazz, it is unlikely that the latter could have received the respectful (if limited) support that it has—recordings, broadcasts, performances—if the idea of 'jazz' had not received the public clarification which it did in the earlier period.

The relative position of jazz in Australian culture continues to be anomalous, though in comparative terms, jazz has enjoyed a more reputable status since the '60s than at any time before that. To begin with, it has achieved a more or less stable public definition. Pop burst the bubble of the 'trad' boom, but it indirectly improved the position of jazz. For one thing, it left the music more clearly defined. It is interesting that, while the fashion for jazz had subsided in the late '60s, *Music Maker* continued to carry regular interviews with jazz musicians, because they were precisely and identifiably that. The establishment now recognized the music, and regarded it with more intellectual seriousness—Don Burrows made a major contribution to this process. Although a less extensive scene than during the boom period, in some ways jazz is now more firmly based in Australian society. Its practitioners are rarely career opportunists, since there is very little money in jazz. Its public, although still often one-eyed, is generally more dedicated and discriminating than the vast majority of teenagers of the early '60s. Jazz has become a recognized component in Australian music, separate from pop and classical. So distinct is it, that, since the late '60s, it has had the confidence to engage in fusion with other musical forms without the fear of being obliterated. It is interesting that after taking so long to be recognized as a distinct musical form in Australia, one of the first impulses it expressed was towards fusion, as Charlie Munro discussed in *Music Maker*, June 1969.

In terms of the other arts, jazz remains the poor relation, very largely because the cultural establishment in Australia still overwhelmingly takes a nineteenth-century European view of the arts. Nonetheless, over the last decade some important gains have been made, and again these are ripples from the big splash of the early '60s. Above all, jazz has situated itself at a new socio-economic level. Even its venues send this signal. The pub venue goes back to the mid '50s when Sydney led the way with extended licensing hours. In 1967, however, it was reported as a significant innovation that Allan English was playing in a 'mod wine bar' in Darlinghurst. This was one of the first signs of the movement of jazz into the wine bars, bistros, restaurants, and up-market hotel lounges, a movement which reflects the lifestyle of the maturing jazz audiences from the trad boom.

For better and for worse, jazz has become predominantly a music for middle- to upper middle-class audiences, with particular appeal to liberal intellectuals with a general interest in the arts. The members of these audiences graduated from their universities throughout the '60s and set about pursuing careers, often in education, arts administration, politics, journalism, the diplomatic corps, publishing. I do not wish to suggest that the corridors of power are crawling with jazz fiends, but that, largely because of groundwork laid with the undergraduate youth of the early '60s, the general ambience in opinion- and policy-making in Australia is more hospitable to jazz than ever before. In 1959 there was talk of introducing jazz studies into conservatoria programmes. Confronted with the idea, the then director of the conservatorium in Brisbane, Dr Lovelock, was quoted as saying, 'I'm a musician, not a dance hall wallah'. In 1973, Rex Hobcroft, director of the New South Wales Conservatorium took up a suggestion by Don Burrows and established the first of the formal jazz diploma courses that have since spread throughout Australia. I have provided some indication of their local impact in regional essays and other entries, but in general it can be said that this innovation has been one of the main watersheds in the history of jazz awareness in Australia.

It has been followed closely by the establishment of several other institutional structures for the fostering of jazz on a national basis. One of these was the inauguration of the Jazz Action Society movement in Sydney in 1974 under the presidency of Mike Williams. The general importance of jazz clubs in Australia has been elaborated elsewhere in this book. The Jazz Action Society is a more recent example of a phenomenon going back to the 1930s. Several things set it apart, however. One is the society's success in attracting government funding for its enterprises, and again this was partly because, by the 1970s, the music has a number of friends, such as Don Banks, in influential positions. The Jazz Action Society of New South Wales also represented a departure from the norm in giving particular attention to sponsoring post-traditional and experimental jazz, in a country where it has usually been traditional jazz that has attracted the kind of *esprit* which generates clubs and societies. Furthermore, the society inspired the formation of kindred bodies throughout Australia, many of which have thrived in areas where formerly jazz activity was uncoordinated. The Sunshine Coast JAS, north of Brisbane, is an example of how effective such a body can be in providing a focus where, previously, jazz was so dispersed as to be almost invisible. In 1986 there were Jazz Action Societies in Sydney and Northern Rivers (NSW), in Brisbane and the Sunshine Coast (Qld), in Adelaide (SA), in Darwin (NT), and three in Tasmania. These bodies are not formally co-ordinated, and indeed any club throughout Australia is at liberty to call itself a Jazz Action Society. Nonetheless, their proliferation exemplifies a general pattern which has developed throughout the country since the late 1970s—that is, the appearance of strong provincial jazz centres. Apart from the Jazz Action Societies, other clubs have appeared in rural centres and in many cases have demonstrated vigour and imagination—the Kiama (NSW) Jazz Concert Committee's presentation of a Bill Evans memorial concert in Sydney in 1985 being a case in point. Other manifestations of this rural jazz energy include the provincial jazz festivals held in Merimbula, Deniliquin, Parkes (NSW), Mildura (SA), Ballarat (Vic.), and York (WA). The increasing frequency with which AJCs are being staged away from capital cities is a parallel development.

The most recent national administrative body created to serve the music is the jazz co-ordination programme established by the Music Board of the Australia Council under the chairmanship of Gordon Jackson in 1983. The objective of the programme was set out as being to foster 'more and better quality jazz activity at the educational, amateur and professional level encouraging in particular innovative activity and widening the base of financial support whether through fees for service or subsidy.' This is a rather abstract and flexible brief, and in practice the state jazz co-ordinators, assisted by advisory committees, have, appropriately, had to improvise according to local conditions. Most of the inaugural co-ordinators came to the position with considerable experience of the jazz scene: Ted Vining (Qld) and Alf Properjohn (Tas.) are both drummers; Paula Langlands (Vic.) is a singer with thirty years experience throughout the eastern states, and had earlier been president of the New South Wales Jazz Action Society. Eric Myers (NSW) is a pianist and jazz journalist who succeeded fellow journalist Dick Scott in 1981 as editor of Australia's only national jazz periodical, *Jazz: the Australasian Contemporary Music Magazine*, and is a forthright critic and defender of the music. Western and South Australia have had several co-ordinators, beginning with journalist Adrian Kenyon and Richard Mallett respectively.* All co-ordinators have attempted to stimulate jazz education through lecture courses, jazz workshops, and in the case of Adrian Kenyon, the establishment of the successful West Australian Youth Orchestra. Different states have presented different problems, and some have proven more amenable to 'jazz co-ordination' than others: funding for Queensland, for example, was not renewed in 1986. It is too early in the life of the programme to evaluate its contribution to the history of Australian jazz, but, at the very least, it embodies an important lobby operating in conjunction with the Music Board which also included jazz representatives Judy Bailey, then Schmoe.

The actions and attitude of these administrative bodies are symptomatic of the fact that jazz in the 1980s has found a place, albeit an uncertain one, in Australia's arts establishment. How far this connects with and determines either the state of jazz itself or the lay perception of it remains unclear. Regard-

---

* In South Australia the first appointee was John Kotow, but he never took up the appointment.

ing jazz itself the jazz co-ordination programme has had little effect on the music actually being played on a week-to-week basis. This is because jazz is essentially a folk art, drawing its vitality spontaneously from a popular base, and no amount of cultural or social engineering from above is going to have much effect on the music unless it is closely allied to that base. Furthermore, as in any art form, major innovations emerge tangentially to the political and academic establishment and by the time that establishment has begun to register an innovation, the latter has begun to harden into brittle mannerism. The position of jazz in the Australian cultural context is equally ambiguous when we consider the lay understanding of it. This emerges nowhere more clearly than in mainstream journalism. In the '70s and '80s the daily press has enjoyed the services of a number of well-informed jazz journalists including Mike Williams, Dick Scott, Dick Hughes, Eric Myers. At the same time, however, there have been writers of an abysmal level of ignorance that would never be tolerated in, for example, sports reporting. The lay public makes no distinction between one jazz byline and another, so that while its members are more aware of the existence of jazz activity than in earlier decades, they have little idea of what in specific terms that means: who are the major innovators, which are the most authentic venues, who are the most determined standard-bearers. If jazz has achieved some visibility and respectability, the fact remains that it operates at two levels. One is highly public, surfacing occasionally in mainstream media like television, AM frequency radio, high-budget recordings, and feature articles in newspapers. The other remains underground, denied general publicity, overlooked by the press and the media gurus, and dropping out of sight in the interstices of arts administration programmes. Ironically, however, it is this 'underground' activity which continues to be the main source of the music's vitality, and is where it continues to express something essential, spontaneous, and unruly about our culture.

Between the Wars

The purpose of this loosely titled essay* is to fill a gap that exists between the introductory reviews of the evolution of jazz awareness in Australia, and the entries on specific major centres, most of which are capital cities. The former category of essay is relatively abstract: it talks about what jazz seemed to mean to musicians and listeners, but without presenting much detail regarding who those musicians were and where they played. The entries on individual towns and cities provide that information, but generally concentrate on the period from the late 1930s to the present. The main reason for this emphasis is that, as far as the surviving evidence suggests, detailed regional distinctions in jazz as opposed to dance music in general did not emerge clearly until (and partly because of) the Second World War.

This overall scheme, however, leaves unanswered the question of who were the musicians setting the standards during the 1920s and 1930s, even if those jazz standards were poorly understood in terms of what was taking place in the United States. This essay seeks to identify some of the important individuals and the conditions in which they worked throughout those two decades. It will, in addition, help to qualify the point made above that regional stylistic differences were post-war phenomena. That proposition remains true to a significant degree, but a review of the '20s and '30s reveals the seeds from which those detailed differences grew. Perhaps the two most pregnant generalizations in this connection are that, first, twentieth-century popular music has always been more progressive, in terms of American developments, in Sydney than in Melbourne; and second, that these two cities have nonetheless generated the important ripples which have spread to other cities at varying intervals and with diminishing amplitude.

When entrepreneur Ben Fuller decided to present a jazz band on his threatre circuit in 1918 he invited Billy Romaine (vln) to assemble the group. This became the band led by Belle Sylvia who shortly left for South Africa to be replaced by Mabelle Morgan. Romaine had arrived from the United States in 1912 and had led a ragtime band for dancing in 1914 and 1915. Following Fuller's pioneering enterprise, Romaine remained active in the dance band industry. Sydney entrepreneur Jim Bendrodt commissioned Romaine to form a band to play at Sargent's Pie Shops, and later to lead the band at the Palais Royal, Moore Park, from its opening in October 1920. This latter event

---

*The material in this essay is based primarily on the work of Jack Michell and Mike Sutcliffe, with a small amount of supplementary factual information added by me.

TABLE TALK

J. C. WILLIAMSON'S MELBOURNE ATTRACTIONS.

HER MAJESTY'S THEATRE.

A Furore of Enthusiasm Created by
THE ROYAL COMIC OPERA CO.
in
"Katinka"
This Sparkling Musical Comedy has taken Melbourne by storm.

THEATRE ROYAL.

MURIEL STARR
in
ANOTHER DRAMATIC TRIUMPH,
"The Man Who Came Back"
With a Brilliant Cast headed by Frank Harvey and Louis Kimball.

Plans at Allan's.

Fuller's Bijou Theatre
Direction .. ... Ben and John Fuller
TWICE DAILY,
Matinee 2.30.   Evening at 8.

PAUL STANHOPE'S
MUSICAL BURLESQUE CO.
in the Up-to-date Musical Revue,
"THIS IS THE LIFE"
and
THE FAMOUS GINGER GIRLS
THE ORIGINAL GRAFTERS
QUARTETTE
FULLERS FIRST IN THE FIELD.
See the Famous
JAZZ BAND
You have never heard anything like it in your life before.
Also
THE FALVEY SISTERS
MABELLE MORGAN,
KELLY & DRAKE
and
UPSIDE-DOWN WRIGHT.

Bargain Matinee Prices, 1/6, 1/- and 6d. Evening, 2/6, 2/-, 1/- and 6d. Box-plan at Pianola till 5 p.m. Afterwards at Bijou office.

Tivoli Theatre

Harry Rickards' Tivoli Theatres L
Governing Director, Hugh D. McInt
Deputy Gov. Director, Edmund Cov
Associate Director .. Robert Gr

CAPACITY
EVERY NIGHT at 8.

MATINEES WEDNESDAY and
SATURDAY.

ESPINOSA'S GREAT REVUE,
"Time Please"

The Most Fascinating Show in Melbourne.

Goes like Lightning for Three Joyo Hours.

COME EARLY OR BOOK

Box Plans at Glen's and Theatre

TABLE TALK
August 3, 1918.

MELBOURNE AMUSEMENTS

King's Theatre
Direction .. .. J. and N. Tait.

EMELIE POLINI

THE CLEVEREST AND BEST-LIKED PLAY IN TOWN.
"De Luxe Annie"
"De Luxe Annie"

The Brilliant Star is Supported by
MR. HARMON LEE,
American Specialist Actor;
MISS GEORGIA HARVEY,
Canadian Character Actress;
MR. CYRIL MACKAY
at his Best;
And by such Favorites as
Mr. CLARENCE BLAKISTON, Mr.
MAURICE DUDLEY, Mr. G. KAY
SOUPER, Mr. JOHN FERNSIDE,
Mr. JOHN DE LACEY, Miss OLIVE
WILTON.

J. and N. Tait are producing this play; hence it is of J. and N. Tait merit. What this means is explained by their great successes of the past.
MATINEE EACH WEDNESDAY.
Box Plans at Allan's and Leading Hotels, and at the King's Theatre every Saturday Afternoon.

WIRTH'S ROLLER SKATING

J. C. WILLIAMSON'S MELBOURNE ATTRACTIONS

HER MAJESTY'S THEATRE.

A Furore of Enthusiasm Created by
THE ROYAL COMIC OPERA CO.
in
"Katinka"
This Sparkling Musical Comedy has taken Melbourne by storm.

THEATRE ROYAL.
LAST NIGHTS OF
MURIEL STARR
in
"The Man Who Came Back"
NEXT SATURDAY NIGHT.
"BOUGHT AND PAID FOR"
FOR SIX NIGHTS ONLY.

Plans at Allan's.

Hoyt's Theatre De Luxe
BOURKE STREET.
and
Hoyt's New Lyceum
Over Prince's Bridge.
ATTRACTIONS FOR NEXT WEEK.
COMMENCING SATURDAY, AUG. 10.
Dorothy Dalton in
"FLARE-UP SAL"
June Caprice in
"UNKNOWN 274"
George Beban in
"JULES OF THE STRONG HEART"
Booking—De Luxe, Ring up 6190;
Special Booking 'Phone for Lyceum,
Central 6133.

A Red Cross and Comforts Fund café chantant will be held at Adams'

Fuller's Bijou Theatre
Direction .. .. Ben and John Fuller.

TWICE DAILY.
MATINEE 2.30.   EVENING, at 8.

ABDY'S BOXING KANGAROO.
A Special Treat for the Children.
FALVEY GIRLS,
Popular Singers.
O'KEEFE and LIVINGSTON,
Sketch Artists.
MABELLE MORGAN and
THE JAZZ BAND,
In the Jazziest of Jazz Selections.
GREAT MINSTREL FIRST PART,
Staged by Paul Stanhope, Entitled
"IN THE PALACE OF THE
BUTTERFLIES."
30 PERFORMERS, INCLUDING
8 END MEN.

Popular Prices. Box plan at Pianola till 5 p.m., afterwards at Bijou Theatre Office, or seats may be reserved per 'phone Central 2251.

PRINCESS THEATRE.
Direction .. .. Ben and John Fuller.

Continued Success of
George Cross, Nellie Bramley,
Austen Milroy,
and the Favorite and Talented
FULLERS' DRAMATIC COMPANY,

SATURDAY, AUGUST 10,
Commencing at Matinee,
Boucicault's Greatest Irish Play,
"Arrah-Na-Pogue"
George Cross as "Shaun,"
Nellie Bramley, as "Arrah."

Tivoli Theatre

Harry Rickards' Tivoli Theatres Ltd
Governing Director, Hugh D. McIntosh
Deputy Gov. Director, Edmund Corell
Associate Director .. Robert Greig

MATINEES WEDNESDAY and
SATURDAY,
NIGHTLY AT 8.

Fifth Week and Unabated Success of
"Time Please"
LONDON'S SMARTEST REVUE.
A Laugh Every Second the Clock Ticks
A Brilliant Cast.
BEATRICE HOLLOWAY, ESPINOSA,
and
BARRY LUPINO
(By Arrangement with J. and N. Tait)
"A Show You are Sure to Like and
See Often."

The Box Plans are on view at Glen's and Theatre, and the Prices are within the reach of all.

THE GOLDFIELDS FETE.
For the
ST. JOHN AMBULANCE.
Grand Monster Ball
PLAIN, POSTER and FANCY DRESS,
ST. KILDA TOWN HALL.
AUG.   TUESDAY   AUG.
27   TUESDAY   27
in AID of
The Golden Nugget Stall,
18 Valuable Prizes.
Di Gilio and Orchestra.
Dancing 8 till 2.
Tickets at Allan's and Ronald
(Secure Tickets Early).

Contemporary advertisements in Table Talk *for the jazz band formed for the Fuller circuit, 1918*

Will James's band at the Wentworth, between 1920–24, with Frank Bull (p.), Will James (bjo), John Warren (vln), Syd Simpson (alt.), Sammy Cope (dms)

ushered in the large dance palais of the '20s, which in turn paved the way for the importation of American bands. The Palais was followed by the Wentworth Ballroom and the Dixieland dance hall at Clifton Gardens. Leslie and Dare's Syncopas Band played at both, but so too, at the Dixieland, was a group with a strong jazz reputation led by Ern Derriman and including Dave Meredith (tbn.) who had been with Belle Sylvia. Will James led a band at the Wentworth, and photographs which show the band wearing party hats and assuming odd poses are a reminder of the contemporary connection between jazz and novelty. James later moved to the Bondi Casino where his band included Len Niven (tpt), Al Hammett (sax) and Frank Coughlan (tbn.). Other suburban venues were the Paddington Town Hall, where Tom Burns's Band was rendering the latest 'jazz' melodies, and Oxford Hall where Violet Morrell's quartet played for cabarets and afternoon tea dances, and featured Jack Nimmo, 'Australia's premier trombone player'. Morrell was by no means an oddity in being a female jazz bandleader; among many others was Minnie Rosenthal who also operated in Sydney's eastern suburbs during the '20s. The Southern Cross Hall was another suburban dance hall with jazz nights, and advertised its instrumental star Ces Thompson as 'the premier traps drummer'. Dancing schools offered instruction in either 'American Jazz, Clod Hop or Tickle Toe' classes, and J. Albert and Sons published at 2s each the song successes of the Tivoli Jazz Teas: 'Darktown Strutters' Ball', 'I'm Always Chasing Rainbows', 'Give Me a Cosy Corner', and 'Jazz Dance', the last named expressly described as 'correct jazz music'.

The jazz craze also overtook Melbourne, and visiting musicians from overseas saw the value of trading on their presumed knowledge of the latest developments. *Table Talk*, 25 Sept. 1919, carried a piece on what it claimed was the only jazz band in Melbourne, an 'aggregate of talent demonstrating— Genuine Jazz!', and 'recognised by the best Jazzers as the Jazz Band Unique'. The band enjoyed exclusive rights to the services of George Arnold (sax), lately arrived from England's jazz halls. St Kilda's importance as an entertainment and recreational centre was strengthened when its seaside facilities were augmented by the earliest Melbourne dance halls, including the Palais de Danse (1913), and Wattle Path, both featuring imported American orchestras throughout the 1920s. Harry Yerkes's Flotilla Dance Band, led by Leon Van Straten (who later worked in top London night-clubs), opened in Wattle Path in 1924; later leaders were Walter Rudolph, followed by Joe Aronson, whose orchestra was made up of Australians apart from himself and his son Clarence on drums.

Neither public dancing nor jazz developed on the same scale in other centres as it did in Melbourne and Sydney during the '20s, but few cities remained entirely untouched, particularly on the mainland. Adelaide's first dance band was probably Tommy Dorling's Original Red Hot Peppers in 1925, but most of Adelaide's dance orchestras were founded in the late '20s—Harry Boake Smith's Palais Royal band, 1929—or through the '30s—Jack McDonnell's Rhythm Boys at the Glenelg Maison de Danse, and a later group led by Dorling, the Good Companions. The advent of swing brought into prominence bands led by Hedley Smith, Jack Barter, and in the early years of the war, Mark Ollington, Ron Wallace, and Stan Hall. The same time frame applies to Brisbane and Perth. Billo Smith opened at the Brisbane Trocadero in the mid-1920s, but with a basically old-time orchestra. Trocadero leaders in the '30s were Linn Smith and Roy Baird, with Billo Smith resuming a long association with the ballroom in 1935. The Carlton Cabaret presented bands led by Carl Wintle and Billy Miller, and Harry Griffiths led groups at the auditorium of Finney's department store from 1936. Carl Wintle had come from Sydney where he had worked with Merv Lyons, and was with the band when it travelled to Perth to play at Temple Court (later the Embassy ballroom) in 1929, augmented by locally based musicians Theo Walters (sax) and Izzy Walters (tbn.). Theo Walters was also in The Knickerbockers (1930), a jazz/novelty band that played the Luxor Theatre. Doreen D'Arcy (p.) led an orchestra in the '30s in Perth, with various personnel including Wally Hadley (bjo) and Bob Gibson. The dance band scene in Perth, however, was dominated throughout the pre-war period by Ron Moyle who led the Temple Court/Embassy bands well into wartime. All of Perth's leading dance band musicians passed through Moyle groups, including many who moved to the eastern states to establish national reputations. Bob Gibson and Cec Mitchell are two examples, touring with Moyle's Westralians in 1933 and staying on after the band returned to Perth.

To continue to trace significant developments in Australian jazz, we must turn our attention back again to Melbourne and Sydney in the 1920s. One category of events which gave those cities their priority was the early exposure to imported American bands playing the latest in hot dance music. Jim Brendrodt had heard Art Hickman's innovative dance band in New York, and was fully aware of how far behind the Australian bands lagged. When Hickman retired, his pianist Frank Ellis took over, and this was the band that Bendrodt engaged at £300 per week for the 1923 season at Sydney's Palais Royal. From the opening on 5 May the band, the Californians, was a local sensation, playing brighter music (including in terms of tempo) than Sydney dancers had been accustomed to. Ellis had brought seven American musicians with him: Frank Rago (tpt), Monte Barton (tbn.), Walter Beban and Loris Lyons (reeds), Bob

Ron Moyle's Temple Court Orchestra, 1932, with, l. to r., Viv Nylander, Ron Moyle, Wally Hadley, Cec Mitchell, Bob Gibson (sax), Arthur O'Keefe (bs)

Frank Ellis's orchestra, 1923–24, with Danny Hogan (dms), Frank Rago (tpt), Loris Lyons (ten. sax), Bob Waddington (bs), Walter Beban (sop.), Monte Barton (tbn.), Frank Ellis (p.), Bob Cruz (bjo)

Cruz (bjo), Danny Hogan (dms). He supplemented this group with Sydney bassist Bob Waddington. The band went on to play at Melbourne's St Kilda Palais in March 1924, then alternated between the two cities until New Year's Eve 1927. Certainly Ellis and Lyons, and probably other band members, gave lessons in Sydney. The band's repertoire was dominated by foxtrots ('Last Night on the Back Porch') with the occasional one-step ('Tin Roof Blues').

Another alumnus of the Hickman band was Bert Ralton, who subsequently led the band at the Savoy Hotel in London. This, the Havana Band, was imported (along with the Savoy chefs and waiters) by Stewart Dawson to open the prestigious Ambassadors night-club/restaurant in November 1923. The band consisted of George Eskdale and Eddie Frizelle (tpts), Fred Saatman (a.k.a. Frederick Roger de Gervaise Saat) on piano, Dave Wallace (bjo), Whitey Higley (dms), and Australian Harry Mehden (tbn.). This band played a sweeter and more restrained style from that of Ellis's. Ralton moved to Melbourne in 1924 and played at Carlyons, where he added other Australians to replace musicians who returned to England. He recorded for the World Record Company in 1925, mostly recent hits from America, but also a 'cover' of the Wolverines' version of 'Copenhagen'. Ralton left Australia, but was scheduled to return to take over from Ray Tellier at the St Kilda Palais. He died of a gunshot wound in Africa, however, in what was reported as a hunting accident.

In May 1925 Ray Tellier (dms) and his San Francisco Orchestra followed Ellis's Palais Royal Californians into the St Kilda Palais, Melbourne, for a season of eighty-two weeks. This all-American band, the second most successful and influential group to visit Australia in the 1920s, consisted of Cyril Collins and Felix Eber (tpts), Carl Voss (tbn.), Ed Rose, Joe Johnson, Garry Fisher (saxes), Alfred Lieb (p.), Eugene Pingitore (bjo), Ed Patterson (bs) and Tellier. The band recorded in both Melbourne and Sydney. Tellier returned to the St Kilda Palais with different personnel from May to November 1928, and later became a booking agent for the Matsons Pacific cruise shipping line, and in the

Bert Ralton's Havana Band, 17/11/'24, with Harry Mehden (tbn.), George Eskdale (tpt), Sam Babicci (ten. sax), Whitey Higley (dms), Bert Ralton (reeds), Fred Saatman(?) (p.), Dave Wallace(?) (bjo)

Ray Tellier and his San Francisco Orchestra during its St Kilda Palais De Danse residency, c. 1925, with (as numbered in photograph), 1 : Cyril Collins, 2 : Felix Eber, 3 : Ray Tellier, 4 : Garry Fisher, 5 : Ed Rose, 6 : Eugene Pingitore, 7 : Carl Voss, 8 : Alfred Lieb, 9 : Ed Patterson, 10 : Joe Johnson.

1950s was active in the San Francisco Musicians' Union. Harvey Ball and his Virginians, and Ruth Varin and her Maryland Maids, shared the bill at the Palais Royal and St Kilda Palais, but enjoyed nothing like the success of Tellier's group and lasted for only six months. Individual musicians as well as bands were also imported, notably Harry Van Dyke and Tom Swift, both American pianists. Van Dyke played at Carlyons ballroom in Melbourne and Swift, who had been with Ben Selvin in the United States, led bands at several night spots in Melbourne (Carlyons, the Embassy, the Green Mill) and in Adelaide.

A number of imported bands arrived in 1928, including an all-female orchestra, the Ingenues, which played the Tivoli circuit throughout the main Australian cities. The Ed Tazer Jazz Band, (also known as Morrison's Five Red Peppers) was a stage band that played in Sydney and Brisbane in June 1928. From Canada came the Canadian Cowboys, who played the Tivoli in Sydney and Melbourne, then moved to the Bondi Casino dance hall and the Brisbane Trocadero. Several musicians from these groups, including Bert Mars (reeds) and Wally Parks (gtr), settled in Australia, as had Eugene Pingitore from Tellier's first band and other musicians from bands that visited throughout the '20s and '30s.

The wave of imported musicians was by no means as welcome to local members of the profession as it was to the public. As early as 1924 Musicians' Union officials were complaining about the visiting Americans, 'drawing crowds to St

Kilda to hear the noise that passes for music'. (*Sydney Mail*, 10 Sept. 1924). Notwithstanding the fact that the Americans were helping to create a demand which Australians would have to help supply, and were providing invaluable models of the latest in dance band repertoire, style, and technique, considerable resistance was mounted to the imported musicians. Accordingly, visiting bands began to be augmented by Australian performers. American pianist Carol Laughner* followed Tellier into the St Kilda Palais on 29 November 1926, with an American rhythm section (Loughner, Frank Fremley bjo, Ritzy Richardson bs, Phil Harris dms) and sax player Chuck Moll, with Australians Len Niven and Jack Coughlan (tpts), Frank Coughlan (tbn.), and reed players Ern Pettifer and Arthur Morton.

Many of Australia's most influential pioneers of jazz or hot dance music served invaluable apprenticeships in the company of visiting American groups. Following a disagreement with Jim Bendrodt, Frank Ellis returned to America and Walter Beban took over his band. To fill vacancies, Beban employed Australian musicians Frank Coughlan, Ern Pettifer, Dave Grouse, and Keith Collins. Under the name the Palais Royal Californians this band recorded several sides for the new Columbia studios in Homebush, Sydney. Two tracks were issued and on one of them, 'Milenburg Joys', Coughlan and Pettifer play solos that indicate that the best of the local musicians were catching on quickly to the new music.

Australian jazz musicians were emerging from a variety of settings in addition to dance halls. Vaudeville theatres and silent picture houses also featured bands in which the dominant figures of the 1920s were gaining their early experience. It is a matter for regret that local recording companies restricted their rosters to working bands from a few top residencies; it seems certain that a number of bands that went unrecorded would have been producing authentic versions of jazz. In Sydney, such bands would include Linn Smith's Jazz Band, Tiny Douglas's Charleston Boys, and Danny Hogan's Frisco Six. Jimmie Elkins's band was one of the few to be recorded, a circumstance which gives us a rare opportunity of gaining at least some idea of what was meant by 'the best jazz band in Australia' as the term was used in 1927 (*Variety*, 6 April) to describe Elkins's group. The band's dance music is comparable in quality with imported recordings, with a credible version of Doc Cook's (a.k.a. Cooke) 'Brainstorm' among its recorded output. Jim Davidson and Jim Gussey were two future bandleaders who worked with Elkins. Members of the Elkins band went into the Ambassadors under Jack Woods in 1928, and stayed on when Davidson took over leadership of this all-star group; these included Gussey, Ray Tarrant, Chic Donovan and Harry Danslow.

The flow of imported bands and musicians dried up from the late 1920s. Apart from the longstanding opposition from union members, which stiffened as the Depression loomed, the cause of imported jazz, and of black jazz musicians in particular, was seriously damaged by the scandal attending the visit of Sonny Clay's Plantation Orchestra from Los Angeles in 1928. At least one critic

---

*Contemporary accounts give different spellings, including this, the most common form.

who heard the band realized as a consequence the difference between black and white approaches of the time: 'jazz as played by Negros [*sic*], and jazz as played by white men, are completely different'. Unfortunately, however, a combination of contemporary morality and sensationalist journalism occluded the band's musical impact on public opinion. Clay's band was part of a touring black revue called the *Coloured Idea*, which incorporated dancers, comedians, and singers including Ivie Anderson. (This was by no means the first black entertainment troupe to visit Australia; minstrel groups had paved the way in the late nineteenth century.) The revue played four weeks at Sydney's Tivoli, then moved to Melbourne where they played for five weeks to good houses. Trouble erupted when vice squad detectives raided a flat and found some of the male revue members with Melbourne girls. The ensuing uproar, played for all it was worth by the local press, forced the band to leave the country. During its stay in Melbourne the group had played only stage shows except for one midnight dance at the Green Mill where it was to have gone on for a week's engagement after the Tivoli season. It was one of the last imported jazz bands during the 1920s, and the last black band to visit Australia for twenty-six years.

Jazz in the '20s was also available in forms other than live performance, a circumstance of crucial importance to audiences outside the major eastern cities. Records provided Australians with the first opportunity to hear jazz. With no major local disc record production prior to 1925, thousands of records were literally 'dumped' in Australia at reduced prices to a public eager to hear the latest song hits. Some idea of the magnitude of this importation can be gained from the fact that the Compo Record Co. of Montreal, Canada, makers of the APEX record, which was sold in Australia by the Sutton's chain of music stores, were shipping to Australia 100 000 records a month in the early 1920s. Quantities of American labels would have been on a similar scale, and although the majority of records were dance bands playing popular songs of the day, many jazz and 'hot dance' records were sold in Australia. Some examples of what a jazz fan could buy over the counter in the early to mid-1920s would have included; Wolverines Orchestra on Gennett, the Cotton Pickers and Mound City Blue Blowers on Brunswick, Original Memphis Five on Banner and the Original Dixieland Jazz Band on Victor and Zonophone. Very few black artist records were imported at the time; those that do turn up from time to time are mainly of Fletcher Henderson's Orchestra on Banner or Domino labels, with an occasional blues singer or instrumentalist on Lincoln, Banner and Gennett. All these labels were sold by music stores and in some cases were sold for half the price of records that were being handled and imported by the well-known His Master's Voice and Columbia firms. From the brief listing above it is apparent that Australian musicians eager to study jazz would have been basing their ideas of jazz on the then popular 'New York white school'. By the late 1920s jazz records were being issued fairly regularly by the record companies, with Parlophone having the best jazz catalogue. Parlophone had access to the American Okeh label, and from that source was issuing excellent jazz records in 1927–28 by Frankie Trumbauer's Orchestra, Joe Venuti and the Dorsey Bros Orchestra. In July 1930 the company started a series of rhythm style records that were listed in the normal catalogue, with the only indication of their being in this rhythm style in the monthly record supplements. Issued in this series were records by Louis Armstrong's Hot Five, Earl Hines, Miff Mole's Molers, Chocolate Dandies and Duke Ellington Orchestra.

Unfortunately, local musicians did not enjoy anything like the same exposure on record. Al Hammett, Jimmie Elkins and Cec Morrison made only a handful of dance band records from 1926 onwards, and it wasn't until Des Tooley began making vocal records for the Parlophone label in 1929 that Sydney musicians at least were given the chance to play on small group sides with improvised solos. Included in these groups that accompanied Miss Tooley were Frank Coughlan on trombone, Abe Romain on clarinet and alto saxophone, Bert Mars on clarinet and guitar, and a so far unidentified trumpeter. These recordings lasted from 1929 until 1933, during which time Des Tooley, known as 'The Girl with the Unusual Voice' and also as 'The Rhythm Girl', recorded over thirty popular titles of time, with many featuring excellent solos and accompaniment.

Radio provided an outlet for Australian musicians, and during the early years of the medium there was much jostling among different musical categories for airtime. Dance bands, frequently with advertised jazz content, came off relatively well. Bands were featured in broadcasts both from the studio and from their residencies, a situation that was to continue right through until the war years. Some examples of broadcasts in 1925/26 were:

Sept. 1925   Saturday 2BL Jazz Night 8 p.m. onwards
              United Distributors Limited Jazz Band

Jan. 1926   Monday 2BL 8–12 p.m.
              The Ambassadors Jazz Orchestra under the direction of Mr Hammett

March 1926   Friday 3AR
              Harding and Burchill's Jazz Orchestra (Miss Burchill–p.)

June 1926   Wednesday 2BL
              Paragon Jazz Orchestra under the direction of Mr Stan Bennett

With the possible exception of the Hammett broadcast, it is doubtful whether any jazz was actually played during these broadcasts. The Melbourne station of 3LO engaged Joe Aronson to lead a 'house' band at the station during 1927, billed as Joe Aronson & His Syncopating Symphonists, the band played for all types of musical broadcasts as well as having their own featured programme every Thursday night and at 11 p.m. every other night. Aronson played the clarinet and had toured throughout the Far East and China before opening at the Wattle Path. After his engagement at 3LO he went to the Melbourne Green Mill dance hall and there he became known as the 'Rajah of Jazz' a title possibly derived from Paul Whiteman's 'King of Jazz' title in America.

The Great Depression affected every area of the entertainment industry. Many dance halls closed, bands dissolved, and several Australian record companies went out of business, leaving the EMI group (HMV, Columbia, Regal-

Zonophone, Parlophone) with a virtual monopoly. Nonetheless, the 1930s saw an upsurge in the recording of local groups, with Jim Davidson's band in Sydney making over one hundred records. Davidson's career is sketched under his own name, but his importance requires that he be referred to in any review of pre-war dance music. Apart from his successes in dance band residencies in Sydney and Melbourne and his recording activities, his national tours as leader of the ABC Dance Band stimulated activity in every city he visited. Although not a specialist 'hot' band, the ABC aggregation included musicians capable of outstanding jazz/swing performances, as indicated on recordings which featured that style.

The Oriental, where Davidson had begun his career as a leader, became the main 'hot' venue in Sydney, and bandleaders who succeeded Davidson included Frank Scott (p.), Dick Freeman (dms), and under its new name, the Ginger Jar, Maurie Gilman in March 1934. Gilman was one of many New Zealand musicians who migrated to Australia in the 1930s; others included Abe Romain (reeds), John Robertson (tpt), Col Bergerson (reeds/tpt), Charlie Lees (gtr), Craig Crawford (sax), Jim Gussey (tpt), Ted McMinn (sax.). The Ginger Jar reverted to its original name in 1938 and Theo Walters and his five piece Personality Band played there for six months before touring New Zealand.

On its return to Australia, the Walters band became the nucleus in Melbourne of a larger orchestra which was led by imported leaders Jay Whidden (October 1938 to February 1939), and Roy Fox (February to May 1939). These were the first overseas leaders to appear in Australian dance halls since Sonny Brooks's ill-timed attempt to reinforce the sweet and sentimental style in 1935 on the eve of the swing craze. Also from overseas was the stage band called the Americanadians. When the group returned home, two members remained in Australia: Jack Carpenter (sax) who joined Davidson, then led his own band, and Sammy Lee (dms) who ran night-clubs in Sydney in the 1940s and 1950s.

The arrival of swing threw several bandleaders into greater prominence, above all Frank Coughlan at the Sydney Trocadero. It was unfortunate that this band recorded so sparsely, with so little 'hot' material, and with inferior sound quality, since it was regarded as the finest swing orchestra in Australia at that time. The Cantrell brothers, Dudley and Pete, who had been with Davidson's orchestra, formed their own swing band with younger brother Bert for an engagement at the Palais Royal in 1937, during which they recorded a convincing version of 'Harlem Heat' for Regal-Zonophone.

In Melbourne from 1931 to 1933 the premier dance hall remained the St Kilda Palais de Danse, where Ern Pettifer's Rhythm Boys played. A featured musician with Pettifer was the legendary Benny Featherstone who later joined Art Chapman and led his own groups. Harold Moschetti succeeded Pettifer at the Palais, and later bandleaders included Tom Davidson and Jay Whidden. Melbourne's equivalent to Sydney's Ginger Jar was the Fawkner Park Kiosk where the house band was initially led by Bob Tough (sax) and later his brother Ern (bs), with Alf Warne, Mick Walker, Neville Maddison, and Featherstone.

In addition to these major venues, both Sydney and Melbourne (and to a lesser extent other cities) boasted other dance halls ranging from ballrooms

Maurie Gilman's orchestra in the Ginger Jar, c. 1934, with Dick Freeman (dms), Frank Scott (p.), Lyn Miller (tpt), Colin Bergerson (alt.), Maurie Gilman (alt.), and Tiny McMahon on ten. sax is obscured

Stan Bourne's band in the Ginger Jar, April 1937, with, l. to r., Stan Bourne, Charlie Lees, Roy Stewart, Jack Baines, Ken Caves, Mick Gardiner

designed as such to local town halls and Masonic temples hired for weekly suburban dances. In Sydney there were the State and Empress ballrooms, the auditorium in Grace Brothers department store, the suburban Albert Palais and Bondi Esplanade, and even a floating dance floor at Luna Park. Melbourne had Leggett's ballroom, and the Wattle Path which was reopened for dancing in 1935, but overall possessed fewer dance venues than Sydney. Except for special occasions, none of the halls or ballrooms was licensed, and in some states of Australia it was an offence to consume alcohol within a certain distance of a dance hall. Patrons would have to repair to a nearby pub, or their parked cars, and indeed often took the precaution of a pre-dance drinking session.

During the 1930s night-clubs sprang up to cater for those who sought a relatively informal evening of dancing with legal and convenient access to alcohol. The law required that, in conformity with 6 p.m. hotel closing, a patron would have to order his drinks (by telephone if necessary) before that hour. Naturally, a law so clumsy was frequently ignored or bypassed on a formality. While the 'society' night-clubs like Prince's or Romano's in Sydney had fairly restrained orchestras, the less ritzy establishments often fostered more enthusiastic swing bands. In Sydney there were the Hayden, the Carl Thomas Club, and suburban cabarets like the Colonnade at Maroubra, Stones at Coogee, and the Golden Key at Bondi. Melbourne enjoyed fewer such establishments, so that in 1936 the New Embassy could advertise itself as the city's 'only cabaret' without stretching the truth beyond public belief. Keith Chew, Bill O'Flynn, Craig Crawford, were successive bandleaders at the New Embassy, with Crawford moving to Prince's when it opened in late 1938 or early 1939. There were in addition short-lived ventures like the Rex Cabaret where Featherstone had led a band in 1934.

The ballrooms and cabarets were obliged to furnish a wide range of modern music, of which jazz or swing would be a small component, diminishing in direct ratio to the dignity of the venue. When swing arrived in 1935, clubs sprang up to provide stronger rations of the music. Radio 2UW in Sydney introduced a Rhythm Cocktail Session late in 1935, which led to the formation of No. 1 Rhythm Club, in conjuction with the magazine *Music Maker*. *Music*

Maker was Australia's first periodical devoted primarily to dance bands and jazz. It grew out of earlier magazines that had covered other areas of contemporary music, particularly brass bands, and first appeared under the name *Australian Dance Band News* in 1932. Edited originally by Eric Sheldon, with art direction by cartoonist Jim Russell, the magazine lasted until 1972, and was one of the two most enduring chronicles of jazz and dance music. The other was *Tempo*, established in 1937, also in Sydney, originally edited by Frank Johnson (not the same Frank Johnson who led the Fabulous Dixielanders in Melbourne), again with assistance from Jim Russell. *Tempo* carried many jazz items, including the highly informed record reviews by Ron Wills, probably the most widely influential reviewer in the history of Australian jazz. Wills was also a stalwart of the Sydney Swing Club, which seemed to grow out of the No. 1 Rhythm Club. The Swing Club met for the first time on 16 January 1936, and Frank Coughlan became its president in March. For a period its headquarters was the Ginger Jar, though the club's emphasis was on recorded music. It lasted into the mid 1950s.

Melbourne also witnessed the formation of clubs in association with the resurgence of jazz interest that followed the advent of swing. The most popular of these was the 3AW Swing Club, inaugurated in 1937, and expanding its membership with such speed that within six months it had moved three times to successively larger premises, including the Embassy, the Victorian Palace Banqueting Hall, and the Commodore. It was as sudden and brief a gust, however, as the initial swing fashion itself, and seems to have passed out of existence by 1940, thus enjoying a mere fraction of the lifetime of its Sydney counterpart. More durable was a club which foreshadowed Melbourne's post-war bias toward traditional jazz. The Melbourne University Rhythm Club, which included Ade Monsbourgh on its committee, was formed in June 1937. This club manifested a preference for the earlier styles of jazz, regarding the commercialized ambience of swing with some suspicion. This attitude was also reflected in the periodical *Jazz Notes* which was inaugurated in Melbourne during the war.

These are appropriate episodes with which to conclude this review of the pre-war history of Australian jazz, since they embody some of the broad generalizations made about the differences between the two major centres of the music, Melbourne and Sydney. In addition, these club activities are a reminder that there was a relationship between Australia's post-war jazz and the popular music of the pre-war period, notwithstanding the inclination of some commentators to deny that such a relationship existed.

Dictionary Entries

A

Abbey Jazz Band (Sydney, NSW)

In the early '70s Barry Pascoe (tbn.) began rehearsing a band made up of fellow staff-members of the Prince Henry and Prince of Wales Hospitals. In 1972, as the Abbey JB (named after Pascoe's place of residence in Johnston St Annandale), they began a residency at the Lord Dudley Hotel, Woollahra, with Pascoe, Bob Haine (tpt), Chris Dixon (reeds), Jim Loughman (bjo), Graham Grant (p.), Peter Paisley (wbd), Alex Watson (bs). The Lord Dudley residency lasted until 1978, by which time Loughman had been replaced by John Cleary (1974), Pascoe by Ken Longman, and Haine by Eric Holroyd. Pascoe moved to Bairnsdale, Vic., and has continued to be active in the Gippsland area, in particular through involvement in the establishment of the annual jazz festival at Merimbula. Holroyd (b. 13/12/38 in Yorkshire, England) had played trumpet in brass and dance bands in northern England, but retired from music from 1961 until resuming in 1974 after immigrating to Sydney. Before joining the Abbey JB he had been a founder member of Tom Baker's San Francisco JB. While with the Abbey he published a chord book which has enjoyed extensive use throughout Australia. Since 1981 he has also led the Triangle JB, but remains also the main tpt player in the Abbey JB, with ex-Brisbane trumpeter Les Crosby frequently sharing duties. Through the late '70s the band played a sequence of residencies at the Golden Sheaf Hotel, Double Bay and when the Lord Dudley finished they went into the Woollahra Hotel for two years. In 1980 the band began residencies at the Marble Bar of the Hilton Hotel and at Red Ned's, Chatswood. The main personnel during this period was Holroyd/Crosby, Longman, Watson, Dennis Tonge (bjo/gtr), John Reid (bjo), Tom Parkonnen (dms) and Graham Spedding. In addition to pub and restaurant engagements, the band played numerous promotions for a building society from c. 1978–83, using Jack Wiard, Paul Baker (bjo) and Wally Temple (dms). Other musicians who have worked regularly with the band incl. Tom Baker (tpt), Mike Hallam, and Rod Lawliss. In 1986 the band was into its third year at the Old Push, with Watson, Holroyd, Spedding, Lloyd Taylor (dms), Hans Karssemeyer and Verdon Morcom sharing the piano chair, and Longman working with the band when economics permitted a six piece.

Acheson, Mervyn Fletcher (Merv)   b. 31/3/22 Sydney, NSW   reeds/ldr

Started on the violin by his father but switched to tenor in 1933 following a sporting injury. By 15 was playing with George Fuller's band at the Cairo Ballroom, Drummoyne. Through the '30s he was involved in the ballroom/night-club scene, and after hours blowing sessions. Following the outbreak of war he enlisted on the understanding that this would give him some choice over the nature of his service. Posted to the 17th Battalion Military Band by 1940, was required to play clarinet and also took up soprano sax (as always, self-taught). Still performing in civilian contexts, Acheson was building a solid reputation as a hard driving, uncompromising jazz player; visiting US musicians admiringly bracketed him with Chu Berry. Following America's entry to the war, Acheson was transferred to 116th Rhythm Ensemble, where, in addition to general musical duties, he led the jazz group and worked under Giles O'Sullivan in the Booker T. Washington Club for black American servicemen and merchant sailors. Other members of the band included Rolph Pommer, Jim Somerville, Marsh and Gerry Goodwin, Billy Weston, 'Slush' Stewart, Ray Price. This unit was broken up in late 1943, and Acheson was posted to the remote training camp at Canungra, Qld. He went AWOL, moved south and attached himself unofficially to a staging camp, then continued to Sydney where, again unofficially, he recommenced playing with O'Sullivan as well as at the US Officers' Club, Bondi, under Les Welch. He resumed his activities with the 2KY Swing Club, with which he had been associated since its foundation in 1943. Acheson had always moved among the underworld circles which intersected nightclub life, and he was at the 2KY auditorium in late 1944 when he became involved in a shooting following a discussion regarding the disappearance of a load of bootleg liquor. Charged with malicious wounding, he was released on bail, to be arrested by MPs, having been AWOL for ten months. After serving 120 days detention, he resumed playing, while awaiting civilian proceedings which led to a nine-month sentence. He formed a band in prison, where he also received his military discharge, and was released early. Subsequently he worked in night-clubs—Ciro's with George Trevare (1945), Maxine's until it was closed in 1946 for violations of the Liquor Act, the Stork Club (1946), the Golden Key (which had been the Bondi US Officers' Club) until Christmas 1947. He also played theatre work at the Tivoli and the State. In 1948 he moved

Merv Acheson, Soup Plus Restaurant, 1981

to Melbourne for domestic reasons, where he played at the Empress coffee lounge. This was followed by three months playing in an illegal casino and other coffee lounges including the Galleone with Stan Bourne. Returning to Sydney, he worked with George Trevare, then, following a brief stint with a rural newspaper, played at the Collonade Club in 1950 until he left after a brawl. His activities expanded with the advent of late night hotel trading in 1955. In addition to beginning nearly ten years with O'Sullivan's night-club, he began the first of many, extended pub residencies: the Balaclava in Alexandria, then the Port Jackson with Doc Willis, and from there to the Criterion in the city where he remained until Christmas 1965. Through this period other engagements incl. the Stage Club, Redfern, the Ling Nam, and various concerts. Most of Acheson's work since the late '60s has been in the pubs and licensed restaurants, with occasional concerts interstate, and opposite Stan Getz with Dick Hughes in 1980. He held close to a ten-year residency at the Windsor Castle, and five years at the Bellevue, both in Paddington; two years at the Mansions in Kings Cross, terminating when he was assaulted on the premises, suffering a broken wrist which necessitated a lengthy layoff. He worked with Adrian Ford at the Old Push. In 1979 he joined Dick Hughes's Famous Five at Soup Plus, remaining until a clash arose with his own band's commitments in 1985. In 1986 he was with Alan Geddes's band at the Bondi Icebergers. At the time of writing he had been forced to give up playing owing to ill health. Extraordinarily, in his long career he has hardly ever been recorded; the best of the few examples is his 60th birthday concert album which shows the full range of his style from a whispered delicacy to the baroque swagger for which he is best known. Acheson has also pursued a parallel career as a music journalist. From a cadetship on the *Labour Daily* in 1938, for which he wrote a regular jazz column, he freelanced for *Music Maker, Syncopation, Tempo* (of which he became editor in 1951), joined the Packer group's *Pictorial Show* in 1952, founded *Staccato*, the organ of the Professional Musicians' Union, in 1960, and contributed autobiographical material to *Quarterly Rag* and *Jazz*. He has served office as vice-president, director, and president, of the NSW branch of the Musicians' Union, and is an honorary life member of the Musicians' Club. He has been a committed jazz player for nearly 50 years, has influenced an extraordinary stylistic range of younger musicians, and continues to appeal to a young rhythm and blues audience. He is probably the only jazz musician ever to win top polls on his instrument 25 years apart: *Tempo* in 1946 and *Music Maker* in 1971. Acheson is the stuff of legends.

Adams, John Charles   b. 29/10/38, Castlemaine, Vic.   p.
First job in 1956 with Lachie Thomson, Graham Bennett (dms) and John Morey (tpt), then to Dave Rankin's band. Played intermission piano at Melbourne New Orleans JB gigs, 1957–58. Met Allan Leake and Keith Stevens (vibes) and developed a more mainstream approach. The most important period of his musical development was with the house trio (Ted Vining, Barry Buckley) for HSV–7's *Cool Cats Show*, with guests including Graeme Lyall, Alan Lee, Brian Brown, Brian Rangott (gtr) and Bruce Clarke. Also played with Bob Barnard's band during this period. From '60s to mid-'70s worked with various groups including some led by Allan Leake, the Driftwood Jazz Band, and the John Foster Quartet on 'In Melbourne Tonight' in 1965–66. During the '70s he was a member of Allan Leake's Storyville band, with whom his work is well represented on record.

Adelaide

The overall pattern in Adelaide has followed that of Sydney and Melbourne, though lagging behind in some respects. In addition to details disclosed in specific entries, there are certain distinctive emphases dictated by regional characteristics. This is not the place to go back to South Australia's origins in free settlement, with an early preponderance of rather austere, often German, Protestantism, though I believe such circumstances have been important in establishing context. Adelaide's distinctiveness emerges in post-war developments, but

the seeds of these lie in earlier soil: a detailed history of jazz in Adelaide would have to recognize the importance of the Adelaide College of Music, founded in 1932 by John Ellerton Becker. Becker imported large quantities of instruments which he persuaded Adelaide parents to buy for their children to learn. His college virtually equipped and instructed the children of the '30s, establishing a pool of musically literate youth unique in Australia. Most young jazz musicians in South Australia in the '40s, including some who were to become most influential nationally, emerged from Becker's Adelaide College of Music. In wartime Adelaide these youthful instrumentalists were able to absorb, and gradually play, swing-inflected dance music at several venues which attracted musicians, as well as American servicemen, with interests in hotter forms of popular music. Particular favourites were the Bondezvous, which had started its own swing club in 1939, and the Astoria, which opened on 11 December 1941 and was owned by Hedley Smith (who had started a swing club in mid-1941), with a band led by Jack Barter.

In the same year the Adelaide Jazz Lovers' Society was formed, its founders including Maurie Le Doeuff, Bill Holyoak and Clement Semmler. The society's interests extended beyond the current fashion for swing, and presented discussion, recordings, and performance which included the earlier Chicago jazz of the '20s, now becoming available on the Columbia re-issue programme. William Verran (Bill) Holyoak (22/4/03–7/4/67), remained one of the most enthusiastic jazz proselytizers in Adelaide throughout the '40s and '50s and not only through his AJLS activities. His broadcasts were often the first contact young musicians had with jazz, and his variously titled swing and jazz shows were the main manifestation in Adelaide of the national concert phenomenon. He also established Memphis records which preserved the history of the Southern Jazz Group.

During the war the local scene was populated by a high proportion of musicians who had gained legitimate training through Becker's college, and whose jazz interest was stimulated through circumstances outlined above. Many found work

The Adelaide College of Music's Banjo Band, 1938. Bill Munro is on the extreme left, and Bob Wright is sixth from right

filling the gaps left by enlistment, thus gaining experience for which they might otherwise have had to wait longer. Consequently, as in most capital cities, war's end found a pool of talent unusually experienced for its age and, in Adelaide, with a high standard of legitimate training. Many of these came under the influence of Alf Holyoak. In the history of Australian jazz, there are few musicians who, in stimulating a generation of musicians, have altered the subsequent history of the music on a national level. Holyoak was one of the few. The post-war period in Adelaide saw a disproportionately large group of musicians who were both progressive in thinking and commensurately sophisticated in execution, and I believe that important factors were Holyoak and Becker respectively. Unfortunately, the cultural soil of Adelaide

was, and remained for decades, inhospitable to the germination of such innovative talent. The public interest simply did not exist on a large enough scale to support its development. Hence there was a post-war diaspora of progressive jazz musicians dispersing to the major centres in the east, Sydney and Melbourne. The list includes some of the major figures in the subsequent development of the music (including all the Australian members of AJQ): Sid Beckwith, Bob Limb, Bob 'Beetles' Young, Jack Brokensha, Errol Buddle, John Bamford, Clare Bail, Bryce Rohde, Ron Loughhead, Laurie Parr, Maurie Le Doeuff (though he shortly returned and resettled in Adelaide). This is a significant concentration of talent for a city as small as Adelaide, and it left the city virtually stripped of an effective body of progressive jazz musicians until

Malcolm Bills' band, c. 1945, with, from l. to r., John Foster (bs), Bill Munro (tpt), Mal Badenoch (tbn.),
Malcolm Bills (p.), Bob Wright (dms), Bruce Grey (clt.)

the early '60s. This in itself cyclically reinforced the city's innate musical conservatism, and accentuated the tendency for traditional jazz to dominate the scene, a situation which has never altered and which can be dated from the mid-'40s.

The traditional movement was therefore assisted in Adelaide by having a fairly clear field to itself; the decline in the vitality of swing throughout the whole country in the post-war period complemented the exodus of the most progressive local musicians to favour the more traditional styles of jazz. Malcolm Bills (b. 30/11/25, Peterborough, SA) arrived in Adelaide as a dental student in 1941. On 12 April 1942 he became the first live performer, playing boogie-woogie p., at AJLS and on 3 Sept. 1944 presented his new band, with Bruce Gray, Bill Munro, Bob Wright (dms), Shirley

Appelt (gtr), Max Dickson (bs). With occasional variations in personnel, including Bob Limb, Mal Badenoch and Eric Hopkins (tbn.) and bass players John Foster and Ron Ackfield, the band remained active until winding down through late 1946 to 1947 as Bills devoted more time to his university work. Upon graduation he moved to Whyalla, returning to take up a dental practice in 1966. Apart from very occasional work and a recording with the Phoenix JB in 1974, he virtually retired from jazz with the dissolution of his band.

That dissolution was accelerated by the involvement of Munro, Gray and Wright in the very important Southern Jazz Group. While Bills's band had been in the contemporary dixieland style of Condon and Bob Crosby, Dallwitz's SJG had been consciously based on earlier forms of traditional, and its dominance in the early years of that movement had an effect on the subsequent history of Adelaide jazz which is still felt. When the SJG broke up at the beginning of the '50s, two divergent streams emerged, one sustained by the revivalist spirit of the now retired leader, the other by the actual band alumni who remained active. The former was embodied in the group that began as the Black Eagle JB which carried the flag of the mouldy movement in Adelaide through the late '50s. The formation of the Adelaide Jazz Society in December 1954 provided a focus for this fraternity; the society's functions at the Wayville Institute from March 1955 and its official organ *Jazz Parade* (edited by Don Porter, who is still a jazz writer in Adelaide) provided a forum for the music. Towards the late '50s the appearance of a new generation of traditional jazz musicians inevitably led to some stylistic broadening, but the AJS continued to constitute the Adelaide chapter of the mouldy movement through the early '60s. In the meantime, Bruce Gray, with Bill Munro and for a brief period Bob Wright, pursued his path down to the present, in some ways finding himself in an anomalous position. Gray was often disapproved of by the more extreme sections of the traditional movement, yet he lived in a city which provided little support for the more progressive styles in which, as a constantly developing musician, he remained actively interested. He therefore found himself occupying, partly by force of circumstance, a mainstream channel which, until a general blurring of stylistic demarcations in Adelaide in the '80s, had little connection with the 'righteous' fraternity, nor with the post-bop/progressive scene that bloomed in the early '60s. Both Gray and Munro have demonstrated the breadth of their musical vocabularies in various contexts, but have had little sustained opportunity to exercise a wide range of their capabilities.

The jazz boom of the early '60s focused in Adelaide on the Cellar and a succession of venues called the Tavern. The latter catered to the traditional dimension. It began as a crowded, smoke-filled room in Grote St in 1962 in conjunction with AJS, largely under the impetus of Roger Hudson

and Dick Frankel. Two newly formed bands provided the music. Dick Frankel's Jazz Disciples (the name inspired by Frank Traynor's Jazz Preachers), with Don Cutting (bs), Norm Koch (bjo), Ernie Alderslade (tbn.), Peter Ubelhor (dms), Roger Swanson (tpt), Hudson and Frankel; and the Adelaide All Stars, with Grahame Eames (tpt), Rod Porter (later Alex Frame) (clt), John Pickering (tbn.), Kingsley Dignum (p.), Mike Rossiter (bjo), John Bradman (later 'Bradsen') (tba), and Ian Bradley (dms). These two bands, with inevitable minor personnel variations, became the mainstay of the Velvet Tavern which opened on 12 January 1963 in Hindley Street. These musicians, augmented and occasionally replaced by newcomers such as reed player Bill Clarke and ex-Melbourne drummer Ken Farmer, also performed at the third such venue in Pulteney Street where the Tavern moved in 1965, by which time Ita and Ted Kennedy, who had come in at Hindley Street, were, for all practical purposes, managing things. The passing of the 'boom' in the mid-60s robbed the Tavern of its base and (after experiments with folk, Afro-Cuban and non-traditional jazz styles) it finally closed in about 1968, victim of licensing laws which prohibited its serving liquor legally and which, from 1966, introduced the competition of 10.00 p.m. hotel trading.

Some activity in modern jazz was proceeding diffusely in the late '50s, involving musicians like Arne Van der Harst (p.), Gordon Latta (dms) and John Bayliss (vibes), but by early 1961 the more progressive forms found a centre in the Cellar in Twin St. As Blinks coffee shop it had featured recorded jazz until two trios were booked to alternate, led respectively by Bayliss and Graham Schrader. On 10 February 1961 a guest appearance by Graeme Lyall with the Billy Ross Quartet (Alan Slater p., Bayliss, Darcy Wright), drew a capacity crowd. At the end of 1961 Joe Richardson, who had bought the lease from Maurie Rothenberg earlier in the year, sold to John Howell, who consolidated its jazz reputation. At its peak the Cellar was one of the most energetic centres of jazz in Australia. Billy Ross, Bob Jeffery and Ron Carson were local stalwarts along with

The Cellar, Adelaide SA, c. 1963, with, l. to r., Billy Ross, Bob Jeffery, Mike Pank (bs), Bob Gebert (p.)

Bayliss, Schrader, Slater and bassists Wright, Mike Pank and Dave Kemp. They were stimulated by visitors residing in Adelaide, often simply to be near the venue: Keith Barr, Bob Bertles, Keith Stirling, locally born Bob Gebert, and later, Roger Frampton. Ted Nettelbeck also participated following his return from England. Howell sold his interest in 1964 to Alex Innocenti who, seeing the jazz wave receding by 1968, gradually introduced blues bands before selling in 1972, after which the premises became a restaurant. Contemporary, post-bop forms of jazz have enjoyed almost no public support in Adelaide since the days of the Cellar. The relevant musicians dispersed into session work, night-clubs, and occasional short-term residencies generally centred upon such veterans of the period as Nettelbeck, Ross, or Schmoe, who had witnessed the cellar era with enthusiasm. Younger musicians emerging in South Australia moved to the eastern capitals if they sought to participate in contemporary jazz forms, many of them like Dave Colton and Bob Johnson becoming established performers in their new base. In the '80s Adelaide has enjoyed some of the increased support that flowed from the national Jazz Action Society movement and the introduction of formal jazz studies courses at educational institutions. The temptation remains, however, to move into session work for survival or to shift to Sydney. In spite of the unqualified dedication of particular individuals, it cannot be said that Adelaide, in the '80s, sustained an actively performing contemporary jazz movement.

As always, traditional forms have fared better. Following the boom of the early '60s, the musicians who sustained the Tavern phenomenon went on to become the nuclei of virtually every traditional jazz band in Adelaide in the '70s and '80s. Following the national slump in the early '70s the level was slowly, if unevenly, rising, assisted by a number of active newcomers, among the more durable of whom is English-born Ron Flack (formerly Slack). In addition to leading his own groups, Flack helped establish the Southern Jazz Club in 1971 under the presidency of Chris Kelsey (clt), using a house band consisting of himself, Bradley, Ubelhor (tpt), Ted Kennedy (tbn./tba), Bill Polain (tbn.), Dave Rigby (ten./vibes), Ken Way (bjo) and veteran of the first SJG, George Browne. The Southern Jazz Club has constituted a virtually uninterrupted forum for bands which, for non-musical reasons such as size, would have difficulty getting bookings elsewhere; Dave Dallwitz's big band is a case in point. Although the chasm remained between mouldy and post-traditional styles, it has narrowed during the '80s as the population of mainstream musicians has been increased partly by the appearance of newcomers like Peter Hooper, partly by the re-emergence of older players like Graham Schrader and Jazza Hall, with more catholic tastes. This new or recycled talent has crossed the borders between the various sub-categories of traditional/mainstream, and has stimulated new activity in some musicians, like Phil Langford, who for many years felt disenfranchised by the blank area between mouldy and modern. There are currently more musicians working in traditional to mainstream styles than during the boom of the early '60s, and Adelaide is also one of the few cities in Australia where those styles are enjoying any significant infusion of young blood.

Allan, Jack Clarence   b. 28/9/29, Sydney, NSW   p.

Began as a child on accordion, changed to piano in his teens and became involved in the modern schools of jazz burgeoning in the late '40s in Sydney. Worked in cabaret, incl. the Dungowan, California, Christy's, Romano's; and in the town hall concerts with, *inter alia*, Wally Norman, Ralph Mallen, Ron Falson, Joe Singer, Ron Gowans. Formed his first band, the Katzenjammers, in 1949 to back visiting American Rex Stewart, with whom it recorded as the Sydney Six: Allan, Clare Bail, Frank Smith, Don Andrews, Frank Marcy (dms) and Reg Robinson (bs). In 1950 the Katzenjammers, with varying personnel that included Keith Silver (clt), were very active through town hall concerts, recording, broadcast, and a country tour presenting a lecture/recital tracing the history of jazz. The group disbanded, but Allan formed a new Katzenjammers in 1953, which made a ten-inch LP with Don Burrows, John Bamford, Andrews, Wally Wickham, Al Vincer (vib.) and Tommy Spencer (dms). Since that time Allan has worked on ABC radio, writing/directing revues, and TV studio work. As an actor, has appeared frequently on TV and in Australian films. Most of his live jazz performance has been in solo to trio work, though there have been occasional reconstitutions of the Katzenjammers. In 1983 Allan formed a duo with John Sangster on vib., playing concerts and festivals. In 1984 he recorded a solo cassette for Anteater.

Alldridge, Raymond (Ray, The Goat)   b. 12/10/44, Brisbane, Qld   p./synth.

Received five years formal tuition, and became professional at 18, having first heard jazz in the form of the Varsity Five, c. 1960. Worked mainly in the commercial field in Brisbane and Gold Coast until moving to Sydney, 1968. In early '70s, played session work with Jack Grimsley, and began deputizing for Col Nolan in Warren Daly's group at Rocks Push. With Warren Daly, 1973–76, and Galapagos Duck, 1976–79. From 1980, freelancing and later with Su Cruickshank's group at the Brasserie. In 1986 freelancing, playing studio work, and occasionally at the Regent Hotel.

Allen, Joseph Reuben (Joe)   b. 17/6/18, Brisbane, Qld   p./vibes/ldr

Began playing piano by ear at age five, then took lessons from Erich John and later from dance band leader Burt Waller. First gig with brother Henry's band, where Les McGrath (clt) gave valued encouragement. From mid-teens was active in dance and cabaret, and led his own group for a long residency at Lennon's Hotel from 1941. From the mid-'40s he worked for the ABC with Jack Thomson, Eric Wynne, Harry Lebler, Neil Wilkinson, Cliff Reese. As a member of the Canecutters and with other bands Allen performed regularly at the city hall concerts during the '50s, when he was also Brisbane correspondent for *Music Maker*. Spent most of the '60s leading a group on Channel 9, leaving in 1967 to work casually with Bernie Hansson. To Sydney in Jan. 1969 where his main band work has been with Mike Hallam, as well as leading groups in club work. Has recorded with Hallam and also with Eugene Wright and John Guerin in Los Angeles.

Andrews, Donald John (Don)   b. 29/5/29, Singleton, NSW   gtr/comp./arr

Incapacitated for normal childhood activities by osteomyelitis, took up ukelele at 10 when the family moved to Sydney, then switched to guitar, taking lessons from Roy Royston and later Ralph and Phil Skinner. Subsequent teachers included John Collins during his visit with Nat King Cole, José Luis Gonzalez (classical), and Pepe Behia (flamenco) in Spain. At thirteen Andrews played for American entertainment units and at sixteen wrote the first of many film scores, under commission from the Commonwealth Film Unit. Has been primarily a studio musician and educator. With the ABC for many years, playing, arranging and directing all forms of music including his own Castillian Players, and his chamber group, the Musique Classique Quartette. Staff musician at Channel 7 with Tommy Tycho. Has published extensively and lectured on classical guitar at the Canberra School of Music and the NSW State Conservatorium; wrote the initial classical guitar syllabus for the Australian Music Examinations Board, for which he was an examiner. Founded the

Academy of Guitar, of which he remained director until leaving Sydney in Dec. 1983 for Toowoomba, Qld, where he was active in the same range of musical activity. In June 1985 he moved to Toukley, NSW. Like George Golla, Andrews plays a seven-string Maton.

Ansell, Anthony Richard (Tony)   b. 21/2/45, Sydney, NSW   keyboards/comp./arr
Began playing 1956, learning from Roy Maling and by listening at jazz venues like the Sky Lounge and El Rocco. Full-time musician from 1971, mainly in clubs and rock groups. In 1976 began playing in Don Burrows groups on various keyboard instruments including keyboard bass and synthesizer, on which he has been among the pioneers in Sydney. Ansell has also worked with, *inter alia*, the Daly-Wilson band and John Sangster, but has increasingly moved into studio work since the mid '70s, writing for TV, films and radio. Has his own sound engineering studio. In 1986 he was president of the Music Arrangers' Guild of Australia.

Arthy, Alan   b. 20/1/29, Brisbane, Qld   reeds
As a teenager Arthy used to listen to Freddy Ailwood's band at Cocoanut Grove, and towards the end of the war he bought a clarinet and entered Perc Garner's Sweet 'n' Swing band, remaining to the late '40s, when he worked with Jim Riley on the Gold Coast. In the early '50s he played casually in quartets which included Merv Boyd (p.) and Maurie Dowden. A regular performer at the city hall concerts, he later played at Cloudland under Tich Bray. A long period of working in non-jazz settings ended with the foundation of the Queensland JAS, for which he became programme director and performs in his own and other groups.

Ashby, Anthony Paul (Tony)   b. 21/6/36, Surrey, England   reeds
Although interested in jazz at boarding school Ashby didn't begin playing seriously until arriving in 1951 in NZ. Resident in Darwin, 1960–62, playing in various groups, notably the Hi-Tones which incl. Ross Anderson (bs). Ashby returned to Auckland, NZ, where he founded the Climax JB with Bruce Haley. Moved to Brisbane in 1968 where he

and Bruce Haley founded the Pacific Jazzmen. This band was important in maintaining a jazz presence in Brisbane during a lean period, and also in becoming the focus of the Brisbane Jazz Club of which Haley was the founding president. The band incl. at various times Harley Axford (tbn.), Denny Olive (bs), Ray Maule (dms), Pat Roche and Rick Purdie (gtr), Percy Cramb (p.), Barry Webb and Len Little (reeds). It continued functioning into the '70s and played support for Acker Bilk and Kenny Ball during their respective tours. Ashby formed the Brisbane Jazz Club Big Band, and led it for its first two years. Secretary/treasurer to the financially very successful 1976 AJC, and shortly afterwards joined Ken Herron at the Melbourne Hotel, taking over the band in the late '70s. Left the group in 1980, and freelanced with every major jazz band in Brisbane as well as playing cabaret and dance work. Although most of his opportunities have been in traditional to mainstream, Ashby also enjoys working in more progressive styles. Currently active with the Brisbane Jazz Club.

Australian Jazz Convention

The jazz movement which gathered momentum in Australia during the Second World War attracted the attention of various radical movements in art and politics. Among the latter was the communist Eureka Youth League, under the aegis of which the Eureka Hot Jazz Society was formed in Melbourne in 1946, with Harry Stein (pron. Steen) as president. With Graeme Bell and other enthusiasts, Stein assembled a committee to hold a jazz convention in Melbourne in December 1946. The venue was the Eureka Hall, North Melbourne, headquarters of the Uptown Club, and the leading radical literary journal of the day, *Angry Penguins* (ed. Max Harris and Harry Roskolenko) gave its tenth issue over to the printed programme, with articles on jazz by Tom Pickering and others. A welcome dinner on the 26th was followed by four days of record recitals, discussion groups, a riverboat trip, a public concert, with music provided by the bands of Graeme Bell, Tom Pickering's Barrelhouse Four (less Cedric Pearce), Dave Dallwitz's Southern Jazz Group (less Lew Fisher), the

ANGRY PENGUINS BROADSHEET

JAZZ: AUSTRALIA

MORRIS GOODE (Harlem, N.Y.)   ROGER BELL

Souvenir Programme
OF
AUSTRALIA'S FIRST JAZZ
CONVENTION

Critics from Three Countries in Articles by
TOM PICKERING, FREDERIC RAMSAY, CHARLES FOX, *etc.*

A DAY WITH DUKE ELLINGTON *by* INEZ CAVANAUGH

EDITORS: MAX HARRIS, HARRY ROSKOLENKO
No. 10   PRICE 1/-   *December, 1946*

Angry Penguins Broadsheet *for the first Australian Jazz Convention, 1946*

Geelong Jazz Group (a contingent from the Dixieland Ramblers), as well as pick-up groups involving these and other musicians attending as individuals, incl. Frank Johnson, Tony Newstead, Doc Willis. Sid Bromley from Brisbane, and Ken Olsen, Eric Dunn, Alan Burton, and Ellis Blain from Sydney also attended. Blain recorded some of the music for an ABC broadcast which Eric Dunn reviewed, noting in particular the outstanding contribution of the Southern Jazz Group. The convention was a major inspiration to all who attended, and the visitors returned to their home states with a sense of national solidarity. Its success led to the planning of a second, and the Australian Jazz Convention has continued annually ever since, now occupying the week from Boxing Day to New Year's Eve.

There have been a number of important developments in its history. As a consequence of some political irreverence at the first Convention Fundraising Concert held at Prahran TH on 10 Oct. 1947, the Eureka Youth League disassociated itself from the enterprise, and the 2nd AJC was held at the New Theatre, Flinders St (though this also had radical connections). The convention was held out of Melbourne for the first time in 1950, the venue being the Sydney suburb of Ashfield, and it has been held in a different city each year ever since, in all states excepting the Northern Territory. In 1955 it was held for the first time outside a capital city—in Cootamundra, NSW. Since 1975, a provincial venue has become the norm, with the 1980 through to 1985 conventions all being held outside state capitals. This shift is associated with the colossal growth of the event, which now attracts around 1000 musicians and non-playing delegates. While such a number is difficult to accommodate in a large city over a vacation period, non-resort country towns are happy to play host to an event that will attract a non-violent crowd who will bring around a quarter of a million dollars into the community in only a week. There is, indeed, some danger of the AJC falling into the control of local chambers of commerce and being promoted primarily as a profit-making enterprise. Checks exist, however, to monitor such a tendency.

Early in its history the AJC began to generate surplus capital, and at first this was passed on in full to the next committee. In 1958, however, some of that surplus was deposited in a trust account, a procedure repeated in 1960. These sums were discussed at the annual general meeting in 1961, and a trust fund established. Originally seen as a reserve to rescue AJCs in financial difficulty (which happened only once), it has been called on to subsidize recordings and publications of books and music. In 1986 it stood at around $20 000.

There has been an increase of public concerts throughout the years, and this has sometimes caused apprehension that the event will be overwhelmed by gatecrashers more interested in the socializing than the music. It is certainly true that on the few occasions there have been unwelcome incidents in connection with the convention, they have usually been instigated by people unconnected in any formal way with the event. Against this anxiety, however, should be set the value of bringing jazz to a wider and younger public, the

An informal street blow at the second AJC, 1947. Among those gathered are Bob Wright (tba), Dave Dallwitz (tbn.),
John Malpas (bjo), Bruce Gray (clt); Bill Munro is the trumpeter between Wright and Dallwitz, and Keith Hounslow
next to Gray. Seated on Hounslow's right is Dick Hughes

latter consideration becoming a matter of particular urgency.

One of the unique features of the AJC is that the musicians appear for nothing and indeed pay to register. It was therefore a significant and controversial innovation when Alton Purnell, the New Orleans pianist, was imported and paid a fee to be a guest at Sydney in 1965 (Ken Colyer had appeared in 1962, but received travel and per diem expenses only). The practice has been repeated at intervals, involving in some instances the payment of Australian musicians, and a debate has proceeded on the matter. On the one hand, it is claimed that the visitor acts as an inspirational centre to the AJC. On the other hand, it is pointed out that Australia is no longer short of visiting jazz musicians, that the imported musician is rarely up

to the high standard of the best of the locals (the exception of Clark Terry should be noted) and that the visitor disturbs the egalitarian spirit of the event.

Although the AJC continues to favour traditional styles, there has been a broadening of interests which now sees big swing bands and mainstream to bop groups appearing. The days when such heretical deviations were ordered from the stage are now gone, but there still exists a core of veterans who regard modern inclinations as a threat to the event; it is undoubtedly true that for many of the delegates the AJC is a welcome retreat into a nostalgic ambience, but equally, it should be noted that at its best the AJCs are seeking to preserve a style of music which has been of incalculable importance in the history of Australian jazz.

By 1986 several events had become permanent fixtures of the AJCs. The original tunes competition, instituted at the 2nd AJC, has continued every year (except at the 8th in Hobart, 1953), with the value of the prize increasing to something far in excess of the symbolic. A picnic, a jazz breakfast, and a street parade have all become more or less perennial events, the last mentioned in particular providing a spectacle which constitutes a major public relations exercise in whatever town or city the AJC is taking place. Riverboat trips and jazz church services have also been held, with a public concert which putatively presents the best of the bands.

Apart from being the world's oldest continuously surviving jazz festival, the AJC has a number of features which distinguish it sharply from the usual commercial festival. Above all, the musicians perform for nothing, and at a time of year when they could be receiving unusually large incomes in their home towns. A band or musician participates in the programme simply by registering: it is open to all interested performers throughout Australia. While this leads to considerable unevenness, it also frees the event from a kind of élitism which implies that only a privileged few have the right to play music on a stage. In addition to the programmed bands, however, the music is dispersed throughout what becomes a temporary community. Bands spring up in pubs, motel rooms, theatre foyers, on lawns, in shop doorways, park rotundas, jetties, cafés, playing among strolling or reclining crowds, members of whom will briefly join a nearby group. Bands at the picnics swell and contract in obedience to some spontaneously organic lines of force. The AJC assumes its own shape around a planned core of events, and the most exciting sessions are rarely foreseen. All of this is made to happen on a completely volunteer basis. Neither delegates nor organizing committees receive any payment. Each committee is autonomous, and although tradition exercises considerable persuasion, the committee of each successive AJC is free to organize the event in its own way.

The AJC has established a national traditional jazz fraternity, and few musicians in Australia have never performed at a convention. Many bands are made up of musicians from different states, rehearsing by correspondence during the preceding year. No one is billed as the 'star' (unless the current committee has imported a special guest), and the permutations of players are incalculable. The atmosphere of the AJCs is unique. Violence, acrimony and (notwithstanding the notoriously massive consumption of alcohol) drunkenness are virtually unheard of. One other characteristic of the AJC which, echoed throughout most of Australia's traditional jazz scene, should give pause for apprehension, is the remorselessly rising average age of the participants. No young musicians are appearing on a significant scale. The AJC is now essentially a middle-age, middle-class recreation. This represents what is currently a very wealthy socio-economic group, and creates the impression of unqualified vigour and prosperity. Suddenly, by the end of the century, this group will be dying. This fact is the greatest question mark hanging over the future of the AJC.

The foregoing was prepared with the generous assistance of Norman Linehan, who is writing the history of the AJC. I wish to record my thanks for that assistance, but also to take full responsibility for the opinions which I have expressed.

Australian Jazz Quartet/Quintet

In 1945 Errol Buddle, Jack Brokensha and Bryce Rohde, based in Windsor, Ontario, began working in Detroit with American Dick Healey. In December, Ed Sarkesian of the Rouge Lounge in Detroit booked them to back Chris Connors, suggesting the name AJQ. Rohde and Healey provided some instrumental arrangements featuring bassoon, vibes, flute, which was to become one of the group's distinctive voicings. This led to a contract with Joe Glaser's Associated Booking Corporation and, with Sarkesian as manager, the AJQ quickly became one of the busiest jazz groups in the country. Its first gig for Glaser's agency was in Washington DC on the same bill as Dave Brubeck, the MJQ, and Carmen McCrae with whom they then toured for some time. From 1955–58 they were among the top-rated groups in the US.

The Australian Jazz Quintet, c. 1958, with, l. to r., rear, Ed Gaston, Errol Buddle, and front, Dick Healey, Bryce Rohde, Jack Brokensha

Travelling constantly, they worked every major venue, often billed above names like Max Roach, Art Blakey, Miles Davis. They were consistently placed in *Metronome* and *Downbeat* polls, and Buddle himself was several times placed as an individual. They played concerts, incl. Carnegie Hall, TV and recorded. Although outstanding soloists, the group's forte was its unusually voiced, cool and composed, ensemble sound, extended in scope by the multi-instrumentality of Brokensha (vibes/ dms), Buddle (ten./bassoon/dms) and Healey (alt./ fl./bs). In about Dec. 1955 they added Jimmy Gannon (bs) leaving Healey to concentrate on horns. Gannon was shortly succeeded in 1956 by Jack Lander, an Australian night-club musician who had moved to Canada in 1950. He was replaced by Ed Gaston in June 1957, and thus constituted, the band was offered an Australian tour organized by Clement Semmler for the ABC in Nov. 1958. Australian audiences were scarcely aware that they were listening to one of the most successful groups in the US and frequently regarded the occasion as simply marking the reappearance of three musicians—Buddle, Brokensha, Rohde—who had achieved local celebrity some years earlier. The response was respectful but not overwhelming, a reflection of Australia's parochial inferiority complex at that time. Following the tour the band broke up. Healey returned to America, followed in late 1959 by Brokensha. Rohde formed a new quartet that included Gaston, and Buddle entered commercial areas. Although the AJQ was primarily active in America, it signifies in Australian jazz, for having established a reputation for this country in the competitive US context, for disseminating some of that progressive spirit which emerged in Adelaide in the late '40s, and for being the first to bring back to Australia the results of first-hand lessons absorbed in the early post-bop environment in America.

B

Bagnall, Leonard Francis (Len)   b. 30/10/38, Melbourne, Vic.   tpt/ldr/vcl/comp.
Bagnall joined a practice group in 1966 then played casually. At the AJC in 1971 with the Sweet Georgians, he received encouragement from Ian Pearce and Tom Pickering, and in 1972 formed the De Quincey Quintet. Established Jazz Menagerie in 1973, operating around Burnie, Tas., and this band became the Emu Strutters, one of the few three front-line groups in Tas. From the AJC in Queanbeyan in 1973 the Strutters toured Qld, and back in Burnie played a residency at the Menai Hotel and backed visiting bands, the Dutch Swing College and Kenny Ball. Shortly before disbanding because members moved elsewhere, the group toured schools on the north-west coast presenting a lecture/recital programme. In 1985 Bagnall was leading Dr Jazz and the Jazzmen (formed in 1981) and through his position as a school teacher, was keeping jazz visible to children. Bagnall has been a major organizing and entrepreneurial force for jazz in Tasmania's north-west since before the advent of the JAS.

Bail, Albert Clare (Clare)   b. 26/5/26, Adelaide, SA   reeds/fl./arr
Began playing at 18 and after three years moved via Melbourne to Sydney to join Reg Lewis's and at the Prince Edward Theatre. In 1949, with Jack Allan's band, recorded with Rex Stewart and has performed with numerous other visitors incl. Artie Shaw, Frank Sinatra, and Henry Mancini. Much of his career has been in big bands and session work with thirteen years conducting and arranging as assistant MD at Channel 7. Has written extensively for film and television, and toured with concert packages, and recorded with Bob Barnard.

Bailey, Judith Mary (Judy)   b. 3/10/35, Auckland, NZ   p./comp./arr
A completely eclectic style, from stride to avant-garde, with influences including Stan Kenton, Lenny Tristano, Horace Silver. Began piano at 10 and at 12 heard Fats Waller records. Associate Diploma in Piano Performance from Trinity College of Music, London, at 16, but thereafter her interest in jazz-influenced music became dominant. Began arranging and composing jazz with the Auckland Radio Band. In Australia in 1960 en route to England or the US but has remained based in Sydney since. On Julian Lee's recommendation she became resident with Tommy Tycho's Channel 7 Orchestra, and involved with session work. Active at El Rocco with, *inter alia*, Stewie Speer, Rick Laird, Don Burrows, Graeme Lyall, Lyn Christie, John Sangster, recording her first album with the last two. In 1974 formed a quartet with Ken James (replaced by Col Loughnan in 1978), Ron Philpott, John Pochée, the nucleus of her jazz activities over the next decade, incl. recordings, concert series, and in 1978 a tour of Asia for the Dept. of Foreign Affairs and Musica Viva. Bailey has been extremely

Judy Bailey Quartet, c. 1980, with, l. to r., Ron Philpott, Judy Bailey, Col Loughnan, Ron Lemke

active in music administration and education. She set up a programme of 'music and movement' in schools. Writes extensively for children's radio, TV and theatre, as well as for film and television documentaries. A founder teacher in the jazz studies programme at the NSW State Conservatorium, and in 1978 was appointed MD for an annual course of jazz performance and lectures at Sydney Opera House. Member of the Music Board of the Australia Council, 1982–86, and inaugural winner of the annual APRA awards for composition in jazz category, 1985. Bailey is one of Australia's leading jazz musicians, and unlike many of her colleagues in administration or session work, has continued to perform regularly at concerts, festivals, and in the regular jazz venues in Sydney.

Baker, John Thomas (Tom)   b. 14/9/52, Oakdale, Calif., USA   tpt/sax/tba/vcl/ldr
Began piano at six, singing in school choir at 12, and bought his first trumpet at 15. His first gig in Australia was with Dave Banham's Northside Jazzmen. Subsequently on tuba with Nick Boston and on trumpet with Ray Price before forming his San Francisco JB which performed to great acclaim at the 1975 AJC. The band toured widely, including the US in 1977, and recorded. Baker later left the group, which still survives in name, though without trumpets, under the leadership of Paul Furniss. Baker toured the US with Helen Forrest and Andy Russell and the Pied Pipers (1978), and since 1979 has been an official guest artist at the Breda Festivals, Holland, in May. With the formation of Groove City in 1981 he turned his main attention to sax, playing in the bop style that has generally characterized his subsequent work. This band also toured in Australia and played support for Anita O'Day and Oscar Peterson shows in 1982. In 1983 formed the seven-piece Swing Street Orchestra. In 1984–85 was leading his own groups and playing bar. sax with the Morrison Brothers Big Band. Revisited the US in 1986.

Ball, Denis Winston   b. 6/9/40, Melbourne, Vic.   clt
Joined the Yarra Yarra JB in 1962 but his evolving style led to departure from the Yarras for Allan Leake's Jazzmakers, a more mainstream group. Subsequently with Frank Traynor, then brief terms with Roger Bell's Pagan Pipers, a fusion group called the Original New Orleans Rock Band, the Australian Dixieland All Stars, and Dick Tattam's Jazz Ensemble. Played in support band for Benny Goodman's Melbourne appearance in 1973. In 1985 was in semi-retirement from music, though still pursuing his profession as an industrial chemist.

Bamford, John (Ocker)   b. 10/7/30, d. 4/10/85, Adelaide, SA   tbn./p./vibes./sax./comp./ldr/arr/ vcl
Born into a musical family, Bamford began piano at four, and subsequently became interested in jazz through the records of Pete Johnson and Albert Ammons. His first paid gig was with Alf Holyoak when the usual pianist failed to appear on New Year's Eve 1945. He participated in jam sessions at the Air Force Association Ballroom on North Terrace, and when Colin Bergerson took over the Palais de Danse orchestra in 1947 he brought Bamford in on tbn. (and Charlie Foster, tpt) as part of an updating programme. Moved to Melbourne in 1948 to join Bob Limb at Ciro's then to Sydney in 1950 to work at Sammy Lee's as part of the vocal group the Tune Twisters, staying on after they broke up to play trombone with Limb who had come at the same time. In early 1951 the two moved to the Colony Club at Sylvania. Bamford also became active in session work, on the concert scene with various bands incl. those of Bob

John Bamford (left) with Stan Kenton, 1957

John Bamford Big Band, c. early 1960s, with, 1. to r., back row: Colin Jones, Mal Pearce, Jimmy Shaw, Jack Iverson, George Thompson, John Edgecombe, Neville Blanchett, Jimmy Stokes, centre row: Arthur Hubbard, Peter Haslam, Norm Wyatt, Wilson Douglas, Cec Regan, front row: Ron Gowans, Errol Buddle, Johnny Green, John Bamford, Ron Mannix, Sid Powell

Gibson, Ralph Mallen and Graeme Bell. He worked in major night-clubs—the Ling Nam, and the Roosevelt (renamed the Sunset Club), and formed a big band which fostered young musicians. TV arranger and musician for Tommy Tycho, Bob Limb, Bob Young, and with Don Burrows, from 1956 to 1981. Won both trombone and big band *Music Maker* polls in 1957, and recorded with other poll winners. Played with numerous visiting performers, incl. Buddy Rich, Ella Fitzgerald, Artie Shaw, Mel Tormé, and in 1957, with the band of Stan Kenton, who offered him a permanent trombone chair. In 1981 Bamford moved back to Adelaide after five months in Perth. There he fostered younger musicians through his work with the SA College of Advanced Education, and as MD for the 5AA Big Band until shortly before his death.

Bancroft, Donald (Don)   b. 19/12/38, Cottesloe, WA   tpt/bs/dms

Began playing jazz in 1956, with the younger generation of musicians who gained prominence in the trad boom of the early '60s incl. Ross Nicholson, John Archer (bjo), and Peter Rumbold (p.) in the Paramount JB (Perth). Moved to Sydney where he worked with the Black Opal JB for Sydney Jazz Club functions, and briefly with the Harbour City band. Returned to Perth in 1968 and joined J.T. and the Jazzmen, for 11 years. Following a brief period with Barry Bruce's Chicago JB at the Ocean Beach Hotel, he was forced by ill health to retire from music for two years. Resumed playing in 1981. At the Ocean Beach Hotel with Nicholson, Viv Booker (el. bs), Duncan McQueen (dms) and Barry Bruce from 1982; worked with Elvie Simmons's (vcl) Ragtime; in 1983 joined the Nooky Hot Six at the Nookemburra Hotel. Bancroft has also worked in non-jazz settings, notably in various brass bands incl. the Fifth Military District Army band (tpt/dms/bs) since the mid-'70s.

Don Banks Boptet, 1949–50, with, l. to r., Charlie Blott, Ken Brentnall, Pixie McFarlane, Betty Parker, Don Banks, Eddie Oxley (sax), Bruce Clarke

Banks, Donald Oscar (Don) b. 25/10/23, Melbourne, Vic., d. 15/9/80, Sydney, NSW
p./tbn./arr/comp

Began piano at five, showing an early awareness of improvisation. In 1938, while at high school he began to play jazz with other enthusiasts incl. Ray Marginson (dms), Keith Atkins (reeds), and some years later, Charlie Blott. In 1939 Banks played trombone with Graeme Bell and made private recordings with him in 1941. Also recorded with Roger Bell/Max Kaminsky, 1943. Resumed music study in 1944 and through association with the technically progressive jazz musicians, became central to the earliest bop experiments in Melbourne, 1946–47. With Errol Buddle on the first Jazzart release, 1948, and formed his Boptet which performed on Melbourne's first public traditional/bop concert. Wrote for the Rex Stewart tour and recorded with Stewart, 1949. In the same year his Boptet broadcast on the ABC and Banks completed his studies at the Melbourne conservatorium. Left for England in 1950.

During the late '40s Banks had successfully synthesized a subtle but distinctive contrapuntal style of jazz which reflected his formal background and foreshadowed his later 'third stream' works. Overseas study with composers such as Seiber, Babbitt, and Pallapiccola led to international recognition and awards, but his continued interest in jazz manifested itself in his arrangements, film music, third stream compositions, and his work in jazz education. In 1970 he was a vigorous spokesman for jazz as Patron of the Jazz Centre Society, London. Returned permanently to Australia in 1973 and appointed Head of Composition and Electronic Music Studies (which incl. jazz studies) at Canberra School of Music. Continued to write third stream work, and became Head of School of Composition Studies at the NSW Conservatorium in 1978, a post he still held at the time of his death. Bank's crucial contribution to Australian jazz has generally been overshadowed by his achievements in other areas.

Compiled by John Whiteoak

Barnard, Leonard Arthur (Len or Sluggsie)
b. 23/4/29, Melbourne, Vic.   dms/wbd/keyboards, occasional comp./arr

Joined his parents' dance band in the late '30s, also the bands of Fred Holland and Harry McWhinney. In 1948 he formed the South City Stompers with brother Bob, Tich Bray, Smacka Fitzgibbon, Bill Frederickson (bs), Fred Whitworth (dms), and Bob Mellis (tbn.), succeeded by Doc Willis then Frank Traynor. The group made an instant impression, particularly with Bob's fiery Louis Armstrong-influenced playing. They started a weekly radio series and a weekly dance at the Palais Royale in 1950, having already established themselves at the Mordialloc Jazz Palais and the Mentone LSC, the latter of which would last until 1955. Recorded for Jazzart in 1949, and in 1951, the first Australian LP (10″). In 1953 Traynor and pianist Greg Clarke (Len had moved to dms) were replaced by Ade Monsbourgh, Graham Coyle (Ron Williamson had earlier replaced Frederickson), for a Sydney record session and SJC concert. When band manager Ross Fusedale began planning a tour of the Eastern states, Monsbourgh, not wishing to travel, was replaced by Willis. The tour began in March 1955, but, defeated by bad weather and clashes with country ball seasons, it collapsed in Brisbane. Tich Bray and Len stayed in Brisbane, playing at the Story Bridge Hotel for Skippy Humphries and doing other casual gigs. Len returned to Melbourne in 1956 and joined the St Kilda Palais orchestra until 1960, in the meantime playing with Alan Lee (Frank May, dms, John

Len Barnard's Jazz Band, c. 1955, with, l. to r., Graham Coyle, Doc Willis, Bob Barnard, Len Barnard, Tich Bray, Ron Williamson, Peter Cleaver

Allen, bs) at the newly opened Jazz Centre 44, alternating with Brian Brown.

During the '60s he recorded, incl. *The Naked Dance* album, played session work for Frank Smith, and in 1962, he joined the Les Patching trio (with third man Ivan Videky) first at the Cockpit, Essendon, then until 1970 at the Southern Cross Hotel. From 1970 worked as a representative for Yamaha, resigning in March 1974 to move to Sydney. From 1974 worked with Judy Bailey, Col Nolan, Ray Price, as well as in touring packages and night clubs. At the beginning of 1977 he joined Galapagos Duck, with whom he has remained except for late 1980 to November 1983, which he spent with Bob Barnard's band.

Len Barnard's career has been an extraordinary spanning of major eras in Australian jazz. Although most prolifically recorded since the late '60s, his first band marks the beginning of a major

new phase in Australian traditional jazz, a second generation with a finely tuned stylistic dedication. Yet he has been at home in almost any company, briefly alighting in the midst of a significant modern movement in Melbourne in the mid '50s, and more recently established in the band which represented much that was new in the '70s.

Barnard, Robert Graeme (Bob)   b. 29/11/33, Melbourne, Vic.   tpt/occasional comp./vcl
Started on cornet at 12 in brass bands; first gig at 14 with his parents' band. Introduced to jazz by his father and brother Len, he joined the latter's South City Six as a founder member, making his first record on his 16th birthday. Except for a very brief association with Geoff Kitchen in 1951, his fortunes were tied to those of Len's band until he returned from Brisbane to Melbourne and resumed at Mentone LSC in 1956. His early poten-

Bob Barnard, 1984

Bob Barnard Jazz Band, Rocks Push, 1976

tial had, in the meantime, been signalled when, in 1952 he unseated Roger Bell and Frank Johnson in the *Music Maker* polls. In June 1957 Mentone finished and Bob spent a year in Sydney with the Ray Price trio (third man, Dick Hughes) and the PJJB. Returning to Melbourne he joined Kenn Jones's Powerhouse group, remaining virtually until moving permanently to Sydney in 1962 to join Graeme Bell. Remained with the Bell band during one of its busiest periods, leaving it in Surfers Paradise and returning to Sydney in 1967. Until 1974 he was mainly involved in studio work of various kinds, though in 1970 he formed what amounted to a recording band for two LPs for EMI—Ed Gaston, Laurie Thompson, Bill Benham, Clare Bail and Norm Wyatt. In 1974 he formed another band, this time for live performance. Its personnel has remained relatively stable: John McCarthy, Chris Taperell, Wally Wickham, John Costelloe,

Alan Geddes. Geddes was replaced by Laurie Thompson, then by Len Barnard (1980), then Thompson again (1983). Costelloe died in early 1985 and has not been replaced, although Bill Howard in Melbourne has joined the band on some tour work.

The Bob Barnard JB has enjoyed considerable acclaim both within Australia and internationally. Its tours have included: Asia (1977), Australia (1977, 1979, 1980), Europe (1980), many of these under the auspices of the Dept. of Foreign Affairs. It has also toured the US (1976, 1978, 1982), on each occasion performing at the Bix Beiderbecke festival where it has received the very highest approbation. The band has also played residencies at the Old Push, the Marble Bar, Red Ned's, and Bob and the rhythm section have played for many years on Saturdays at the Orient Hotel. Although traditionally based, Barnard's versatility has placed him with a wide range of company, incl. the Daly-Wilson Big Band, of which he was a founder member. His playing could be regarded as an extension of the style of Bobby Hackett, but expanded in instrumental and harmonic sophistication, infallibly poised and swinging, always able to revitalize an idiom sometimes felt to have exhausted itself. It is heard with particular clarity on quartet albums from 1982 and 1983, and on a session with strings recorded by the ABC. He has been an inspiration to and model for some of the best players of a later generation, notably Grahame Eames and Mal Jennings. Bob Barnard is regarded internationally as one of the greatest trumpet players in the world in his broadly mainstream style.

Barr, Keith   b. 1927, London, England; d. 2/6/71, Sydney, NSW   reeds, incl. oboe/ldr/arr/comp
Started piano at 10, tenor at 14. Enlisted in the services at 18 and took up clarinet and studied arranging while posted in Europe. Broadcast with his quartet in Vienna. Following discharge, 1947, worked in dance and theatre bands incl. Jack Parnell's and Basil Kirchin's. Study with Eric Gilder led to increasing interest in experimentation. Won *Melody Maker* poll on tenor, 1955, then worked with John Dankworth and made several visits to

the US. Brought to Australia by entrepreneur Lee Gordon, 1960, Barr later spent a period in Adelaide, playing with Bob Bertles at the Cellar, 1963. Back in Sydney he was very influential at El Rocco, and in the late '60s played in pop/rock settings incl. Nutwood Rug, and Heart 'n' Soul. Played with Bob Gebert, John Sangster, Jack Dougan at the Cell Block Theatre, 1970. Following hospitalization in 1971, he died after falling through an apartment window while in a diabetic coma.

Bartolomei (a.k.a. Bart), Michael   b. 30/9/56, Perth, WA   keyboards/comp./arr
Began piano at seven and subsequently moved to Sydney to study at the conservatorium. After work in the commercial field joined Kerrie Biddell. In late '70s also played duo keyboards with Julian Lee. Studied in New York, 1981–82, returned to Sydney, rejoining Kerrie Biddell. Also works with John Hoffman's Big Band. In 1983 awarded first prize in the JAS of NSW composition competition.

The Basement (Sydney, NSW)
In 1973, Bruce Viles and Tom Hare, with the help of other members of Galapagos Duck and Horst Liepolt, began renovating a derelict basement print shop in 29 Reiby Place near Circular Quay, opening it in August as The Basement, with Galapagos Duck as the house band six nights a week. From November, other bands were booked for seasons, incl. Kerrie Biddell, Judy Bailey, Dave Fennell's Power Point, and Brian Brown from Melbourne. Through the '70s The Basement was the focus for the first concerted contemporary jazz movement since the closure of El Rocco. Major groups formed to play the venue incl. The Last Straw, Out to Lunch, and Jazz Co-op. It also played host to shorter-lived experimental ventures and various big bands incl. those led by Craig Benjamin and Dick Lowe. The Basement provided something of a public workshop for students attending the jazz studies courses at the conservatorium, presenting the student big band (June 1975) and the Music is an Open Sky Festival (May 1975 and April 1977) as a tribute to the course director Howie Smith.

There was some abatement of the intense experimental and innovative energy towards the end of the '70s, though the jazz policy was continuously maintained, with the Duck, Errol Buddle, and Kerrie Biddell's Compared to What being mainstays. There have been occasional presentations of more traditionally oriented bands, incl. Graeme Bell, the Harbour City, and Tom Baker's San Francisco JB, but The Basement has always been primarily a venue for mainstream to contemporary and fusion styles. In the '80s there has been more emphasis on the latter, notably with Crossfire and more recently with the big band called Supermarket, as well as overtly blues/rock artists like Lonnie Mack, and blue grass performers. The Basement has also been the main non-concert hall venue for visiting musicians both from within Australia (Vince Jones, Bob Sedergreen, Schmoe) and from overseas, incl. Art Pepper, Georgie Fame, Chris Hinze, Toshiko Akiyoshi and Lew Tabackin, Freddie Hubbard, Johnny Griffin, the Modern Jazz Quartet, Mark Murphy, Mike Nock. It continues to provide a forum for otherwise economically constrained big band performance—John Hoffman, Barry Leef, Espirito. The Basement is probably the best-known jazz venue in Australia, and played a particularly significant role during the mid-'70s burgeoning of contemporary groups. In the '80s it has been less uniformly adventurous—The Benders had been in existence for more than four years before appearing at The Basement. Experimentation is less common, and there has been a lean towards established names and a flavour which some would call an extension of, and others a deviation from, jazz, moving into a pop area. The Basement has moved up from the underground and has become something of the jazz establishment. It continues, however, to play an important role in maintaining a broader public awareness that jazz is a major music in Sydney.

Bates, John Herbert Malcolm   b. 5/1/36, Sydney, NSW   tbn./tba/bs/vcl/comp
Inspired by jazz at local dances and concerts, began trombone with lessons from Harry Berry, 1957–62. Played in Bexley Brass Band and sang in

church choirs in late '50s. Joined a .dance band reading dixieland orchestrations in 1959 and began sitting in at Sydney Jazz Club functions. Through the '60s worked with Nat Oliver's NO jazz band, and for shorter periods with the Society JB, and Mick Fowler. Active in the early '70s with Adrian Ford's Big Band and with Nick Boston. A founder member of Tom Baker's San Francisco JB, 1975, and in 1986 was still with the band under Paul Furniss. He has also sung in gospel groups, is an AJC original tunes competition winner, and a director of the Sydney Jazz Club since 1983.

Batty, George Philip   b. 15/11/26, Toodyay, WA
reeds
Early member of West Coast Dixielanders in Perth. Left Australia, April 1954, as a merchant marine radio operator. Played in Wellington, New Zealand, listened to jazz in New York, and during a period in London (May 1958–June 1960) played with several 'trad' bands sometimes providing support for the bands of Humphrey Lyttelton, Nat Gonella, Dick Charlesworth. Travelling again (June 1960–March 1962), heard veterans in New Orleans. Returned to Perth, 1962, and resumed working in bands with brother Phil. From 1964 his day job took him to Esperance, WA and Darwin, NT, where he played casually. Returned to Perth 1975 and has since freelanced and is a member of the Corner House JB.

Batty, Philip James (Phil),   b. 28/6/31, Toodyay, WA   tpt
Became interested in jazz through his elder brother George's records and began cornet in 1947. Played in the Alvan St Stompers and then at the 6PR sessions, becoming a member of the West Coast Dixielanders which resulted. Following two years in England in the mid-'50s he worked in the Riverside and Westport groups, the latter his last long-term band commitment. He subsequently worked casually, incl. with Westport alumni under the name of Red Notes, and in deputizing situations, but since 1980 has virtually retired from music.

Bell, Graeme Emerson (Gay) b. 7/9/14, Melbourne, Vic.   p./ldr/comp
His parents were performers, his mother, Elva Rogers, a noted contralto who toured with Melba. Graeme studied piano from age 11 under Jessie Young, and was later coached in dance music by band leader George McWhinney. Converted to jazz by brother Roger, the two played their first gig (Roger on dms) for a dance at the Deepdene Scout Hall in 1935 as the Hot Air Men. This duo was augmented with Bill May (bs), who later manufactured Maton guitars, and a succession of reed players incl. John Weston, Ivan Arthur and, for about three years, Tom Crowe. The Bells were increasing their jazz knowledge from records, the Sunday afternoon sessions at Fawkner Park, and later under the guidance of Bill Miller. In 1939 they began making private records, the first on 14 September as Ding Bell and his Belfry [sic] Bats, with Ding [Roger] Bell (cnt), Spadge Davies (clt), Gay Bell (p.) and Plunk [Ade] Monsbourgh (gtr/jug). That Christmas the Bells worked with Bill May, Russ Murphy (dms) and Hadyn Britton* (clt) for a company function on the paddle-steamer *Weroona*, and again the following year with Sam Dunn (bs), Don Banks (tbn.), Murphy, and either Britton or Pixie Roberts, whom they had met at Saul's coffee lounge in Feb. 1940. That Christmas they also played at the Nepean Hotel, Portsea, as the Portsea Four, with Murphy and Roberts, with Ade Monsbourgh regularly sitting in. In 1941 the band was stabilizing and Graeme was taking frequent initiatives in organizing opportunities to play jazz. On March 2 the brothers started at Leonards café, advertised as Graeme Bell and his Jazz Gang, 'the only band playing real jazz', with Murphy, Roberts, and Stan Chisholm (bs) and vocalists. The band's popularity led to a second weekly night, Wednesdays, though the gig folded after three months. In Aug. Graeme gave a history of jazz concert for the Victorian Jazz Lovers' Society, which he had formed. In Oct. the band played for the third annual exhibition of the Contempo-

---

* Variously spelled in written accounts also as Haydn, Haydon, and Britten, in various permutations.

rary Art Society. Apart from the affinity felt between two groups (the Bell coterie and the CAS) in
rebellion against commercial philistinism, Graeme
himself had studied art under Max Meldrum.
Graeme was also playing with other groups, incl.
those of Mollie Byron (tpt/vcl) at the Swing In coffee lounge, and Gren Gilmour for dances. In May
1943 Graeme left to join Claude Carnell's dance
band in Mackay, Qld, stopping off in Sydney
where he jammed at the Booker T. Washington
Club with Giles O'Sullivan's band. From Mackay
he moved to Brisbane where he played with Freddie Ailwood at the Cocoanut Grove and met Max
Kaminsky, who roomed with him. Arriving back
in Melbourne in Nov., he took over leadership of
the band from Roger, who had a day job. The band
was now resident at the Heidelberg TH and the
Palais Royale, using different drummers incl.
Laurie Howells. Bill Lobb, Charlie Blott, and
Murphy. Monsbourgh had gone into the RAAF
and was replaced by Peter Law, Cy Watts, and
Harold Broadbent in succession. In June 1944 the
Heidelberg residency ended and Graeme began
teaching. The band was also playing occasional
dances for the communist Eureka Youth League,
and when the League started its Hot Jazz Society
in Nov. at Chapel St, Bell played the opening
(Graeme, Roger, Roberts, Murphy, and Bud
Baker, gtr/bjo). In Oct. 1945 the Palais residency
finished, and in Nov. the band recorded *Alma St
Requiem*, nominated by George Avakian, when it
was subsequently released, as one of the 10 best
jazz records of the year. In 1946 the band was
playing the Eureka Hot Jazz Society functions,
now in Queensberry St, North Melbourne, and
began running its own cabaret on the premises
under the name Uptown Club. Beginning on June
29, the band consisted of Graeme and Roger,
George Tack, Lindsay Motherwell (dms), with
Will McIntyre playing intermission. During the
year the band developed into a seven-piece consisting of Graeme and Roger, Ade Monsbourgh
(tpt/clt), Cy Watts (tbn.), Lou Silbereisen, Pixie
Roberts, Sid Kellalea (dms) later replaced by Murphy; later again Jack Varney (bjo) was added. In
Dec. of that year, Graeme was one of the prime

*The Uptown Club, North Melbourne, 1947, with, l. to r.,
George Tack, Lindsay Motherwell, Roger Bell, Graeme
Bell*

movers in the inauguration of the AJC, at which
his band also performed.

By 1947 jazz and the Bell band were developing
a considerable following. The Uptown Club continued, and in Feb. they started at the Manchester
café, using Monsbourgh or Tony Newstead on
trumpet while Roger was recovering from an
appendectomy. In March they played in Sydney,
sharing the stage with the PJJB, then returned to
Sydney in April to make three records under contract to Columbia. Roberts, immobilized after a
motor cycle accident in the Uptown Club, was
replaced on this occasion by Geoff Kitchen. Released on Regal Zonophone, these records sold at
the extraordinary rate of over 50 000 each, and
established the band's name throughout the country. Back in Melbourne, Graeme was invited by
Harry Stein of the Eureka Youth League to take
the band to Prague, Czechoslovakia, for the World
Youth Festival. The band sailed on the *Asturias* on
3 July: the Bells, Monsbourgh, Roberts, Varney,
Murphy, Silbereisen. Following the festival the
band played concerts in Europe, then based themselves in England with Mel Langdon as their manager. In Feb. 1948 they opened their own Leicester
Square Jazz Club, advertising jazz for dancing.
Hitherto, traditional jazz in England had largely
been the preserve of rather scholarly enthusiasts
who absorbed the music through records, discussions, and in a passive concert format. Although
regarded as something of a heresy, Bell's idea of a

jazz dance revolutionized the perception and social function of jazz in England, made it the centre of entertainment for the young, and gave direction and impetus to the 'trad' revival of the '50s and '60s. They returned to Australia in Aug. 1948, having been preceded by Russ Murphy who was replaced for the final period in England by local drummer Dave Carey. The Bell band had been the first major cultural export from Australia to Europe. In many parts of Europe it was the first jazz band heard live, and it inspired the formation of bands and jazz clubs in its wake. It left behind the enduring conviction that its loose, exuberant approach (presented through an extensive library including much original material) constituted a distinctive Australian jazz style, and it was acclaimed as one of the world's most convincing jazz groups, surpassing many American bands. If Bell's career had finished at this point he would already have been the most influential figure in the history of Australian jazz.

On its return in 1948 the band severed its connections with the Eureka Youth League and embarked upon a three-month tour for the ABC, which had been organized from England. Charlie Blott and Ian Pearce were now with the band and in early 1949 Varney was replaced by Bud Baker, and Jack Banston came in on drums. Monsbourgh moved to second trumpet, with Ian Pearce succeeded by Johnny Rich on trombone. In the same year Bell founded the Swaggie record label, issuing its first record at the end of the year. Swaggie has since become the most prolific and important jazz specialist label in Australia, and has international distribution. Graeme sold Swaggie Records to Nevill Sherburn in 1954. In 1949 Graeme imported American cornet player Rex Stewart, the first American jazz musician to tour as such for decades (not incl. US service musicians during the war). The tour attracted attention beyond the jazz fraternity, and aspects of it were covered by the daily press and women's magazines. Stewart stayed on for a period after the tour and worked with other groups in Sydney.

In 1950 the band began its second European tour, arriving in England in November; Bud Baker (gtr/bjo), Deryck 'Kanga' Bentley (tbn), and John Sangster (dms) were newcomers, with Roger and Graeme, Roberts, and Monsbourgh. On this tour the band was fêted. They played concerts, BBC broadcasts, recorded with Humphrey Lyttelton, gave royal command performances, toured and recorded with Big Bill Broonzy, playing to capacity houses through England and the Continent. Apart from being a musical success, the trip was a triumph of logistics: with families, the party consisted of 18 people, with three children and two more born during the tour. The band returned to Australia in April 1952 to ABC concert tours at which members received the hysterical adulation later accorded rock stars. In addition to awards given to other members, the band won the jazz band category, and Graeme the piano category, in the *Music Maker* polls of 1952. After seven years of continuous touring, the band broke up in June, with only Graeme and Sangster remaining professional, though all the others continued a part-time involvement with jazz. Bentley returned to Adelaide and a later association with Dave Dallwitz. Norman 'Bud' Baker played with bands in Melbourne, where he died on 14/2/83. From the first tour, Russ Murphy joined Tony Newstead briefly, won the *Music Maker* poll for Drums, 1952, and worked casually around Melbourne before settling in Qld.

This was the end of a major phase in Graeme's career. From 1954–55 he toured Korea and Japan for Combined Services Entertainment, with John Costelloe, Sangster (tpt), Yolanda Bavan (vcl), Jack Baines (reeds), and Banston. In 1955–56 he based himself in Brisbane playing at the Story Bridge jazz cabaret and working in night-clubs with Sangster. Moved to Sydney, 1957, for lack of work in Melbourne, and through the rest of the '50s worked in various musical settings, and opened an art gallery in Double Bay. At the onset of the 'trad' boom, Harry Harman suggested that Graeme form a new jazz band. In June 1962 he led the first of the All Stars at Yarra Bay Yacht Club, where he had earlier been leading a group with Graham Spedding, Mick Martin (tpt) and Jim Shaw. Norm Wyatt, (b. 2/6/19, Bristol, England),

Graeme Bell and his Australian Jazz Band, 1952, with, l. to r., back row: *Don 'Pixie' Roberts, Bud Baker, Lou Silbereisen, Graeme Bell,* front row: *Roger Bell, John Sangster, Deryck Bentley, Ade Monsbourgh*

Alan Geddes, and Bob Barnard, whom Graeme had invited up from Melbourne, were founder members of the All Stars. Thus began a second very active jazz phase which incl. a recording contract with Festival, tours and concerts, and his own TV series for which shooting began in August 1962, with Geddes, Harman (bjo), John Allen (bs), Wyatt, Barnard, and Laurie Gooding (reeds). Later replacements incl. Laurie Thompson, Ken Herron, Spedding, John Bartlett and ex-Dutch Swing College bassist John Van Oven. In April 1964 the All Stars began a two-year residency at the Chevron Hotel where they gained considerable celebrity, reinforced by the band's prolific recording activity. Bell took the band to Surfers Paradise Hotel for an eight-month residency in 1966, after which the group broke up.

Bell visited England again in 1967, playing with Terry Lightfoot's band, and on returning to Australia in Sept. 1968 formed a band for the club circuit. He formed a new All Stars group in 1973, incl. Cliff Reese, Paul Furniss (who had already been with Bell in the clubs), Ken Herron, Ed Gaston, and Geoff Proud (dms), and has kept an All Stars group together ever since. Subsequent members have incl. the following: clt: John McCarthy, Dave Ridyard, Jack Wiard; tpt: Bruce Johnson, Russell Smith, Bob Henderson; bs: Harry Harman, Richard Ochalski, Stan Kenton, Dieter Vogt; dms: Laurie Thompson, Ken Harrison, Alan Geddes; tbn.: 'Kipper' Kearsley, Don Thomson. Kearsley (b. 29/1/43, Leeds, England) had been involved in traditional jazz in Perth before moving to Sydney where he also worked with the San Francisco JB, the Harbour City JB, and toured with *One Mo' Time* and Bill Dillard's Blues Serenaders. Graeme Bell's All Stars groups have played numerous residencies in Sydney, incl. Rocks Push,

Marble Bar, Crows Nest Hotel and the Purple Grape at Homebush. They have produced a succession of recordings, worked on TV, toured for arts councils and the ABC, played festivals incl. the Breda festival in Holland in 1981 and 1983. In 1983 the All Stars went into the Don Burrows Supper Club and in 1984 began a two-year residency at Bondi RSL.

Graeme Bell has also been active during that period on an individual basis. In 1974 he was MD for a TV show, and was a founder committee member of the NSW JAS. In 1975 he travelled to the US to study jazz conditions under a grant from the Arts Council. In February 1976 he was flown to Cairns, Qld, to 'christen' a Grotrian-Steinweg piano at a mayoral function. He established a new record label, Sea Horse, in 1977, and appeared as a guest soloist at the 1980 Breda festival. In 1977 he was awarded the Queen's Silver Jubilee Medal, in 1978 was the subject of a *This Is Your Life* TV show, and was awarded the MBE. In 1986 he is still performing, maintaining a hectic tour schedule, and working on his autobiography. Bell is the most influential jazz musician Australia has produced, in terms of participating in enterprises which have altered the history of the music. His involvement in the foundation of the Uptown Club, the AJC, Swaggie records, his national and international tours, his early alliances with other artistically and politically radical groups, his high visibility during three successive boom periods for jazz—any one of these would make him significant. The total makes him a giant in the history of Australian jazz.

Graeme Bell, c. 1950   *Graeme Bell, c. 1986*

Bell, Roger Emerson   b. 4/1/19, Melbourne, Vic.   tpt/dms/wbd/comp./ldr

Louis Armstrong and Clarence Williams records on radio drew him to jazz in 1932, an interest shared with school friends Ade Monsbourgh and Spadge Davies. Drum lessons from Ron Crane, 1935, then played local dances with brother Graeme, whom he gradually converted to jazz. Listening to records with the guidance of Bill Miller, and Sunday afternoons at Fawkner Park Kiosk, consolidated Roger's jazz interest. Began trumpet, 1938, with two lessons from dance band musician Frank Arnold. From 1939 his career was parallel with Graeme's until the latter left for Qld in 1943. Roger led the band which opened in Heidelberg TH in July, the first opportunity to play undiluted three-piece front-line traditional/dixieland jazz in an extended residency. With Roger were Ade Monsbourgh (tbn.), Pixie Roberts, Laurie Howells, and to begin with, Gerry Watson (p.) and Arthur Knight (bs) from the dance band with which Bell's group alternated. Subsequently Don Banks, Jack Varney and Cy Watts in succession replaced Watson, and when Lin Challen (bs) was unable to join because of other commitments, Bud Baker came in on bjo. On 23 Sept. 1943, the band began at the Palais Royale, with Watts (p.), Bill Lobb (dms), Jim Buchan (bs). When Graeme returned from Qld in November Roger handed over leadership because of the distractions of his day work in engineering. He remained in Graeme's hands however, until the break up of the group in 1952. Roger had occasionally worked in other groups incl. Benny Featherstone's Dixielanders in the Manchester café in 1943.

Roger took a year off after leaving Graeme's band, then in 1953 resumed playing by joining Max Collie for several years. In the mid-'50s he worked at the Mentone LSC with Fred Parkes, Frank Traynor, and Graham Coyle, and in 1958 he joined the house band of the Melbourne Jazz Club, remaining well into the '60s. In 1963 he recorded *The Wombat* with his own Pagan Pipers (he had first led a recording group of that name in 1949), and this marked the beginning of a series of Pagan Pipers albums using various musicians incl.

Roger Bell, c. 1985

The Benders, 1985, with, from l. to r., Jason Morphett, Chris Abrahams, Lloyd Swanton, Andrew Gander

Ade Monsbourgh, Mal Wilkinson, Neville Stribling, Rex Green, Peter Cleaver, Lou Silbereisen, Bud Baker, Len Barnard, and featuring his own highly melodic compositions. During the '70s he freelanced and also led his own groups at different venues. Revisiting Europe in 1976 and 1981 he played with Claude Luter, whom he had met during the earlier tours with Graeme. Roger has continued to play into the '80s with his own groups and is a regular performer at AJCs.

The Benders (Sydney, NSW)

In November 1980 Dale Barlow (sax) introduced his quartet at the Paradise Jazz Cellar, Kings Cross, with Chris Abrahams (p.), Lloyd Swanton (bs), and Tommi Parkonnen (dms). This band became The Benders in December, with Andrew Gander replacing Parkonnen, who returned to Finland, in 1981. Barlow was originally coached by his father Bill, also a sax player, with later lessons from Col Loughnan. An alumnus of the Young Northside Big Band and a student in the NSW Conservatorium jazz programme, he had also worked with Bruce Cale, David Martin, and Phil Treloar. Abrahams (and Swanton and Gander) too had also come through the Northside band, and had earlier absorbed the music of Phil Treloar, The Last Straw, and Jazz Co-op, at the Pinball Wizz. Gander had worked with Kerrie Biddell and Bruce Cale before joining The Ben-

ders, as well as freelancing in the young contemporary scene. Lloyd Swanton graduated from the jazz studies course at the conservatorium in 1980, when he also worked and recorded with Jenny Sheard (vcl). Swanton has been the most diversely active member of The Benders, freelancing and working regularly with Roger Frampton's Intersection and Bernie McGann's groups at Jenny's and on tour. Like other members of The Benders, he had also been involved in the activities of Keys Music Association (KMA) since its inception in 1979. The Benders defined their intentions as 'A serious attempt to fuse modern improvisation with slapstick comedy'. The latter consideration is an important component of their energetic contemporary style, but is often overshadowed in reviewers' eyes by the members' genuine and often awesome musicianship. They were resident at the Paradise for two years, and also played occasional concerts, sometimes, still using the name the Dale Barlow Quartet. Barlow, one of the most creative jazz musicians to emerge in Australia in the '80s, played his last gig with the band on 23 Sept. 1982 before leaving for Europe, where he established an impressive reputation. For the farewell concert, he was joined by the equally prodigious Mark Simmonds who shared with Barlow a background in KMA. Simmonds's style drew broadly upon soul and free jazz, with influences ranging from the inescapable John Coltrane to local sax players Paul

Furniss and Merv Acheson. He established his reputation in the late '70s primarily as leader of the Freeboppers, which also incl. Gander and Abrahams. Simmonds has subsequently been less visible in explicitly jazz settings, moving sideways into other contemporary forms. He was replaced by about December 1982 by Jason Morphett, who had studied in the so-called 'classical' field at NSW Conservatorium. The Paradise residency finished in January 1983, and The Benders have since played in relatively short-lived venues incl. Jenny's Wine Bar and at concerts both in Sydney and in other capital cities. Their first performance at The Basement was in March 1985, and they have also played concerts for the Sydney Improvized Music Association. They have produced three albums (not incl. a solo LP by Chris Abrahams, and tracks on a KMA double LP), with Barlow on the first and Morphett on the later two. These not only document the band's performances, but also the composing talents of its members. In July 1985 the band embarked upon an extensive overseas tour embracing the Northsea and Montreux Festivals sponsored by departments of the Australian government. The Benders represent the most cogent, continuously active, and creative outcome of the upsurge in jazz activity which Australia has experienced since the late '70s.

Benham, William Michael Francis (Billy)
b. 27/10/28, Dubbo, NSW   p.
Began piano at nine. Played drums in high school jazz band and at country dances, where the pianist coached him in chords before persuading him to replace her when she retired. Left Dubbo for first professional engagement of 12 months at Katoomba, then moved to Sydney in 1952. Active in the concert scene and in the mid '50s worked with Ron Gowans at Segars ballroom Sunday jazz functions, Warren Gibson at Parramatta Rivoli, and at the Sunday jazz sessions at Phyllis Bates Ballroom. In 1957, as a foundation member of Club Eleven, was in the house band with Dave Owens (sax), Fred Logan (bs), Don Osborne (dms), and placed second in piano section of *Music Maker* poll. In 1960 recorded with his own quartet (Ed Gaston,

George Golla, Colin Bailey dms). Less visible on the jazz scene through '60s and '70s, but active in club and session work incl. recording with Bob Barnard, TV work with Jack Grimsley, and radio. In demand for backing visitors (Mel Tormé, Buddy Morrow). In the '80s, has played a long solo residency at Manor House Restaurant, Balmain, and freelancing as a sideman.

Bennett, Lawrence (Laurie)   b. 21/12/40, Sydney, NSW   dms
Began listening to jazz at age 14, took up drums at 15. In 1961 was with John Speight and Ford Raye at the Mocambo; later worked with groups at El Rocco, and established the pattern of extensive freelancing which has characterized his subsequent career. With Don Burrows at the Wentworth Supper Club in the early '70s, and in the late '70s and '80s led his own groups for a lengthy seasons in the Soup Plus. With Bob Bertles's Moontrane (1977), Richard Ochalski's Straight Ahead (1978–79), and led a quartet at the Orient Hotel, George St (1978). Has played concerts for the JAS, incl. sessions which he organized and led. Bennett is an outstanding example of the Sydney musician whose versatility and taste are so respected that he is able to survive on a freelance basis in both jazz and commercial areas. There are few musicians in Sydney with whom he has not worked, ranging stylistically from Noel Crow's and Nancy Stuart's bands, to Bernie McGann and Peter Boothman.

Bentley, David Keith   b. 31/8/43, Brisbane, Qld   keyboards/comp./vcl/ldr
His first major jazz activity came after moving to Sydney in 1962, when he joined the Riverside Jazz Group. Two years in Melbourne as a journalist, then with pop groups, which took him to London where he studied with Peter Ind (bs). Returned to Brisbane in 1972, and although backing visiting musicians at the Cellar Club, incl. Phil Woods, Richie Cole, Milt Jackson, Mark Murphy, most of his work was in a more commercial vein. Formed a jazz trio (Lach Easton bs, Geoff Proud dms) which played Qld JAS concerts backing visitors incl. John Sangster and Errol Buddle, and toured with Mark Simmons and Eddie D'Amico (tpt).

Bertles, Robert Anthony (Bob)   b. 6/3/39, Mayfield, NSW   saxes/clt/fl./some dms/p. comp./arr

Began playing professionally in 1957, and was involved with the modern scene that grew out of venues like the Mocambo in Newtown and found its focus at the El Rocco. Active in clubs, TV and the burgeoning local rock scene, including tours with Johnny O'Keefe and the Dee-Jays. To Melbourne, mid-1963, and co-led a quartet with Keith Barr (with Brian Fagan* bs, Barry Woods dms) at Fat Black Pussycat, Toorak, then to Adelaide where Barr and Fagan rejoined him to form a sextet (Keith Stirling, Billy Ross or Trevor Frost, Bobby Gebert) to play at the Cellar. Returned to Sydney late '64 and, apart from casual gigs took, over the band at Sammy Lee's Latin Quarter until 1966. In 1967 joined Jeff St John's The Id, then after three months began five years with Max Merritt and the Meteors, which took him to London where he became extensively involved in session work with, *inter alia*, Cliff Richards, Cilla Black, Allan Price, and also joined Ian Carr's Nucleus, a leading fusion band. Forced by conflicting commitments to choose between Max Merritt and Nucleus, he chose the latter, with its hectic schedule of touring, festivals, recording, TV work, and continued with this band even while based in Germany, 1975–76. Returned to Sydney, June 1976 and joined Col Nolan, remaining to the end of 1977. Started teaching in the jazz studies programme at NSW Conservatorium, formed Moontrane with Mike Bukovsky, Tony Esterman, Jack Thorncraft and Laurie Bennett, and joined John Hoffman's Big Band. In Sept. 1980 left for seven months in Europe on a study grant; on his return in 1981 he rejoined the jazz studies programme and went into the orchestra for the long-running stage show *Chicago*. Has subsequently continued to follow a varied and extremely busy schedule of concerts, festivals, session work and touring, not only in the more commercial areas but also maintaining a strong presence in strictly jazz performance. The beginnings of Bertles's career and

of a major development in the modern movement in Australia coincided. He has remained an essential figure in that movement by virtue of both his commitment and his unsurpassed musicianship. His extensive discography includes recordings with his own Moontrane, Richard Ochalski's Straight Ahead, and a duo album, *Misty Morning*, with Paul McNamara.

Biddell, Kerrie   b. 8/2/47   vcl

Studied piano for 12 years, but self-taught as a vocalist, with Don Burrows an important inspiration. First professional work was in commercials and backing Dusty Springfield. Joined the pop group The Affair, 1968, and with them in England as part of a contest prize. Joined Daly-Wilson Big Band, 1970, leaving in 1971. Travelled to US and Canada, working in TV, cabaret, and studio settings. Returned Sydney, 1973, active in session and concert work, had her own TV programme and released an LP which was acclaimed as the best female vocal album of the year. In 1974 played Expo 74 in the US and worked in Los Angeles and Las Vegas. Formed the band Compared to What, 1975. With fluctuations in personnel this has been her support group ever since. Credits incl. a long residency at Red Ned's (from 1977), Australia's first digital album, concerts with Dizzy Gillespie, Don Burrows, Judy Bailey, the Auckland Jazz Festival (1979) and the Perth and Adelaide Festivals (1981). In the '80s Biddell has also formed a female vocal group and taught at the NSW Conservatorium.

Black Eagle Jazz Band (Adelaide, SA)

Following the demise of the Southern Jazz Group in 1951, a new generation of dedicated traditionalists completed their apprenticeship in various short-lived bands before converging in the Black Eagle JB in about 1955 in conjunction with the functions begun at the Wayville Institute by the Adelaide Jazz Society. The core of the band was Alex Frame, John Pickering, Dick Frankel, Col Shelley (p.), Bob Wright, Glyn Walton, with Roger Hudson playing washboard on occasions. The raw, hot, rolling two-beat style was the main

---

* Also spelled 'Fagen' and 'Fagin' in the literature of the period.

embodiment in Adelaide of mouldy fygge music in the late '50s, and younger musicians who entered that fraternity were drawn into it by the sound of the Black Eagles. In 1959 the band changed its name to the St Vincent JB in order to shed any possibility of a motorcycle gang image, the new name being taken from the gulf that washes Adelaide's beaches. They moved into the Duke of Leinster Hall, running functions under the St Vincent Jazz Club banner, and began using drummers more frequently, John Day becoming the most regular. They played the opening of the Storyville Jazz Club in Nov. 1960, but gradually became inactive thereafter, although they regathered for the 1961 AJC with John Bradman (tba) and Peter Ubelhor (dms). Most alumni joined later bands; Col Shelley withdrew from jazz in the '70s, and died in September 1984.

Bland, Geoffrey Brian (Geoff)   b. 1928, Melbourne, Vic.   p.
Took lessons 1934–40, and was school pianist at Melbourne High School where he met Max Marginson, founder of MHS Jazz Society, Geoff Kitchen, Nick Polites, and came under the influence of the MHS music director Ray Fehmel. A founder member of Frank Johnson's Fabulous Dixielanders in 1945, leaving the band, and music in general, in 1951. Returned to jazz at the urging of Nick Polites whom he joined, with Graham Bennett, in the Atlantic Trio, formed to launch the VJC in 1968. Subsequently worked with Roger Bell, Tony Newstead, Ade Monsbourgh. Since 1980 has been a full-time musician, concentrating on teaching both privately and at Melbourne CAE.

Blott, Charles (Charlie, Porky)   b. 21/1/25 dms/tpt(?)
Mother played piano to some extent and at age six Blott began to play the drum in the Caulfield State School (drum and fife) band. He was given no formal musical training but was strongly influenced by the family collection of imported records, particularly those with solos by Bix Beiderbecke. In 1939 Blott was also influenced by the 'golden age' jazz of W.H. Miller's 3UZ session, and also liked

Charlie Blott, c. 1946, with Don Banks's Boptet, with, from l. to r., Lin Challen, Don Banks, Charlie Blott, Splinter Reeves, Doug Beck (?)

other styles of jazz. Soon after the war began Blott began to associate with Melbourne jazz lovers such as the Bells, Don Banks and the Atkins brothers. In 1941 played with Banks as a duo in association with Graeme Bell's band and later began to do small casual jobs with Jack Varney. He was also becoming influenced by the Gibson band and musicians such as Frank Coughlan, Bob Tough, and Ben Featherstone. In 1942 Blott took over as drummer for Coughlan and by 1943 had recorded alongside Coughlan, Featherstone and Tough. The same year Blott was deeply impressed by Artie Shaw's drummer, Dave Tough. Blott also played for the 1943 Kaminsky session alongside future 'progressives' Splinter Reeves, Don Banks, and Linton Challen. After 1943 Blott gravitated increasingly towards technically progressive music and musicians but maintained strong links with traditional/dixieland jazz. War-related events which had forced Blott to leave Coughlan after approximately six months led Blott into a variety of casual engagements but by 1947 his progressive attitude and enthusiasm saw him at the centre of the first Melbourne experiments with bebop and the setting up of the Modern Music Society jazz sessions. Blott also worked with Bob Clemens to establish Jazzart label and appears with the Errol Buddle Sextet on the first releases, recorded February 1948. The same year Blott toured with the Bell band, helped to organize Melbourne's first Stan Kenton-style band, the Fred Thomas Orchestra.

Blott also made time to rehearse with the newly formed Banks Boptet which, in November, featured at Melbourne's first public bop/traditional concert, Jazz Parade. This successful concert was organized by Blott and Graeme Bell. During 1949 Blott continued to play or record with leading modern line-ups such as the Boptet, Splinter Reeves's Splintette, Rex Stewart and His Jazzartists, and the Palm Grove Orchestra. He also recorded with the Bell band. During the early 1950s he remained a leading Melbourne drummer in both modern and traditional jazz, playing a variety of club, concert and radio engagements, personally organizing groups and working to promote concerts such as the Rhythm Festivals and the Downbeat concerts. In 1954 Blott played in the Melbourne bands which supported various overseas stars including Frank Sinatra, Frankie Laine, Nat King Cole, and on one occasion traded solos with Gene Krupa. Blott won the *Music Maker* poll in 1955 but with the advent of TV and the decline of Melbourne's concert era in the late '50s fewer opportunities occurred for playing jazz. In c. 1958 he played with Frank Johnson's band at 431 St Kilda Road and between 1969 and 1974 worked with Smacka Fitzgibbon. At present working with young musicians in '20s revival band, the Cairo Club Orchestra. Blott has been admired as a professional musician and remarkably versatile jazz man for decades but he was also responsible for greatly enriching Melbourne's cultural history by the role he played in generating the exciting jazz events of the late '40s and early '50s.

Compiled by John Whiteoak

Bloxsom, Ian Charles (Blocko)   b. 11/12/38, Mt Morgan, Qld   dms/perc./vibes
Began piano at five and drums at 15. To Brisbane in 1957 to study law; worked with the Varsity Five plus Two and joined Stan Walker, by whom he was much influenced, as also by Lloyd Adamson. To Sydney in 1965 and worked with Doug Foskett's Wentworth Hotel band for two and a half years. With Sydney Symphony Orchestra through the '70s, and also working with various groups incl. Crossfire. Recorded on John Sangster's *Lord of the Rings* project. Since 1979 has freelanced, mainly in session work.

Boardman, Don   b. 29/1/39, Melbourne, Vic.   dms
Began drums in 1955 and worked with various groups until joining Sny Chambers and the Bayside JB. With Frank Traynor and the Jazz Preachers, 1962–79. Withdrew from music for two years to develop his work in sound engineering. Returned to jazz as drummer with Dick Tattam's band.

Boothman, Peter Ralph   b. 2/9/43, Sydney, NSW   gtr/comp./ldr
Initially self-taught, then lessons from George Golla, Don Andrews, and classical tuition from Antonio Losada. Began working commercially in clubs in the late '50s, proceeding to jazz and pop/rock oriented groups in the late '60s, incl. the band of Jeannie Lewis. Extremely active as a freelancer, and formed his Ensemble in 1970. In addition, during early '70s was working regularly with other bands, running a guitar shop, teaching at the NSW Conservatorium, and touring with commercial packages incl. Kamahl. In 1973 resident for periods of both Old and New Rocks Push. Withdrew from music and lived out of Sydney, 1975, briefly returned and made a recording, Dec. 1975, before going to the US where he became involved in a counter-culture lifestyle. In the late '70s, returned to Sydney and resumed work in the contemporary jazz scene, playing with Don Andrews, Roger Frampton, and Ken James's Reunion band. Played Sydney Jazz Festival, 1979, and has subsequently concentrated on freelancing although led his own quartet for seasons at Soup Plus, 1985. Boothman's adaptability places him in the broadest range of stylistic settings, and since 1984 he has worked with Roger Frampton's Intersection, the bands of Joe Lane, Dave Ridyard, and Bruce Johnson. In 1984 he broadcast a solo performance of his jazz suite 'The Village', and joined Eclipse Alley Five, of which he was still a member in 1986.

Boston, Nicholas Charles (Nick)   b. 20/2/40, Ashtead, England   cnt/vcl/arr., some p.
Began playing in England during the jazz revival of the late '50s. Arrived in Brisbane in 1964 and formed a band with Marty Mooney who per-

suaded Boston to follow him to Sydney in 1966. Boston formed a band with Mooney, Pat Qua (p.), Des Bader (bjo), Neil Macbeth, Bob Learmonth (tbn.). In the early '70s formed his Colonial JB which brought together a nucleus of musicians who were to form the basis of Tom Baker's San Francisco JB incl. John Bates, Hans Karssemeyer, and Baker. Subsequently directed his interest away from two-beat approach to the NO idiom represented by George Lewis, with a band formed c. 1978 with Chris Williams (tbn.), Barry Wratten, Peter Gallen (bs), Lloyd Taylor (dms) and Graham Kellaway (bjo). Although there have been personnel changes, in spirit and flavour this, the Nick Boston NOJB, has continued to 1986, playing mostly pub work. In the mid '80s it is one of the very few bands (and certainly the longest surviving) keeping NO jazz alive in Sydney.

Bray, Earl Lloyd (Tich)   b. 22/12/24, Geelong, Vic.   reeds/fl./ldr/comp./arr
Began tinkering on piano at five, joined the Mitcham State School Fife Band in 1933, and then took formal tuition on a flute given him by a relative after Bray won a talent quest. In the army he switched back to piano and following discharge, bought piano and drums, hoping to start a dance band. Moved to Mordialloc where he met the Barnards and became a founder member of Len's band until it broke up in 1955 in Brisbane, where Bray settled. Worked at the Story Bridge Hotel with Rae 'Skippy' Humphreys until late 1957, and following a period of casual cabaret work, replaced Vern Thomson as leader of the Cloudland Ballroom band in 1960. He also formed his Mainstreamers, incl. at various times Lloyd Adamson, Stan Walker, Alan Arthy, Errol Clyde, Neil Wilkinson, Bill Boone, John Bosak. Left Cloudland after refusing to accommodate pop, and apart from occasional work such as with the Varsity Five and a two-year period at the National Hotel with Robin McCulloch, gradually withdrew from performance. The AJC at Brisbane in 1976 rekindled his interest and he returned to jazz at the Adventurers' Club in Oct. 1977. He sat in with Ken Herron at the Melbourne Hotel and worked with Mileham Hayes at the Cellar Club before working

briefly with the Caxton St Band in early 1981. In October he shifted to the Vintage Jazz and Blues Band with whom he was still playing in 1986. In 1984 he began teaching music for the state education dept.

Brentnall, Kenneth John (Ken)   b. 1/2/25, Melbourne, Vic.   tpt
Took up trombone and cornet as a child and played in brass bands in Daylesford and Shepparton, Vic. At 13 joined Ned Tyrrel's orchestra at the Regent Theatre. Worked in the Melbourne clubs and at American Red Cross until joining the army in 1942, serving in the Entertainment Unit headed by Jim Davidson. On discharge, returned to Melbourne and played the opening of Sammy Lee's Storklub, 1946. Became a central figure in the Melbourne bop movement of the late '40s. Moved to Ciro's with Bobby Limb, then in 1949 joined Jack Brokensha with whom he toured extensively, playing the flourishing concert scene. Moved to Sydney and worked in clubs and concerts until joining Jim Gussey's ABC orchestra. In 1956 retired from music to work as a director/cameraman for TCN 9, but returned to playing in 1959. Played lead trumpet with Tommy Tycho's Channel 7 orchestra until its disbandment in 1965. Since leaving Brokensha he has been primarily a studio and club musician.

Brinkman, Alan Ernest (Brinky)   b. 24/8/17, Glen Lusk, Tas.   reeds
Moved to Hobart as an infant and started at age nine on a clarinet on loan from a church group. Learned from Alex Caddy, whose stress on the importance of harmony provided a basis upon which Brinkman later established his reputation as an improviser when few professionals understood its principles. His first work was for church functions, magic lantern and silent movie shows. Introduced to jazz through the records of dance band musician Sid Collins (sax), whom Brinkman recalls as 'the first to play jazz in Hobart'. In c. 1932 Brinkman began to play dances with blind pianist Cecil Foster. The church objected, took back its clarinet, and Brinkman bought his own, which he still plays. Active in dance bands incl. those of

Tate's coffee lounge band, Melbourne 1942, with, from l. to r., Al Royal, Alan Brinkman, Billy Hyde, Wally Nash

Alan Brinkman, 1977

Brian Horne and John Knowles, and in 1936 joined the leading local band led by Ron Richards. In the late '30s and early '40s, Brinkman was also a regular at the jam sessions at Fouché's Stage Door, until he moved to Melbourne where he joined Stan Bourne at Tate's Tea Rooms in 1941. Frank

Coughlan took him into the Trocadero, and upon Brinkman's departure in 1942, guaranteed him a chair in the orchestra should he return to Melbourne. Having joined the army, Brinkman was posted back to Tas. where he played in a service unit, the Tasmaniacs, before being posted to Qld, then New Guinea where he contracted Dengue fever and malaria, for which he was returned to Hobart in 1945 for hospitalization. Resumed participation in the Stage Door sessions where Melbourne saxophonist Sel Chidgey introduced him to the harmonic structure of the blues. In 1947 he was back with Ron Richards, and in 1948 with Stan Huntington (tpt) at the Belvedere and Don Denholm at Wrest Point. In 1949 he worked with the ABC orchestra, and continued to be ubiquitous in Hobart's pit, cabaret, and dance orchestras through the '50s. His reputation as a jazz improvizer predated the war, and in the '50s he began sitting in with Tom Pickering's band at the town hall dances, subsequently joining Ian Pearce's sextet. Since the '60s he has regularly visited Japan where he has become well known through frequent sitting in and occasional engagements. On one of these visits in the '80s he added soprano sax to his array of reeds. Although less active musically since 1965, he continues to guest with Pearce/ Pickering and plays for JAS functions. Brinkman shows complete poise through jazz styles ranging from traditional to West Coast; he is certainly one of Tasmania's most durable and accomplished jazz musicians, and it is incomprehensible that he has never been commercially recorded.

Brisbane

Up to the Second World War the history of jazz activity in Brisbane followed the pattern evident in the larger centres albeit at a lower level. The first wave of jazz consciousness can be dated from the tour of Billy Romaine's 'jazz band' in 1917, to the period of the nation-wide slump during the Depression. Swing arrived in 1936, bringing the formation of the Brisbane Swing Club, and within a few years, the war stimulated more extensive mingling among Australian musicians and contact with the Americans. Brisbane's comparatively small size and its greater proximity to the Pacific theatre of war meant that the effect of the US servicemen was more marked than in the south. So radically did they alter the character of local entertainment that some public dance halls, incl. Trocadero and the Cocoanut Grove, were unable to reclaim the more conservative local populace and, at least in part for this reason, closed down as the Americans left. In the meantime, however, the visitors had created an opportunity for many of the local musicians to work in a hotter style which encouraged improvisation. The Cocoanut Grove and the Bellevue Hotel were two venues where young musicians like Keith Bentley (reeds), Jack Kenny (tpt), Harry Lebler (dms) and Cliff Reese were able to establish reputations as up-and-coming hot players. As in other parts of Australia, the end of the war found a pool of young musicians, who had had unusual opportunities to develop, and of demobilized veterans of the dance band industry, some of whom had had their progressive interests stimulated by wartime contacts. Perc Garner became an important mentor of the former, coaching many of the younger men and giving them regular opportunities to work in his Sweet 'n' 'Swing' bands, which he kept together throughout most of the '40s. Many of the older musicians working today in a swing/mainstream style served their apprenticeship under Perc Garner. Their jazz interests were also shared by a small proportion of other musicians whose main income was provided through dance band and cabaret-style work. The opportunity to express those interests under that billing in regular public performance had to wait until the advent in Brisbane of the jazz concerts. Until then 'jamming' was something smuggled inconspicuously into cabaret or an activity pursued in private parties for which musicians like Joe Allen, Darcy Kelly (bs), Jimmy McLaren (reeds), and Harry Lebler would sometimes hire a hall. The exception to this was the work of the band named the Canecutters, organized by Sid Bromley. Consisting of professional musicians with highly developed jazz abilities, they played an essentially contemporary style compounded of swing and dixieland.

These three overlapping groups—the established dance/cabaret musicians, the younger generation coming into their ranks, and the Canecutters—found their main forum for jazz in the concerts which began in about 1950 as ripples from Sydney and Melbourne. Jack Brokensha's opening at Theatre Royal, 8 April 1950, to great acclaim, inspired more experiments in the promotion of local jazz concerts modelled on the 'band battle' style in the southern capitals. The most usual format was built around imported big- and small-band leaders like Brokensha, Wally Norman, Bob Gibson, Bob Limb, Billy Weston, Jack Allan, with support from the more jazz-oriented locals. Many musicians who might otherwise have remained categorized as dance or nightclub performers thus gained experience and built reputations as jazz musicians. In addition to names already mentioned, Eric Hall, Errol Clyde (tbn.), Lloyd Adamson (tpt), Maurie Dowden (p.), Stan Walker (p.), Alan Arthy (reeds), and Eric Wynne (bs) were among the local stalwarts of the concert era. They performed under various concert titles and in conjunction with different promoters incl. Limb, McColl, Ron Brown and Ron McCune, and latterly Jim Burke under the usual billing of 'Public Command Jazz Concert' or some variation. As such, the jazz content began to grow stale and to diminish in favour of variety presentations. By the late '50s the jazz concert movement faded out, as it was doing throughout Australia. Apart from the concerts there was little jazz activity in Brisbane during that decade except for the activities generated by Sid Bromley, and the regular jazz policy instituted at the Story Bridge Hotel. Organized by Rae 'Skippy' Humphreys* (tbn.) in mid-1955, the Story Bridge jazz cabaret, called Storyville, was advertising jazz five nights a week at its peak, using a succession of musicians that incl. Tich Bray, Len Barnard, and established locals like Frank Chernich (bs), Billie Blackmore (p.), Bill Townsend (bjo), and Maurie Dowden. Following reductions in the number of nights through the late '50s,

---

* Alternatively spelled in contemporary written reports also as Ray and Humphries, in various permutations.

Humphreys was finally replaced by a continental band in 1959.

Towards the end of the '50s the old dance hall tradition, which had been largely maintained by Billo Smith, entered the national decline, and professional musicians were forced to concentrate on session and night-club work, but were also able to find some outlets in the extension of live hotel music that was taking place. Brisbane musicians have also found some compensation for the comparatively limited local opportunities in the proximity of the large resort areas clustered along the Gold Coast. Although it has never sustained a dedicated and explicit jazz scene, the Gold Coast has over the years attracted many musicians from various parts of Australia who had established jazz reputations—Ron Gowans (reeds) and Rick Farbach (gtr) being two notable examples—and provided employment also for Brisbane musicians living only an extended commuter distance away. Other musicians, however, saw their futures elsewhere, and Brisbane suffered the same emigration of many of its progressive talents as other smaller capitals have witnessed. Some drifted south to larger centres, or took work on shipping lines. Nonetheless, jazz activity continued, even though deprived of the regular concert forum. Desultory episodes involving progressive jazz styles were witnessed at some of the night-clubs, like the Si Bon and the Primitif, some of these activities continuing through into the next major phase in Brisbane's jazz history. Again, Sid Bromley was an indefatigable force for jazz during this period, organizing club functions, broadcasting, and performing. His involvement with medical school functions in the late '50s provided the tenuous thread that linked his dixieland/mainstream activity with the more specifically traditional movement which began in the early '60s. Brisbane is distinctive in not having developed the kind of purist mouldy fraternity which was evident through the '40s in other capitals. The formation of the Varsity Five in 1960, from musicians who had been involved with the medical school dances, was in part the expression of a revivalist spirit that had been evident two decades earlier in Melbourne. Various members of this band at different times were veteran professionals, but essentially the group represented the beginning of a new stream. With its undergraduate and upper middle-class audience, its deliberate orientation to early traditional styles, and its strong amateur component, the Varsity Five became the focus of Brisbane's equivalent of the 'trad boom'. As elsewhere, the boom saw a sudden spread of interest into more remote areas. Clubs and bands sprang up in more northerly centres like Cairns, Townsville, Rockhampton; in the capital itself, the Varsity Five helped spawn other groups.

The quiescent period in the late '60s affected all jazz styles in Brisbane. The progressive scene, which had always been relatively weak, enjoyed a brief surge of energy in the mid '60s in the form of the Queensland Jazz Art Society which featured the playing of most of the city's mainstream and bop-tinged musicians—Adamson, Arthy, Clyde, Terry Widdowson (p.)—and brought together a big band directed by Arthur Hubbard. By 1970, however, there was virtually no focus for more modern styles in Brisbane, and they returned to night-clubs both local and on the Gold Coast, where the music was scarcely perceived publicly as having anything to do with jazz. The thread of more traditional forms was also broken during this period, and the next important group, the Pacific Jazzmen, was formed by musicians who had had little if any involvement in the Brisbane boom of the early '60s. The organizers of the Pacific Jazzmen also fostered what would become Queensland's most durable formal jazz society, the Brisbane Jazz Club, in which there was some mingling between the city's amateur and professional musicians, and which continues to be extremely active. The foundation of the Pacific Jazzmen in 1969 represented the beginning of a new growth in traditional jazz in Brisbane, accelerated by the formation of the Vintage jazz group, by the energetic lobbying conducted by Mileham Hayes, and in particular by Brisbane playing host to the 1976 AJC. Although this last did comparatively little to irrigate the channels of more modern jazz styles, it helped swell the minor flood of traditional jazz activity. Since that time the main standard-bearers of traditional jazz have been the Vintage group under its various names, Hayes's groups, and the Caxton St JB. Mention should also be made of the Sugar Daddies, partly because of the larger contributions made by its leader, Ray Scribner. In a career spent in Newcastle and Brisbane, Scribner has served jazz as a musician, member of jazz clubs and AJC committees, as a journalist, historian and lecturer. Mileham Hayes has continued to extend and broaden local awareness of the music through his involvement with the Cellar clubs, his vehement publicizing of the jazz cause, and promoting festivals which have enabled the local populace to hear, and local musicians to work with, a wide stylistic range of American and interstate musicians of the first order.

In the '80s Brisbane has enjoyed the benefits of the nationwide increase in institutional recognition of jazz. The activities of the Brisbane Jazz Club are both ambitious and successful. It instituted an annual jazz carnival in 1980, organizes concerts in conjunction with the city council, has begun to establish satellite clubs in other parts of the state incl. the Gold Coast, and in the meantime maintains a big band and promotes the balls and other regular functions which constitute the usual activity of most state or city jazz clubs. Its patronage extends to all styles of jazz, so that it provides a platform for relatively free intermingling of musicians who formerly tended through force of circumstance to pursue separate paths. The Jazz Action movement arrived in Brisbane following initial meetings in 1979, and although its charter specifies modern jazz as its concern, it has stimulated considerable activity, particularly in styles which have enjoyed less favour in the past. The cause of more contemporary forms of jazz has also been aggressively championed by the first Queensland state jazz co-ordinator, Ted Vining. Various jazz studies courses have created a pool of young talent. In addition to school orchestras such as those coached by Roy Theoharris and ex-Sydney entrepreneur Greg Quigley, a tertiary diploma in jazz studies was instituted in 1984. The spread of community FM radio to Brisbane in the form of 4MBS-FM has vastly increased the amount of jazz broadcast locally. As elsewhere, it is too early to assess the long-term effects of this vigorous and multi-faceted jazz activity. On the one hand, it is providing a medium through which many veterans of the cabaret and night-club scene are able to receive overdue recognition for their jazz capabilities. At the same time, young blood is being introduced to the local jazz movement. While both of these are causes for satisfaction to those interested in the music, Brisbane manifests the symptom common to most Australian jazz centres: the new young talent who will carry the music into the next century is almost wholly committed to post-traditional jazz forms, and this city, like most others, prompts the question as to whether or not the end of era of the earlier styles of jazz which dominated the music in this country is intimated in the rising average age of its exponents.

Broadcasting

Whatever may have been its shortcomings, in quality and quantity, the broadcasting of jazz in Australia, over some 50 years, has been a major factor in the national perception of the music as a cultural force in the mid-1980s.

It is true that jazz on radio (and on TV in recent years, to the extent that it does occur) has been almost exclusively, except in the very early years of broadcasting, confined to the ABC. But granted that, it is equally true that jazz programming on the ABC, in continuity and regularity, has depended on the presence within the ABC's power structure of those who believed that the music deserved a place in its programmes, or, if that was not the case, on sufficient pressure being exerted influentially from the outside to force the ABC to make concessions to jazz listeners.

The broadcasting of jazz had its origins in Sydney. The momentum it achieved was due entirely to the efforts and enthusiasm of Ron Wills. A Sydney commercial station, 2UW, began importing Brunswick jazz records from England in 1935 and inaugurated two 15-minute sessions at 9 p.m. each week. When Wills found that the station had exhausted its supply and was repeating the programmes he wrote and offered to make available records from his own collection. He then took over the presentations and out of this was born the 2UW Swing Music Club (later the Sydney Swing Music Club.)

The ABC, as Wills recalls* was a 'reluctant starter'. He had approached its management early in 1935, suggesting a programme, but was told that the ABC saw little merit in broadcasting jazz when it had so much classical music at its disposal.

It was not till 1938 that the ABC showed interest. Wills had been corresponding with the jazz critic and writer, Leonard Hibbs, in London and as a result, Hibbs offered the ABC a series of scripts called *The History of Jazz*. He told the ABC that Wills had most of the recordings on which the scripts hinged. This led to Wills being commissioned to prepare the scripts for a national (i.e. relayed through its Australian network) series. It went to air in May 1938 as the first nation-wide broadcast of jazz in Australia.

Wills was not invited back to the ABC until August 1942 when he began a series called *Swing Notes*, using as his theme the opening of Ellington's 'It Don't Mean a Thing', with Ivie Anderson as singer. This series continued spasmodically during the next three years, during which Wills was consolidating his reputation as a leading jazz critic with his reviews and articles in the music magazine, *Tempo*.

Meanwhile, in Victoria, another pioneering jazz authority was well to the fore. He was William H. (Bill) Miller, a lawyer, Oxford-educated, who had returned to Melbourne in 1938 with a fine collection of jazz recordings and an unquenchable enthusiasm for the music. In 1939 he began a weekly programme on commercial station 3UZ called *Jazz Night* using his own record collection. Fledgeling jazz musicians, like Graeme Bell and his brother, Roger, Ade Monsbourgh and others came to know Miller through this programme and he exercised a formative influence on their careers, as he did on other local jazz musicians, from this point.

Radio jazz was by this time actively flourishing in South Australia. I note that Andrew Bisset in his

book *Black Roots White Flowers* credits me with having begun 'what was probably the first jazz programme on ABC radio in South Australia.' But the honour belongs to Kym Bonython who, though still only in his teens, had somehow persuaded the ABC to let him present occasional local jazz programs in 1938 (following on the national Wills series) which he continued till he joined the RAAF in 1940. Kym, as he recalls in his autobiography, *Ladies' Legs and Lemonade*, was hooked on jazz as a St Peter's College schoolboy and bought a drumkit that he played as an accompaniment to his jazz records.

In mid-1940 I was involved in school broadcasts as a freelancer (I was a high school teacher at the time and one of my English students was Bryce Rohde) and, having by this time got to know ABC programmers fairly well, I suggested that I might conduct a regular weekly jazz programme using my own records to supplement the few that existed in the ABC library. An 'experimental' series was agreed upon. So to my theme, Ellington's 'Drop Me Off at Harlem' (which was also the title of the series), the show was launched and ran, in fact, till the end of 1943.

By this time I had joined the ABC staff, in charge of educational broadcasting, but the free-and-easy and somewhat confused atmosphere of the ABC in those early days left me at liberty also to dabble in entertainment broadcasting (admittedly no one else was much interested in it). By this time also, along with William V. Holyoak (a Wills-Miller of early SA jazz), Maurice Gerdeau (who had the best collection of Bix Beiderbecke records I have ever seen), Maurice Le Doeuff (a brilliant jazz clarinettist) and a few other kindred spirits, we had founded the Adelaide Jazz Lovers' Society, which brought me into contact with local jazzmen who came to hear records and also for occasional jam sessions. These included Dave Dallwitz (with whom I had shared a jazz friendship since our Adelaide Teachers' College days in 1936).

So towards the end of 1943, to replace my weekly records show, and with the limited money I could squeeze from the local budget (I fear some of the educational funds went into it too), I began a

---

*In letters and conversations I have had with him

monthly series of live jazz broadcasts—which certainly were the first of their kind in Australian radio. They featured groups led by Le Doeuff, Dallwitz, Alf Holyoak, Bobby Limb (who with the pianist 'Beetles' Young was playing with Harry Boake-Smith's Palais Royal Band in North Terrace at the time), Bruce Gray, and others. These broadcasts were continued sporadically till I left Adelaide for Sydney in 1946.

Simply because there was no one going to bat for it after that, ABC jazz in SA went into decline, a situation seized on by Bill Holyoak who began a weekly jazz show on commercial station 5AD. His show ran for 15 years—and would certainly rank as the most noteworthy contribution yet made to jazz by any Australian commercial radio station.

Though my official position with the ABC in Sydney was federal script editor for educational broadcasting, I soon found light entertainment activities dumped in my lap. Indeed for a time I had a dual role and acted as director of Light Entertainment.

I found the weekly *Swing Notes* programme—I think it was then called *Swing Show*—almost dead on the vine. I decided to give jazz on the ABC a facelift and began by organizing *Thursday Night Swing Club* (modelled on a jazz show I had heard in the USA a couple of years before). Relayed to all states, it alternated a compered show with records with 'live' jazz groups. Allan Saunders, Ron

Wills, Ellis Blain** (one of the most vigorous proselytizers for ABC jazz in its history), Wally Norman (a fine musician of the day), Eric Dunn and others took turns as comperes. The live bands, which came from various states in turn, included Graeme Bell's Dixieland Band (its initial broadcasts were from the first, 1946, Jazz Convention in Melbourne which I sent Blain down to record), Dave Dallwitz's Southern Jazz Group from Adelaide, Sid Bromley's Brisbane Canecutters, the Pearce-Pickering outfit from Hobart, Kevin Ryder's Harbour City Six (which included Don Burrows), Ken Flannery's PJJB (with Ray Price) and a number of others. Thursday Night Swing Club, with various changes in format, ran for 13 years till 1959 when its title was changed to *Swing Club* and its day to Friday; in 1961 it was re-named *ABC Jazz Club* restored to Thursdays, and it ran till 1965. I should mention that in 1949 I had persuaded a rather stuffy ABC Concert department to tour Graeme Bell's band on its concert circuit after the band's triumphant return from its first overseas visit. Since the ABC then, by law, was required to broadcast ('live' or by recording) every concert performance, the 1949/50 *Swing Club* was pleasurably loaded with Bell performances.

There is no question that the 1950s and 1960s were the golden years of jazz on the ABC. It would be false modesty for me to deny my hand in this since, during this period, I had assumed sole control of all ABC programmes and I had determined to implement my belief that jazz should have an important place in the national broadcasting service.

I heard of a Brisbane announcer, ex-RN, who was trying to have local jazz programme launched without much success. I went up to see him, found he had a wonderful collection of records and arranged for him to take over the Saturday morning show with the title *Rhythm Unlimited*. We chose the theme jointly 'String of Pearls'—I forget whether it was the Goodman or Miller version. It began on 1 March 1952 and was an instant success; listeners immediately took to Eric Child's relaxed, low-key style (for he it was). Through various little changes (*World of Jazz; Jazz on a*

**In his book, *Life with Aunty* (1977) Blain claimed to have begun jazz broadcasts in 1938 in Sydney, soon after his arrival from Hobart. However there is no ABC archival evidence, in the material generously made available to me, to support this claim. Blain joined the RAAF in 1939 and it is possible he did occasional jazz broadcasts on local Sydney radio before this: he certainly was a regular and accomplished jazz broadcaster on his return, as a staff announcer, to the ABC in 1946. When he returned to Hobart in 1950 as Tasmanian ABC programme director he put his knowledge and enthusiasm into practical effect by launching a national series with the English jazz pianist Arthur Young who had then settled in Hobart and who had made many recordings in London with musicians such as Danny Polo, George Chisholm and Tommy McQuater and as accompanist for the Australian jazz singer Marjorie Stedeford. Blain also organized a splendid series of local jazz shows under the title of *After Dark*, featuring Cedric and Ian Pearce, Alan Brinkman, Tom Pickering and other Hobart jazzmen of the time.

Saturday) it ran until February 1983, and over 31 years was thus the longest-running jazz show in the ABC's history.

Kym Bonython, whom I had not forgotten, was next on my jazz list. It seemed to me appropriate, in decentralizing the ABC's jazz output, that there should be a programme from Adelaide, and Bonython began *Tempo of the Times*, one of the best jazz shows the ABC has ever had, each Friday night in peak time (8.30 p.m.) from September 1953. It ran as a weekly show till the end of 1971, when, for various reasons it was made to alternate with another programme Eric Child had begun on Monday nights, in 1965. Bonython's session concentrated on the latest releases—he imported huge quantities of records from America—it was indeed a jazz cornucopia for listeners all over Australia.

I left the ABC in 1976; it was no coincidence that Bonython's programme was rudely stopped at this time—rudely because he was given only a few days' notice of its cessation and received no letter or expression of thanks for his magnificent contribution to ABC jazz over so many years.

Eric Child's Saturday morning show was so successful that the next step was to find an evening spot for him. In September 1957 he began *The Late Night Show* which followed *Tempo of the Times* and made Friday night a paradise for jazz listeners. Its title was changed to *Workshop on Jazz* in August 1963 and ran as such into the 1970s.

I referred above to a Monday night Eric Child show. This was *Jazzography* which began on Thursday nights in mid-1965, replacing *ABC Jazz Club*. It was then moved back to Monday nights where in 1966 it became *Discourse on Jazz* until it began alternating with Bonython's *Tempo of the Times* in 1971.

Eric Child took over the Friday night spot on a weekly basis in 1976 after Bonython's contract was terminated. It was clear that the change was dictated by parsimony; Child had now left the ABC staff; he was retained on a package deal, and it was hoped thus to save money by dropping Bonython (who was only paid a relative pittance anyway). Furthermore, Friday night now only had one jazz programme instead of the former two, because

Child's late night show was subsumed into this single spot, renamed *Jazz on a Friday Night*. Not that Child was immune from ABC bureaucratic anti-jazz thinking which was manifest at this time; when he retired from the ABC in the previous year (1975), it was decided to cancel his Saturday morning programme to reduce the fees now payable to him as a freelance broadcaster. But at least on this occasion, a vociferous outcry from jazz listeners and some press support, persuaded the ABC to change its mind.

However, at least in 1965 (my final year as programmes head), my last move for ABC jazz was to contract Arch McKirdy (whose work on commercial radio I had long admired—his jazz shows on Sydney's 2UW and later 2GB were the last vestiges of the music on commercial radio) to begin, in August of that year, his *Relax With Me*, nightly from 10–12 p.m. This was not only one of the most notable sequences of continuous jazz ever presented on the ABC, but it was also in content a brilliant compilation that covered the whole spectrum of recorded jazz. When McKirdy joined the staff of the ABC in 1972 in an administrative capacity, the programme was handed over to Ian Neil (who had had considerable experience as a jazz presenter in *Thursday Night Swing Club* and its successors) and he continued it with impeccable jazz taste until 1983, when he too retired from the ABC.

Although the programme still continues, compered by Ralph Rickman and others, it has a tendency occasionally to stray into non-jazz fields. Nevertheless Rickman is a polished presenter and is especially effective in publicizing national jazz activities.

Unfortunately, listeners to jazz on the ABC are now, in the mid-1980s, in a position where they must be thankful for small mercies, because it is sad but true that from the late 1970s jazz has taken a nosedive on ABC radio. Whereas in the palmy days of the 1950s and 1960s, there were at least three jazz shows a week nationally broadcast in prime time (between 8 and 9.30 p.m.), now there is but one. Apart from being almost wholly relegated to the listening ghetto of the late night hours, jazz has been reduced since the end of the 1970s by some 50 per cent of air-time on ABC–AM stations. On a state (i.e. local) basis, jazz is virtually non-existent except for a weekly programme in Queensland conducted by Garry Ord.

On FM where the ABC could and should give jazz a reasonable proportion of prime time (since ABC–FM is largely devoted to music), the position is disgraceful. There are but two weekly periods, from 10.30 p.m. to 12.30 a.m. on Mondays and Tuesdays, and jazz listeners who could well enjoy the superior reception of their music that FM offers, but do not wish to stay up so late, can, in effect, lump it.

Not surprisingly, therefore, the many so-called community FM stations which have been licensed in the last decade, have capitalized on this instance of the ABC's dereliction of its responsibility to provide 'adequate and comprehensive' programmes. These FM stations, especially in the Sydney and Melbourne areas, present a wide selection of the best in recorded jazz at mid-evening times, notably 2MBS–FM (in Sydney) which pioneered this move, on five nights a week. There are bi-weekly or weekly programmes on at least a dozen other stations throughout Australia. But unfortunately the restricted reception areas of these stations prevent the greater part of the Australian listening population from enjoying this jazz alternative. And of course, except for isolated and infrequent instances, jazz is ignored by commercial stations.

There are two postscripts to this general account of the history and present situation of jazz in Australia. The first is that the ABC, which maintains perhaps the largest concert circuit in the world, has shown little interest in the place of jazz in that circuit.

The other postscript is that, mainly through the hard lobbying of Don Burrows (and, I suspect, the clout he carried through his teaching activities at the NSW Conservatorium of Music), ABC television began in a sporadic series of television jazz shows under the title of the *Don Burrows Collection*. This has not only featured, to splendid musical effect, some of our best-known jazz groups, but has also had access to some outstanding overseas jazz packages. But here again, the programme is invariably relegated to a late hour, in the vicinity of 11 p.m. Again the question can be fairly asked— why not an earlier time, and a regular weekly feature, at a time in its history when jazz musicianship in this country has never been at a higher nor more versatile level?

Clement Semmler

Brokensha, Jack Joseph (Jazza) b. 5/1/26, Adelaide, SA   dms/vibes/ldr/arr./comp
Began on xylophone, playing his first paid gig at six. Took piano lessons, then learned drums from his father. Broadcast on the ABC at nine, and at 14 was percussionist with the ABC, and the Adelaide Symphony Orchestra in which he played under Dorati and Beecham. Entered the RAAF in 1944, serving in No. 3 Mobile Entertainment Unit and becoming interested in jazz. An American serviceman coached him on the brushes, and in 1950 he was to create a mild sensation at the Galleone playing out in front of his band just with brushes and snare. Left the RAAF in 1946, went to Melbourne and played in Claridges, then returned to Adelaide later that year, where he rejoined the ASO, played the Cairo Club (formerly the 400 Club), then, with his own quartet (Bail, John Foster bs, Ron Lucas p.) opened a coffee lounge in Glenelg. He took the band for a brief season in Perth on Fuller's circuit, then to Melbourne in 1947–48 where, initially under the name the Rockets, its after-hours jamming became a focus for the early bop experiments, even though the quartet's usual style was a more commercial progressive swing. He also played the Galleone coffee lounge and the New Theatre concerts of 1947. His quartet, now under his own name, boasted Buddle, Foster, with Lucas being replaced by Ron Loughhead early in 1948, and Edwin Duff and Marion Morley (vcls). Putting together a show which presented a mixture of jazz, light classics, and novelty items, he began frenzied touring through the eastern states, establishing a national reputation as well as suffering a breakdown which left him recuperating in Adelaide for the first half of 1949. He resumed concert work in Adelaide, then moved to Sydney in late '49

where he played at Golds' and the Roosevelt, and with Loughhead's imaginative compositions and arrangements, assembled large concert orchestras, one of which recorded with Rex Stewart. Following another interval in Melbourne, he embarked on a tour to Brisbane in 1950 with Ken Brentnall replacing Buddle and John Mowson replacing Foster. The group disbanded in Brisbane and Brokensha stayed on until early 1951 playing concerts, ABC work, and with Jack Thomson's Cascades quartet. Back in Adelaide, he teamed with Bryce Rohde, who accompanied him to Canada in March 1953 in response to Buddle's invitation. The three worked at the Killarney Castle, Windsor, Ontario, prior to the formation in Dec. 1954 of the AJQ, with which they remained until the band broke up in Australia in 1959. Brokensha then worked in pubs and TV for ten months in Sydney before returning to Detroit where, after nearly 10 years in TV, he established his own music production company, in the meantime continuing to play and record in small band settings.

Bromley, Sidney Joseph (Sid)   b. 7/6/20, Brisbane, Qld   dms/clt

As a youth began buying records, guided by Ron Wills's reviews in *Tempo* and *Music Maker*. Took up clarinet in 1940 and started a local jazz club, followed by the establishment in 1941 of a new Brisbane Swing Club. A combination of council regulations and wartime enlistment ended the club and Bromley, who had been drafted in 1940, spent the rest of the war in various postings. In Mel-

Sid Bromley, 1958

The Canecutters, c. late forties, with, from l. to r., Joe Allen, Maurie Dowden, Cliff Reese, Jack Thomson, Darcy Kelly, Eric Wynne, Max Humphreys

bourne he met Bill Miller and heard local musicians incl. Bill Cody and Benny Featherstone. In 1944 in Bougainville he associated with Tony Newstead, Will McIntyre, Don Reid (dms) and formed the Bougainville Jazz Appreciation Group which promoted record recitals, jam sessions and issued a magazine. Discharged in May 1946, Bromley re-established the Brisbane Swing Club, renamed the Brisbane Hot Jazz Society in 1947. On 24 March 1947 he organized a broadcast to publicize the club, with Jack and Vern Thomson, Jock McKenna (sax), Cliff Reese, Roy Theoharris, Gordon Peake (bs), Jack Hawkins (p.), Max Humphries. This led to a request from the ABC to assemble a regular broadcasting jazz band and Bromley formed the Canecutters with the basic personnel of Jack Thomson, Reese, Humphries, Eric Wynne, Maurie Dowden, Joe Allen (vibes), and later Arthur Howard (ten./tpt/tbn.) and at various times Darcy Kelly (gtr). Bromley managed the Canecutters throughout its lifetime, recruiting various musicians incl. Gordon Benjamin (sax), Jack Brokensha, and drummers Neil Wilkinson, Stan Penrose, Billy Johns and Harry Lebler. Reese left for Sydney in 1951 and was replaced by Lloyd Adamson. The Canecutters made numerous broadcasts, incl. the first national programme by a Brisbane group on 3 Aug. 1950, and also played the early City Hall concerts. In 1948 it won *Tempo*'s award for top Australian group. The band ceased to operate in around 1952, at about the same time the Brisbane Hot Jazz Society went into abeyance. In 1955 Bromley started at the Story Bridge Hotel with Len Barnard (p.) and Tich Bray, and in 1957 organized the Brisbane Jazz Club, based at La Boheme night-club and publicized through his weekly jazz broadcast on 4KQ. The club's brief success prompted similar activities at the Si Bon and the Primitif, but by 1958 it went into recess for lack of support. Bromley reactivated the club in 1959 with a house band consisting of himself (dms), Terry Hickey (p.) and Lachie Thomson. With the addition of Bruce Dodgshun and Bruce Hyland (bs), this became the usual band playing medical school functions at which members of the Varsity Five gained early experience sitting in, in 1959–60. Bromley established another short-lived jazz club at the Grande Hotel in 1961 and then reopened the Brisbane Jazz Club at Si Bon on 4 March 1962. (It should be noted that this succession of clubs was not connected with the Brisbane Jazz Club later established by Tony Ashby and Bruce Haley.) Bromley led a trio that incl. Rick Henry (bs) at the Oxley Hotel in the '60s, but since that time his jazz activity has continued to be mainly organizational, in particular through the Qld JAS of which he has been a committee member since its inception.

Brown, Brian   b. 29/12/33, Melbourne, Vic.
reeds/fl./comp./arr/ldr
Briefly played trumpet before entering National Service in 1951. Took up tenor in 1952 and was playing gigs within a few months, leading to work in John Fordham's quintet and William Flynn's

Brian Brown, with Barry Buckley (bs), probably during the seventies

Orchestra, Caulfield TH, then the Trocadero under Mick Walker. In late 1953 he left for 14 months in London, where he studied with Eric Gilder and came under the influence of post-bop, most notably, the music of Sonny Rollins. Upon his return to Melbourne he briefly led a group at Studio 1, then formed a quintet with Stewie Speer, Keith Hounslow, John Shaw (p.), Dave Anderson (bs), the last two soon replaced by David Martin and Barry Buckley. This was notable for its hot, aggressive approach at a time when the contemporary scene was dominated by the cooler sound of Brubeck. They played at St Kilda LSC, recorded for the Score label, and in 1957 began Sunday afternoons at the Katherina Club under the imprimatur of the newly formed Jazz Centre 44. During this period Brown took up the flute. Following a successful and influential residency, with Keith Stirling added towards the end, the quintet broke up in 1960 (apart from a brief reunion in 1961 for the Basin St Jazz Centre), with Brown temporarily withdrawing from jazz, moving into studio, TV, and pit work, while studying architecture.

In 1965, Adrian Rawlins booked him, with Buckley, Vining and Gould, for Sundays in the Fat Black Pussycat. Brown led a group here until 1969, playing more exploratory material than in the earlier quintet. Following this period, work was sporadic, and through the early '70s Brown worked with Dave Tolley and Mike Murphy at the prospect Hill Hotel, with Ted Vining, Ian Lawson (p.) and Bob Sedergreen. In 1974 he recorded *Carlton Streets*, subsidized by a Music Board grant. He opened the Commune in Fitzroy where, from the mid-1970s to 1980, he worked with various musicians including Bob Venier, and many of the young players who are likely to emerge as important avant-gardists of the '80s, including David Jones, Jeremy Allsop, Virgil Donati, Alex Pertout. During the late '70s he played in Sydney with different bands involving such musicians as Buckley, Vining, Sedergreen, Tolley, Dure Dara, and in 1978 he toured Scandinavia to great acclaim.

The '80s have seen the consolidation of his work as a jazz educator. He had become a lecturer at the Victorian College of the Arts in 1978, and subsequently director of its full jazz studies programme. He established the independent record label AIJA, led his Australian Jazz Ensemble, which appeared at The Basement in August 1985, and recorded a new suite based on Holst's *Planets*. There have been two major waves of innovative jazz in Melbourne: the first, the bop movement of the late '40s to early '50s centred on Splinter Reeves, Don Banks, Charlie Blott et al., and the second, dating from the late '50s, in which Brian Brown's work has been a central and almost unbroken thread.

Brown, Wesley (Wes)   b. 24/9/22, Melbourne, Vic.   dms
Began on cornet in the Carnegie State School Band in 1932, and continued in civilian and army brass bands until 1947. Taught himself drums and played his first dance with Will McIntyre in 1938, followed by work with Roger Bell. Most important association was with Frank Johnson's Fabulous Dixielanders, replacing Ken Thwaites in 1946, remaining until the band's devolution in 1956, when he took up clerical work with the State Electricity Commission from which he retired in 1970. A motorcycle racer with the Hartwell Motor Cycle Club until 1983, he led the Hartwell Hotshots JB. Joined Nick Polites's NO Stompers in 1979, and appeared in the film *Jazz Scrapbook*.

Browne, Allan Vincent   b. 28/7/44, Deniliquin, NSW   dms/ldr/comp.
Browne's early jazz career was with the Red Onions JB. Before the break up of the group in the mid-'70s he was extending his stylistic range. Began working with Ray Martin and Bob Sedergreen, 1974. Joined Dave Rankin's fusion group, Rankin File, 1976, and Peter Gaudion's Blues Express, 1978. Studied drums with Graham Morgan, 1980–83 and during this time formed Onaje as a vehicle for his own ideas and those of Dick Miller and Sedergreen. At the same time, he played with Geoff Kitchen, Ken Schroder, Odwala, Frank Traynor and Paul Grabowsky, as well as irregularly with Brian Brown. In 1984 he was leading his own trio (Geoff Kluke, Jex Saarelahrt) at the Renown Hotel and preparing a new repertoire for Onaje.

Bruce, Barry Robert (Bazza) b. 4/9/38, Cottesloe, WA p./vcl
Began playing in the early '50s, initially dance work. In the late '50s formed the Westport JB, with which he played at the Melbourne AJC in 1960. In 1961 he joined the Riverside Jazz Group, which became his sole activity until moving to Dubbo for six months in 1963. Returned to Sydney to rejoin the band, now under King Fisher's name, and following its demise joined Ray Price until 1969. Returning to Perth he worked in non-jazz settings throughout the '70s until he started his Condon-style Chicago JB at the request of the manager of the Ocean Beach Hotel, where the group played from c. 1978–80. Has subsequently played for Perth Jazz Society functions, leading a band called Bazza's Jazzers or Vintage Jam, with Denise Dale (vcl). Bruce has a broad stylistic range which places him in demand to back visiting musicians and bands, incl. for the Festival of Perth. He also teaches in the Mt Lawley College jazz courses.

Buckland, Laurence John (Laurie) b. 17/2/44, Violet Town, Victoria bs
Began bass in a boarding-school orchestra in Melbourne, 1960, and played in private jazz sessions with fellow students. Took lessons at conservatorium, joined a big band organized by the Musicians' Union, and sat in at Jazz Centre 44, in 1961. Joined Liam Bradley's band in 1963, and while studying at Univ. of Melbourne was active in the jazz club and the university jazz band with other students incl. Evan Thompson (p.) and Duncan McQueen (dms) who later settled in Perth. In Darwin, 1967–68, returned briefly to Melbourne, then settled in Canberra. Worked casually with local groups incl. Greg Gibson's and the Fortified Few, and joined Neil Steeper's Clean Living Clive group, 1969–72, with a brief interruption in 1971 while he was at Monash. In 1974 co-founded the Black Eagle JB with Terry Kennedy (tpt) and rejoined Steeper, also worked with Alf Harrison's big band. Treasurer of Canberra Jazz Club, 1976–77. Through 1978–80 was mostly occupied in orchestral and pit work. Since 1980 has worked at Bogart's, the Contented Soul, played with the Jer-rabomberra JB, and casual deputizing, incl. with the Black Mountain JB. A regular performer at AJCs.

Buckley, Barry b. 10/9/38, Melbourne, Vic. bs
Buckley picked up an interest in music from his father who was a drummer, and began bass in his teens, joining a trio with David Martin in 1955. With the Brian Brown Quintet, 1956–60, with Ted Vining's TV group, and for a short time with Keith Hounslow at the Katherina. In US studying dentistry, 1961–62, then freelanced among the Melbourne modern jazz fraternity, rejoining Brown for the Fat Black Pussycat residency in the mid '60s. Joined the Ted Vining Trio 1976 and with Brown, 1976–79, during which they toured Scandinavia. Withdrew from regular work for a period, but in the early '80s he joined Odwala, Bob Sedergreen's Blues on the Boil in 1985, and in 1982 recorded a highly acclaimed album with Ted Vining's trio.

Buddle, Errol b. 29/4/28, Adelaide, SA reeds/fl/bassoon/oboe/ldr
Began on banjo-mandolin with the Adelaide College of Music, moving on to sax at eight, playing his first gig within a year at the Prince Edward Theatre in Sydney. At 16 successfully responded to an advertisement for a sax player at the King's ballroom. His first experience of live jazz was at the Astoria, where he was particularly impressed by the work of Bob Limb. This drew him towards the tenor, which he began in 1946. When Limb moved to Melbourne in the same year, Buddle took over most of his work. In Easter 1947 he joined Jack Brokensha in Melbourne at the Plaza, and became involved in the after-hours sessions which incubated early Melbourne bop. Following six months in Adelaide while Brokensha was laid off, Buddle rejoined him at Golds' in Sydney, to be replaced by Ken Brentnall when he returned to Adelaide after six months. Buddle came to Sydney in mid-1950 and joined Bill Weston at the Gaiety ballroom, then Bob Gibson's newly formed Sydney orchestra. By late '51 he was working every night at Chequers, and, via the work of Stravinsky,

Errol Buddle, 1982

formed AJQ in December 1954, Buddle remaining with it until it disbanded in Australia in 1959. For the next 14 years he was primarily involved in commercial work, incl. 10 years as lead alto with Bob Gibson. Apart from occasional stints in El Rocco, the recommencement of his jazz activity was in 1972 when he joined Col Nolan's Soul Syndicate at the New Push. This was followed by guest spots with Bob Barnard at the Rocks Push, then the Nolan-Buddle Quartet with Dieter Vogt and Laurie Bennett (followed by Warren Daly). Since then he has worked in concert, at the Rocks Push and Soup Plus with his own groups, toured Russia, London and the US with the Daly-Wilson big band in 1975, recorded extensively, incl. with John Sangster's *Lord of the Rings* project and an over-dubbed multi-instrumental *tour de force* called *Buddles Doubles*. In March 1978 he went to the USA for 15 months on a grant from the Australia Council and in 1982 he toured SE Asia for Musica Viva with his own funk or fusion oriented group: Phil Scorgie (bs.), Mark Isaacs (p.), Sunil de Silva (perc.), Dean Kerr (gtr) and Rodney Ford (dms.). In 1985 he moved back to Adelaide. Buddle has been a major virtuoso jazz performer throughout most of the post-war period, a pioneer of double reed instruments in jazz, and winner of numerous awards beginning with *Music Maker*'s Musician of the Year in 1952.

Bukovsky, Miroslav (Mike)   b. 5/12/44, Czechoslovakia   tpt/p./comp.
His father led a jazz band, and Bukovsky began piano at age seven and trumpet at 12. Began playing in dance bands, 1959, and conservatorium study in 1961. Worked with small bop groups and in a big band, playing festivals in Europe. To Sydney, 1968, and played feelance in jazz and rhythm-and-blues groups as well as in a quartet made up of fellow Czech expatriates. With David Martin Quintet, 1971. Studied under Howie Smith and later Bill Motzing at the NSW Conservatorium, where he went on to teach. Toured with Marcia Hines, Renée Geyer, and Daly-Wilson Big Band through the late '70s, and played with Bob Bertles's Moontrane, 1977. In 1980 his style developed

had become interested in the bassoon, which he began studying first at the NSW, then the SA conservatoria. In August 1952 a letter from Don Yarella enthusing over the US jazz scene drew Buddle to the States where, following an orgy of listening to the music, he based himself in Windsor, Ontario, playing bassoon in the symphony orchestra. He began working in Klein's at Detroit when the management asked Buddle to take over the band after hearing him sit in, and, with changes in personnel, this group came to incl. Elvin Jones, Barry Harris, Major Holley, Pepper Adams. When an agent suggested that an Australian group might have appeal, Buddle wrote, first to Terry Wilkinson who failed to answer, then to Brokensha, inviting him over. Brokensha arrived with Bryce Rohde, and with local musician Dick Healey they in a group with Dale Barlow, Steve Elphick, Phil Treloar and Roger Frampton, taking him into more contemporary areas. In US 1981, on a Music Board grant, studying with David Baker, David Liebman, Randy Brecker, Woody Shaw, et al. Since returning, has played theatre work, toured with Australian Crawl, the Eurogliders, and Renée Geyer. Formed nine-piece band Major Minority playing his own compositions for the Sydney Festival, 1986, and involved in John Pochée's Ten Part Invention, for which Bukovsky and Roger Frampton write the material.

Bull, Geoffrey Randolph (Geoff, Alby)
b. 26/5/42, Sydney, NSW   tpt/vcl/ldr
Introduced to jazz records by Geoff Holden in the late '50s. Began trumpet and attended Sydney Jazz Club workshops in 1959. With the Melbourne NOJB, 1961; back in Sydney formed his first Olympia JB, 1962, with Holden, Peter Neubauer (clt), Dick Edser (bs) et al., and played jazz club functions and the Brooklyn and Orient Hotels. To England and New Orleans, 1966–67, recording with Alton Purnell, Barry Martin, Cap'n John Handy. On returning to Sydney, resumed playing

in the Orient, and recorded, 1969, with Chris Williams (tbn.), Barry Wratten (clt), Gary Walford (p.), Holden, Don Heap, Viv Carter. Continued to lead New Orleans style bands under the name Olympia, at Unity Hall Hotel Balmain, 1971–72, and at Captain Cook Hotel, 1973. Returned to New Orleans, 1974, and has since divided his time between there and Sydney, where he has played long residencies back at Unity Hall and at the Cat and Fiddle Hotel, Balmain. Has recorded with a number of New Orleans veterans, and organized tours in Australia by Alton Purnell and Sam Price. 1984–85 ran Sweet Emma's, a creole food restaurant, but sold out, and in 1986 was planning another trip to New Orleans. Bull has been the focus for the NO-style movement in Sydney.

Burrows, Donald Vernon (Don)   b. 8/8/28, Sydney, NSW   reeds/fl./comp./arr/ldr
Began on flute in Bondi Public School band, becoming captain of the Metropolitan Schools Flute Band at 12. As a child he listened to and later participated in jam sessions on Bondi Beach and at Palings music shop. Studied at the conservatorium in early '40s and began to get work in 1942 filling gaps created by enlistment. Principal clarinet with the Sydney Studio Orchestra of the ABC in his mid-teens. By 1944, with Bert Mars at the Roosevelt and broadcasting with Monte Richardson on 2UW. In 1945 was with Reg Lewis, played on the George Trevare Regal Zonophone sessions,

Geoff Bull Olympia Jazz Band, c. late 1960s, with, l. to r., standing, Chris Williams, Don Heap, Barry Wratten, Geoff Bull, Gary Walford; kneeling, Viv Carter, Geoff Holden

Don Burrows with the Australian All Stars, c. early 1960s, from l. to r., Freddie Logan (bs), Don Burrows (reeds), Terry Wilkinson (p.), Ron Webber (dms), Dave Rutledge (reeds)

and joined the ABC Dance Band under Jim Gussey, remaining for five years. During this period also worked with Kevin Ryder's (p.) Harbour City Six, and the bands of Wally Norman, Bob Gibson, Col Bergerson. In 1950 he won the *Music Maker* poll and visited Canada, the US, and England. Returned Sydney, 1951 and resumed night-club/dance band activity, incl. with Colin Bergerson at the Trocadero. In the mid-'50s he began his long-time association with George Golla; won *Music Maker* poll and led a quartet (Ted Preston (p.), Ron Hogan (bs), Joe Singer) at André's in 1957. Over next decade became increasingly active in all forms of studio/session work, in concert performance (incl. touring New Zealand with Oscar Peterson, 1960, and Katoomba Jazz Festival, 1964), led groups at El Rocco and was with the Australian All Stars at Sky Lounge until 1966. In the mid '60s, Burrows worked in a range of settings, recording with the Noel Gilmour Players, with strings scored by Bob Young, as well as more conventional jazz recordings for APRA (1967). He helped initiate the Cell Block Theatre concerts in which his quartet (Golla, John Sangster, Ed Gaston) alternated with the New Sydney Woodwind Quartet. In Canada for Expo '67 he participated in the world's first satellite telecast. Began a long residency at the Wentworth Hotel Supper Club, 1968 with Gaston, Golla, and at different periods and different nights, Julian Lee, and Laurie Bennett or Jack Dougan (dms). This engagement helped to make Burrows's name one of the most familiar in Australian jazz, and in the '70s his presence was increasingly taken as a seal of jazz authenticity. In 1970 played for the Aboriginal Theatre Foundation in Darwin, gave a lecture/recital to the Victorian Flute Guild, played a concert in Tasmania, presented a Cell Block Theatre concert with Nigel Butterley of the Contemporary Music Society, and travelled to Osaka as MD for the Australian Expo '70 contingent. His quartet played New Zealand's Tauranga Jazz Festival, 1972. Awarded MBE 1972, appointed to the Australia Council and instigated the jazz studies programme at the NSW Conservatorium from 1973. His activities are too

diverse and numerous for anything but a sketch which outlines their breadth. He has become the most travelled jazz musician in Australia; in addition to regular ABC tours within the country, he has played through SE Asia for Musica Viva and the Dept of Foreign Affairs (1974), at the Indiana Univ. flute convention (1975), in New Zealand, Hong Kong, South Korea, New Guinea (1976), Brazil at the invitation of the Brazilian government (1977), Egypt, Iraq, and Europe (1978), India and the Jazz Yatra in Bombay (1980), Hong Kong Festival (1981). He has accompanied scores of visiting musicians ranging from Blossom Dearie, Stephane Grappelli, to Lee Konitz, Dizzy Gillespie. He is one of the very few Australians to boast 'gold' jazz records, the first jazz musician to receive a Creative Arts Fellowship at Australian National University (1977). In 1977 he was awarded the Queen Elizabeth Silver Jubilee Medal, in 1978 he spent three months in Adelaide with the SA Arts Council and the Sturt CAE. Has been director of jazz studies at the NSW Conservatorium since 1980 and has presented several series of his ABC jazz programme, *The Burrows Collection*. His main base as a public performer in the '80s has been the Supper Club at the Regent Hotel, where he also presents other local and visiting musicians in association with promoter Peter Brendlé. As a musician, Burrows's mastery of his mainstream/bop-based style is unsurpassed, but his importance to Australian jazz extends beyond his musicianship. He has been central in the acceptance of jazz as a musical form to be treated with the same seriousness accorded other performing arts. He is one of the great popularizers of what was for so long a wholly underground music. No other Australian jazz musician could be pictured, un-named, to advertise a daily newspaper as he was in 1985. In 1986 he was awarded the Order of Australia. As well as being the most travelled, Burrows is also the most widely known Australian jazz musician internationally, and at home he is virtually the focus of the jazz establishment, with unparalleled influence at that level. For most Australians he is, with Graeme Bell, the embodiment of local jazz.

Bursey, Brian (Burz)   b. 10/12/31, England
bs/perc.

Learned drums from the drum major of the Royal
Marine Band in Chatham. Bursey's first paid work
was at the Star ballroom in Maidstone in 1947, and
in 1949 he moved to Perth and studied bass under
John Mowson. Through the '50s he worked in
dance bands (incl. Sam Sharp's at the Embassy)
and in pit work, often with other jazz musicians
like Jim Beeson, Bill Clowes, Terry Ingram (p.). In
Sydney, 1962–65, where he worked at Sammy
Lee's Latin Quarter in a group with Graham Morgan. Returning to Perth he worked in TV and with
the WA Symphony Orchestra, with some jazz
activity during the Shiralee/the Hole in The Wall
period. In the early '70s he taught bass with WA
Education Dept and worked with the Will Upson
Big Band. In 1985 he had taken a BEd, was
teaching and working with Gary Lee (vibes) and in
a restaurant with his wife Olwyn Thomas.

Burton, Roy Victor   b. 28/4/36, Rochester, Kent,
England   tbn.

Began on piano and moved to trombone at age 16,
playing along with Graeme Bell records on the
farm in Victoria where he lived at that time. Returned to England at age 18 and joined his first
band after completing National Service. Since
arriving in Perth he has worked in various groups
but his most durable association has been with the
Corner House JB which he joined in 1979. His
style, one of the most fully formed and forceful
among traditional trombone players in Perth, was
much developed under the influence of Corner
House leader Leon Cole.

Cale, Bruce   b. 17/2/39, Leura, NSW
bs/comp./ldr
Studied violin from age nine, and began professional career on bs in Sydney, 1959, working studio, night club, TV, radio and concerts. Closely associated with Bryce Rohde in the early '60s, he also worked with other leading jazz musicians incl. Judy Bailey, John Sangster, George Golla and Errol Buddle, before leaving Australia in 1965. Worked with major figures such as Tubby Hayes and Ronnie Scott in UK before taking up a scholarship in the US where he was based for 11 years. He led his own group, as well as playing with major figures across a range of styles: Bobby Hackett, Phil Woods, Zoot Sims, Toshiko Akiyoshi, John Handy et al. In 1974 he composed a work for the Los Angeles Philharmonic Orchestra woodwinds, which led to more commissions. Under a National Endowment for the Arts grant in 1976 he completed a work for voice, jazz ensemble and small symphony orchestra dedicated to John Coltrane. Residing in Australia since 1978, he has concentrated on teaching and composition. His 'Land of the Aborigine' was performed by the Melbourne Symphony Orchestra for the ABC in 1979. In 1980 his quartet performed Cale compositions at the Adelaide Festival of Arts, and his music was incorporated into the film *Notes on a Landscape*. Teaches both privately and institutionally, with particular reference to the Lydian Concept of Tonal Organisation, and in 1981 studied with its major spokesman George Russell in the US on an Australia Council grant. Probably the most wide-ranging and ambitious jazz-based composer in Australia, Cale's recent projects have included the performance of his Concertino for Double Bass and Orchestra, op. 33, with the Melbourne Symphony Orchestra (1981), a series of concerts and recording sessions with his 10-piece jazz orchestra (1983), and a concert with his quartet for SIMA's Winter Fire Series in 1985.

Campbell, David Walter (Dave)   22/9/43, Ballarat, Vic.   p.
After some classical training in childhood, became interested in jazz at high school and joined Ballarat Jazz Messengers in 1964–67. Moved to Melbourne 1968; with Storyville band 1969–74. Freelanced until joining Dick Tattam's Jazz Ensemble in 1977. Also with Peter Gaudion at the Victoria Hotel for periods in 1978–79. Records with Storyville and Tattam.

Canberra (Australian Capital Territory)
While most major capital and provincial cities in Australia manifest parallel patterns in their jazz development, Canberra* in many ways fails to conform to that broad picture. Standing between the two most important centres, Sydney and Melbourne, it has been fed to some extent from both directions. Its status as a city built around politics and the public service, however, combined with other aspects of its geographical position to create certain unique characteristics in its entertainment patterns. There is almost certainly no city in Australia of similar size and importance with less public night-time recreation. The impact of what little night life there is in Canberra is further dissipated by the decentralized design of the city: it has no particular area in which its music venues are concentrated, and indeed, at night there is no particular sense of being in a city at all. The reasons

---
*I wish to acknowledge gratefully the painstaking and clear-sighted assistance of John Sharpe in preparation of this entry.

for the dearth of night life doubtless relate to the fact that the population includes an abnormally large number of people at an age when attention is devoted to the establishment and consolidation of career and family, so that most evening recreation consists of family and career-related private parties. Many are also based in Canberra in temporary postings, and are more than usually likely to spend weekends in Sydney or Melbourne. Furthermore, the snowfields and the coast provide weekend attractions during both summer and winter seasons, and these therefore tend to occupy the attentions of the more affluent. For demographic, geographical, and reasons simply to do with the indefinable spirit of a city designed around an exclusively political function, Canberra has not spontaneously evolved the same kind of recreational history in general as other cities, and this has inevitably affected the character of its jazz scene.

There is little evidence of any noteworthy jazz activity in Canberra before the '50s, and a visit by the Port Jackson JB under the leadership of Jim Somerville in 1949 was reported as the city's first public jazz concert. Throughout the 1950s broadcaster and drummer/saxophonist Bruce Lansley was most continuously involved in whatever jazz activity was going on, in association with, *inter alia*, Sterling Primmer and John Hamon (bs), both of whom are still active. Further impetus was provided by the arrival of Greg Gibson in 1957, and in the same year the establishment of the first Canberra Jazz Club, with pianist Gordon Reid a prime mover (not to be confused with drummer Gordon Reed who leads JB and the Jazzmen during the '80s). By 1960 jazz was enjoying locally some of the national upsurge. Lansley was leading a group called the Presidents (with Primmer, Terry Wynn, Trevor Holgate (bs), Cees Kruithof (dms), Gordon Reid was leading a quartet at the Civic, and John Hamon was working in a mainstream band led by Alex Powell (p.) at the Canberra Club. Ex-Melbourne drummer Jim Vallins was at the Fiesta coffee lounge in 1961, where jazz club functions were also held, and the club was operating successfully enough to be able to sponsor an annual band competition.

The return to Australia of Alan Pennay (p.) from England in 1961 introduced an energetic spirit into the local picture, and in the same year Ray Price presented what was reviewed as Canberra's first public jazz concert since 1949. Pennay worked in a quartet with Gibson, then in mid 1962 opened the Pendulum coffee lounge at 18 Garema Place, Civic Centre, with a trio (Vallins, Gerry Gardiner bs). The Pendulum at its peak was presenting jazz five nights a week, and regularly featured guests from Sydney like Roy Ainsworth (sax). As the brief ripple of the jazz boom of the early '60s died down in Canberra, the Pendulum became the main, then the only, focus for jazz, particularly after the demise of the Canberra Jazz Club in 1963. The Pendulum continued with a jazz policy (sometimes with excursions into folk music of the kind becoming fashionable in the mid '60s) until 1966. There had been other activity during those years, incl. a well-attended concert by Don Burrows, Errol Buddle, Jack Grimsley, Judy Bailey, John Sangster and George Thompson (bs) in March 1964, and the establishment of the Australian National University Jazz Society at Ray Price's suggestion in 1964. It was the Pendulum, however, which maintained continuity until 1966 when a new thread was spun. Pennay went on to work in the snow resorts, and later moved to Sydney. Vallins (who was replaced by Melbourne drummer Paul Davis) and Gardiner ultimately moved to Sydney also.

The Fortified Few was founded in the year the Pendulum ceased its jazz activity. As elsewhere in Australia, traditional jazz tended to draw bigger audiences than more modern styles. A 'progressive' night at the Canberra Jazz Club in 1962 featuring Roy Ainsworth playing a John Coltrane-inspired approach was reported to have caused considerable bewilderment in the audience. The Fortified Few (many members of which had been in the Cavaliers Dance Band) was firmly in the traditional mould, and built up a large following. Neil Steeper and John Sharpe, founder members of the band, were also important in the re-establishment of the Canberra Jazz Club. Unfortunately the early files of the club have proven impossible to locate, but it seems to have made a fitful start through 1968–69, and was firmly established by 1970 with Sterling Primmer as president. Its constitution was based on that of the Sydney Jazz Club, with a brief extending beyond the traditional styles in recognition of the comparatively small overall jazz scene in Canberra. Steeper was also an important contributor with the formation of his Clean Living Clive's group in 1968, and with, in 1976, the Clean Living Clive's jazz venue, which enjoyed a brief but enthusiastic life during which it presented local and imported musicians. The Fortified Few, the Steeper enterprises, the re-established jazz club, were the main manifestations of jazz activity in Canberra through the late '60s and through the '70s, with, in addition, the groups put together by Greg Gibson during his postings back to the city. Other short-lived or occasional projects were big bands assembled by Jim Latta (dms), Alf Harrison and John Carrick, facilitated by the proximity of a pool of formally trained musicians working in bands in the Duntroon military academy: this reservoir continues to provide Canberra with section musicians to the present.

By the mid-'70s jazz in Canberra had settled to a steady and continuous activity. The Fortified Few provided an unbroken thread to 1983. In addition there was the Antiquity JB, the Clean Living Clive's operations, Gibson's Mood Indigo, Alf Harrison's occasional concerts, Alex Powell's mainstream group, and the activities of the jazz club which included the importation of musicians and bands from Sydney and Melbourne. In the late '70s the Southern Cross Club, an organization connected with the Roman Catholic church, began its jazz concerts series which have continued into the '80s, during which time Canberra has seen more jazz activity than during any previous period. The Jerrabomberra JB led by Jim Hilson (tbn.) evolved from the Antiquity followed by the Phoenix JBs from the mid '70s, and worked with reasonable regularity until disbanding upon Hilson's departure in 1985. Gordon Reed formed JB and the Jazzmen, incl. ex-Brisbane sax player Gordon Benjamin (father of Craig), John Hamon, and ex-Adelaide pianist Ron Lucas who had been associated with Jack Brokensha in the '50s. This mainstream group also played regularly for jazz club functions, at the Boot and Flogger in Green Square, Kingston, and continues to perform as opportunities present themselves. Pierre Kammacher's Hot Four has included at different times Fortified Few alumni Tony Thomas, Tex Ihasz, John Sharpe, with Rene Koppas (bs) and leader Kammacher on reeds, with a residency at the Contented Soul. Carl Witty (dms) and Marylyn Mendez (vcl) formed various jazz-based groups for restaurant work in Canberra before they moved to Sydney in 1983, and Ross Clarke has led numerous small modern groups for a number of years. The most venerable residency of them all has been the Federation Lounge in the Hotel Dickson which began its jazz policy with the Fortified Few in 1966, continuing until 1986, at which time the residency was held by Tony Thomas's Black Mountain JB (formerly the Double T JB).

Although Canberra is distinctive in the way in which its jazz scene has evolved, it shares with most other centres an ambiguously related traditional/post-traditional division. In the '80s, traditional jazz continues to dominate the public venues. At the same time, the average age of the musicians and audiences continues to rise: little or no new blood is entering this area of jazz. The national boom from the late '70s which has echoed in Canberra has resulted in an expansion of talent primarily in more modern styles. The young musicians who are entering the music are doing so mainly through the educational system. Canberra has an energetic and extensive music education programme in its schools, both primary and secondary. These produce big bands with a distinct jazz inflexion, and the Canberra School of Music has also introduced a jazz studies course. As in most other Australian music education programmes, the emphasis is upon late-mainstream and post-bop approaches, and the long-term prognosis for 'modern' jazz is far more favourable therefore than it is for traditional, notwithstanding the continued energy and the educational endeavours of the vigorous Canberra Jazz Club.

Capewell, Derek   b. 20/8/38, London   bs
Began piano at seven, trumpet at 14. Moved to Australia in 1957. Took up bass and drums in 1958.

Returned Melbourne, Oct. 1961. Played The Embers with Ted Nettelbeck, Alan Turnbull, and also Ted Preston, then (The New Embers) with the Tom Davidson big band. From the mid '60s with, *inter alia*, Frank Smith (sax), Don Burrows, Bryce Rohde, Brian Brown, John Sangster and Billy Hyde, concentrated mainly in session work in recording and TV studios; also backed numerous visiting musicians, including Phil Woods, Dizzy Gillespie, Mel Tormé, Clark Terry. Toured with Herb Ellis, Barney Kessel, Henry Mancini, Carmen McRae, Buddy de Franco, Terry Gibbs, and Cleo Laine and John Dankworth, with whom he made one of his many recordings. In 1984, resident in Melbourne, playing with Peter Gaudion's Blues Express.

Carter, Vivian James (Viv)   b. 2/5/35, Springvale, Vic.   dms/wbd
Began drums in 1949 and drawn to jazz through Southern Jazz Society functions. To England, 1955, and active through the trad boom, with Trevor Williams's band (1955), Ken Colyer's Omega Brass Band (1955), Mike Peters's band (1955), Cy Laurie's band (1956). Joined Acker Bilk (1957–58) and with the Bob Cort Skiffle Group. To Germany with Pete Deuchar JB, then with Dick Charlesworth (1959–61), playing concerts, broadcasts, touring and recording. Following work with Colin Sims, Llew Hird, Mickey Ashman, returned to Melbourne, 1964, moving to Sydney, 1965, to join Geoff Bull until the late '60s, interrupted by a tour in Vietnam with the Hirds, 1967. In addition to periods with Chris Williams JB, the Unity Hall JB, and Neil Steeper, Carter's main work since the '70s has been with the Eclipse Alley Five (joined 1970) and the San Francisco JB (1979), with which bands he was still working in 1986.

Caxton St Jazz Band (Brisbane, Qld)
Started in July 1977 by Andy Jenner as a New Orleans-inspired group, with John Reid (bjo), Jack Connelly (tba), John Hall (p.), Peter Rex (tbn.), Ian Oliver (tpt), Bob Mair (wbd). Personnel changes have been as follows, the last named

on each instrument being the current player: bjo—Robbie Robinson 1977–79, Peter Ransom; bs—Peter Freeman 1977–85, Brian Eydmann; tbn—Dick Rigby 1977–79, Tom Nicolson 1979–85, Kirk Jaress; reeds—Tony Ashby 1979–81, Tich Bray 1981, Barry Webb 1981–82, Col Wharton. Hall was replaced by Bernice Haydock in 1977, and Oliver by Graham Duffin in 1979. Bob Mair, who had inherited leadership from Jenner, left in 1985 to be succeeded by ex-Sydney drummer Geoff Allen, and Haydock became leader. The band began at what was known as Caxton Hall in Petrie Tce, but its acoustics were not fashionable enough to draw a crowd and they moved to Knights Disco in July 1980, running their own Caxton St Jazz Club functions on Thursdays. The most successful residencies began in 1981 at the Caxton Hotel and the Barn in Waterloo Hotel. In the same year the group began playing at the Cellar Club for the Qld Jazz Club. Following a year break in 1983, the band resumed at the Barn, where it was still in residence in 1985. Highlights of its history have incl. performing at jazz festivals organized by Mileham Hayes, at one of which, in Oct. 1980, it received a Yamaha award. With the Vintage band, the Caxton St JB has become one of Brisbane's most important traditional-based groups.

Chambers, Alan (Sny)   b. 12/6/32, Carlton, Melbourne, Vic.   tpt/vcl/ldr
First gigs in the early '50s on the Yarra riverboats with the Melbourines, which incl. Don Standing (bjo), Laurie Gooding (clt) and Harry Price. Joined Smacka Fitzgibbon's Jazz Seven, and worked with various other groups until forming his own Bayside JB in the '60s. This band played Downbeat Club and Melbourne TH concerts for many years, and made a 7" EP. Chambers is currently working only occasionally, residing in Melbourne where he works as a motor mechanic.

Clarke, Alfred Bruce (Bruce)   b. 1/12/25, Melbourne, Vic.   gtr/bs/vibes/synth./comp./arr/ldr
At 17 Clarke began guitar at Buddy Waikara's

Hawaiian Club. In the mid '40s he was exposed to jazz and was later inspired by the work of Doug Beck, who gave Clarke coaching. Began participating in jam sessions at the Katherina, organized by Charlie Blott with the Melbourne musicians who generated the local bop movement of the late '40s. Clarke was a founding member of Splinter Reeves's Splintette. Played nightclubs 1949–52: Ciro's with Bela Kanitz and later Neville Maddison, the Copacabana with Geoff Brookes, the Adelphi with the Splintette. During the same period he broadcast with Don Banks's Boptet, and played the Sunday jam sessions for the Boposophical Society in the Galleone. By the early '50s Clarke was broadcasting organizing concerts, writing for shows, and still performing both as leader of his own groups and with other bands incl. that of George Dobson (tpt). With the advent of TV he went into HSV 7, 1956–63, and in 1962 established his Recording Workshop which became a base for a jingle production company, with the involvement of Dick Healey and Frank Smith. During this period he organized numerous recording sessions with local and visiting musicians, some of which was later released on his Cumquat record label, and he played backing or in support bands for many visitors incl. Billy Eckstine and Dizzy Gillespie in the Lee Gordon packages. In 1960–61 he established a rehearsal group incl. Billy Hyde and Frank Smith to experiment in areas defined by Ornette Coleman and Lennie Tristano, and some of this work appeared on the first Cumquat release, *Looking Back*, in 1973. In the late '60s his interest in experimentation took him into electronics and aleatory concepts. He became president of the Melbourne branch of the International Society of Contemporary Music and performed on synthesizer with the electric group Tully. Clarke resumed more conventional jazz performance in 1972 and formed his jazz quintet in 1984. Appointed to the Music Board of the Australia Council, and the Kenneth Myer Music Fellow to the Victorian College of the Arts where he was involved in the inaugural jazz studies programme in 1977. Toured Yugoslavia 1979 with the avant-garde ensemble Australia Felix and began 18 months at the Victoria Hotel. In the early '80s he

formed a quartet with sons Marshall (dms) and Bradleigh (bs), and Vince Hopkins (gtr), and worked with Geoff Kitchen at Potter's Cottage restaurant, Warrandyte. Influential as teacher, with his own guitar school founded in 1976 and as author of several instructional books. Still very active in session work for film, TV, and advertising, in which areas he has won many awards. Further information in Mike Williams's *Australian Jazz Explosion* and James Murdoch's *Australia's Contemporary Composers* (Sun Books).

Cleaver, Richard Peter (Peter)   b. 17/4/30, Mildura, Vic.   bjo/gtr
Began in 1958, joining a band led by Martin Finn (tpt), and incl. Frank Traynor and, later, Graham Coyle. With Len Barnard in 1950–55. From 1955–60 worked in bands led by Max Collie, Bob Barnard, Frank Johnson, and Ade Monsbourgh. With Kenn Jones's Powerhouse Band 1960–65. Led his own band, Old Faces, at AJC in Melbourne, 1966, with the personnel later recorded on Swaggie albums *Hot Tuesday* and *The Mountebank*. Since his first session with Len Barnard on Jazzart, has recorded extensively including the famous *Naked Dance* session, and from 1962–73, with Frank Johnson, Roger Bell and Ade Monsbourgh. In 1985 was resuming playing after a period of retirement.

Clowes, William (Bill)   b. 19/6/30, Perth, WA p./vcl/ldr
Took lessons from Henry Cochran and later Harry Bluck, and in 1949 won an Amateur Hour competition. In the early '50s worked with Jack Harrison's band and also at the Adelphi Restaurant. In 1952, at Bluck's urging, Clowes, who had been blind from infancy, moved to Melbourne to study braille music. Before returning to Perth in 1956, he had also worked with a number of musicians based in Melbourne incl. 'Splinter' Reeves. Back in Perth, Clowes became active at jazz concerts, but most of his work was of necessity in cabaret. In the early '60s he assembled a succession of small groups incl. Brian Bursey, John Milne (dms), Colin Scott, which played in the

Shiralee, the Melpomene, and the Hole in the Wall during the brief flourishing of the modern movement at that time. In the late '60s he broadcast on the ABC, incl. with a group formed in 1968 which incl. Theo Henderson. From 1969 to the early '70s he worked in the pop area on electric piano with Murray Wilkins and Barry Cox, but withdrew from music until the mid '70s because of commitments with the Guide Dogs Association and at his employment at the Vox Odeon music store. Since the late '70s Clowes has enjoyed renewed opportunities for jazz performance, both on the pub scene and with jazz clubs, notably the Perth Jazz Club/Society. In the face of what was at the time a relatively unsupportive public climate, Clowes established one of the earliest reputations as a bop-oriented jazz pianist, and his work has maintained a driving edge which has continued to attract respect among other musicians.

Colborne-Veel, John   b. 21/2/45, Melbourne, Vic.   tbn./ldr/comp./arr
Started on piano accordion, violin, and guitar as a child, settling on trombone at 14. Left Merchant Navy, 1963, to take up music professionally, playing dances around Melbourne. In National Service, on active duty in Vietnam, 1967–69. Returned to full-time music, 1970, joining Eclipse Alley Five until 1972. Studied at NSW Conservatorium, and through the early '70s mainly active in commercial areas, incl. pit work, arr./tbn. for South Sydney Leagues Club, and freelance writing. Settled in Deniliquin, 1974–76, commuting weekly to Sydney for club commitments. Returned to Sydney, formed a concert big band, and joined Graeme Bell, 1977–79, then Ray Price, 1979–80. During the '80s has composed prolifically in jazz, chamber music, ballet, concerti, opera, incl. a jazz mass which has been frequently performed, and the 'Toad Hall' jazz suite; and has organized several concerts devoted to Australian jazz compositions. Also freelancing as player, arranger, and touring with concert packages. MD for Sydney Business and Football Club, 1982–83. In 1986 was leading his quartet at the Purple Grape Wine Bar at Homebush, where he had been since 1979, and composing.

Collins, Ross   b. 28/1/33, Brisbane, Qld   p.
Started piano at age nine. Began playing in wartime concert parties in 1943. To Melbourne, 1949, for conservatorium study, but changed his mind and returned to Brisbane where he became interested in jazz. Settled in Melbourne c. 1956, attended the AJC and won its original tunes competition. Freelanced with the local jazz fraternity, and joined Derek Phillips's Port Stevens JB, which incl. Trevor Rippingale (reeds), a driving and versatile player who later moved to Sydney. Collins moved to Sydney c. late 1964, and entered the commercial field. He has continued, however, to appear in jazz settings, excepting a brief sojourn back in Qld in the early '80s. His outstanding technique and versatility place him in considerable demand, and he remains essentially a freelance player, although in 1986 he had been regularly associated with Ted Sly's East Coast JB.

Conley, John Lancaster   b. 22/12/53, Melbourne Vic.   gtr/el.bs/comp.
Learnt piano, then violin as a child, switching to guitar at 12. In 1970 was living in Canberra playing local jazz clubs with, *inter alia*, Craig Benjamin (reeds). To Sydney (1972) to study at NSW Conservatorium and played with Benjamin's Big Band at The Basement. In 1974, with the Craig Benjamin/Mark Simmonds Boptet, then in the mid '70s worked with the important contemporary/experimental group Out to Lunch (with Benjamin and Barry Woods). Worked with groups led by Serge Ermoll and Mark Simmonds, 1976–77, then with Col Nolan and Harry Rivers (dms) at Soup Plus, 1978. Also worked with Rivers, and Ray Alldridge, at The Basement, then Errol Buddle at Soup Plus. Joined Galapagos Duck, 1981. Conley has also played in commercial areas, incl. about five years as touring musician with Winifred Atwell (p.), in session work, incl. for films. In 1986, still with Galapagos Duck.

Conlon, Grahame John   b. 23/1/48, Newcastle, NSW   gtr/el.bs/comp./arr
Began playing in pop groups, 1963. To Sydney, 1968, where Freddie Payne (ex-Daly-Wilson, now

resident in Adelaide), converted him to jazz. To Adelaide, 1970, mainly in session work, playing composing, arranging, incl. for films. Led the group Onions which incl. Sue Barker (vcl), Payne, Geoff Kluke, Schmoe, Jim Shaw. To Sydney, 1980, again primarily involved in session work and touring with visiting commercial packages. With Marylyn Mendez (vcl) at Rose Bay Hotel, and Freddie Wilson at the Cat and Fiddle, Balmain. In 1986 playing jazz on a freelance basis.

Conlon, Keith Andrew   b. 30/4/44, Adelaide, SA   dms/vcl
Began sitting in with the University Jazz Group from 1962, ultimately replacing Kent Fuller after it had become the Campus Six, and remained with the group until it disbanded. Joined Bruce Gray's All Stars in 1978. Has played with various Dallwitz bands since 1977, incl. recordings. Active in broadcasting, incl. jazz projects such as Don Burrows Workshops in 1976 on 5UV, of which he was foundation manager.

Connelly, John Keenan (Jack)   b. 5/2/26, Bendigo, Vic.   clt/bjo/gtr/bs/tba
Started on clarinet at age 13, banjo at 18. Joined Frank Johnson's Fabulous Dixielanders on bass, and later other instruments, in 1947, remaining until c. 1954. Played with Len Barnard and began the tour of 1955 but left the band before it broke up in Brisbane. Settled in Sydney, and played with the Paramount JB, 1956–c. 1961, then returned to Melbourne briefly. Back in Sydney by late 1962 and replaced Bob Cruickshanks in the Black Opal JB. In the mid-'60s played with the Harbour City Jazzmakers. Most of his subsequent career has been freelancing in Sydney, with periods in Melbourne. In the early '80s he suffered from ill health and had retired from music by 1985.

Cook, James Alfred (Jim)   b. 31/12/43, Perth, WA   reeds/fl./arr.
Began sax in c. 1952 and within two years was sitting in with his aunt who played at local tennis club dances. In 1958 in a practice band which incl. Don Bancroft and working dance engagements

Jim Cook, 1985

with Will Upson and in the early '60s with the Traditionalists. When Keith Stirling arrived in Perth he encouraged Cook further, at the same time broadening the young man's style. Cook joined Stirling and worked at The Hole in The Wall and the Shiralee, also playing with Bill Clowes in the same venues. From January 1965 he spent 12 months in England, listening avidly but not playing. Back in Perth he entered the night-club/restaurant scene, first at Romano's and in the late '60s at La Tenda. Resumed jazz performance from the mid-'70s with the fusion/contemporary group Collage, Uwe Stengel's Manteca, and the Will Upson Big Band, with which Cook recorded. Although still a busy studio/session musician, Cook is one of the more visible post-traditional jazz musicians. In the '80s his major jazz activity has been in the Mt Lawley College courses and in 1985 he was also performing with Gary Ridge's Nova Dreams.

Costello, Gary   b. 7/6/52, resident in Melbourne, Vic.   bs
Began electric bass in 1967, working first in pop. In the early '70s he studied acoustic bass with Marjan Brajsa with jazz coaching from Murray Wall. In the late '70s worked with Suzy Dickinson at the Grainstore. With Onaje, 1982, and also worked with Ken Schroder. Joined Vince Jones, 1983, and the trombone group McCabe's Bones. Has backed visiting musicians incl. Mal Waldron,

Richie Cole, Bobby Shew, and Australians Don Burrows, John Sangster, and Kerrie Biddell. In 1985 Costello was teaching at the Victorian College of the Arts.

Costelloe, John Eric (Cossie)   b. 8/10/30, Cootamundra, NSW; d. 31/1/85, Sydney, NSW   tbn. and other brass/ldr/comp./arr/vcl
Began on violin at seven, and on trombone in the Cootamundra City Band, of which his father was bandmaster. Played drums in The Modernists, 1947, a trio with father Eric (tpt) and John Ansell (p.). Moved to trombone when the group evolved into a six piece. This band became the Cootamundra JB, one of the most successful rural jazz bands in Australian jazz history, and unusual in having an inspiring effect on city bands of the period. The personnel stabilized with Costelloe, Ansell (ldr), Greg Gibson, Lloyd Jansson (tpt), Kevin McArthur (dms), Bob Cowle (bs). Through the '50s the band was active in performance and broadcasting in the Riverina area, and in 1956 played for the Sydney Jazz Club and made its first record. It played at, and its members served on the committees of, AJCs held in Cootamundra in 1955 and 1959. Through his work with this band Costelloe came to the attention of Graeme Bell with whom he toured in Japan and Korea, 1954–55. Moved permanently to Sydney, 1960, and joined Ray Price, the PJJB, and the ABC Show Band. His TV work incl. 10 years with Bobby Limb on *The Sound of Music*. Also very visible in jazz settings, freelancing with most traditional groups (incl. Graeme Bell's). Foundation member of Daly-Wilson Big Band and of Bob Barnard's JB, the latter becoming his main public performing activity. At the time of his death from cancer Costelloe was with Barnard, working in a club orchestra, and leading his own quintet.

Coughlan, Francis James (Frank)
b. 10/9/04, Emmaville (formerly Vegetable Creek), NSW; d. 6 or 7/4/79, Sydney, NSW
tpt/tbn./ldr/arr/vcl
Played in Glen Innes District Band, of which his father was the bandmaster; brothers Tom, Jerry,

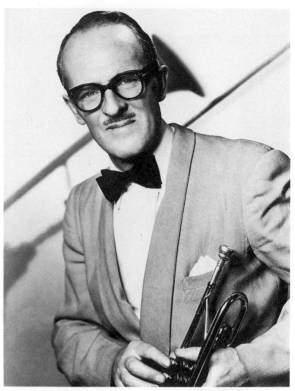

Frank Coughlan

Charlie, Jack, were also musicians from childhood. To Sydney 1922, as a trainee teacher, and heard records of Miff Mole, who revolutionized his attitude to music. With Bill James at Bondi Casino, 1923, then played in Melbourne with Frank Ellis and his Californians, 1924. Joined the Californians under Walter Beban's leadership at the Palais Royal, Moore Park Sydney, 1925, then back in Melbourne began a season at St Kilda Palais de Danse, November 1926, with Carol Laughner's orchestra. With Linn Smith on the Fuller theatre circuit, Sydney 1927–28. To England, arriving Dec. 1928, and joined Arthur Rosebury at the Kit Kat Club. Played with Jack Hylton on the Continent, with Fred Elizalde (whose band incl. Adrian Rollini), Al Collins at Claridges, and Al Starita at the Piccadilly Club, before leaving England in December 1929. Back in Sydney, joined Al Hammett at the Palais Royal, May 1930 to Feb. 1931, then to Brisbane to lead the Carlton Hotel band. In mid-1932 he was back in Sydney with Harry Whyte at Romano's. To Melbourne at the end of 1932,

Frank Coughlan's dixieland group in the Sydney Trocadero, probably during the late 1940s, with, l. to r., Reg Robinson, Jim Riley, Frank Coughlan, Dennis King, Bobby Bell, Ron Murray, Les Dixon, Les McGrath

where he joined Ern Pettifer at the St Kilda Palais, then Don Rankin, and in 1934 played with Art Chapman at the Embassy and with Benny Featherstone at the Rex Cabaret. In early 1935, back in Sydney with Sam Babicci at Romano's, then leader at Bondi Esplanade. On April 1936 he opened with his newly formed band at the new Sydney Trocadero. The band, most of the members doubling on other instruments, consisted of Dick Freeman (dms/deputy ldr), Norm Goldie (bs), Jack Moore, Colin Bergerson, Keith Atkinson, and Frank Ellery (reeds), George Dobson and John Robertson (tpts), Geoff 'Dutchy' Turner (tbn.), Frank Scott and Stan Bourne (p.), Charlie Lees (gtr), with Cardru Llewellyn (vcl). Coinciding with the arrival of swing, and led by one of the most accomplished and informed 'hot' musicians in Australia, this event established the reputations of the young musicians, of the venue, and of Coughlan, throughout the country and beyond. The band's performance at the Trocadero was included in the internationally distributed feature film *The Flying*

Doctor. Coughlan wrote a series of highly perceptive articles on the history of Australian jazz in *Music Maker*, and band members were regularly featured subjects in professional journals, and as featured performers at various swing club functions. The music was a broad spectrum of modern dance music, incl. dixieland segments, and with the later addition of vocalist Barbara James (with her husband Reg Lewis), it boasted one of the most authentic hot singers in the mould of Adelaide Hall that Australia has produced. Coughlan finished at the Trocadero on 31 Aug. 1939 and went on to the Bondi Esplanade. From Feb. to April 1940 he led the band at the New Romano's, then at Prince's in Brisbane. In Nov. 1941 he took over leadership at the Melbourne Trocadero from Stan Bourne. Joined the army in June 1943, serving as a bandleader in the 9th Division. His discharge was reported in *Tempo*, Sept. 1945, and from March to Sept. 1946 he led the band at Rose's in Sydney. He returned to lead the Sydney Trocadero band in Oct. 1946, and remained until July 1951.

Following several months as leader at Christy's night-club, he went as leader to the Melbourne Trocadero in January 1952. Returned again to the Sydney Trocadero in September 1954 and stayed until its closure, 31 Dec. 1970, after which he retired except for occasional club engagements. Coughlan was arguably one of the first authentic jazz musicians Australia produced. His importance is not simply in this historical priority however. He was an influential propagandist for jazz in all its pre-bop forms, not only leading some of the country's most accomplished swing groups, but encouraging the traditional movement also, as in his support in the foundation of the Sydney Jazz Club. He was involved in the significant Fawkner Park Kiosk sessions in Melbourne, and recorded a dixieland session which reveals a strong, Roy Eldridge-influenced trumpet player. Along with Clarrie Collins (tbn.) who worked with Paul Whiteman in the film *King of Jazz*, Coughlan was one of the first of many Australian hot dance or jazz musicians to establish a reputation at the highest professional level overseas. A discography and biography by Jack Mitchell (upon which much of the foregoing is based) was published in *Quarterly Rag* no. 13 (New Series), together with an appreciation by Graeme Bell, who called Coughlan 'the father of Australian jazz'.

Coulson, Gordon   b. 10/12/37, Easington, Durham, England   tbn./ldr
Began trombone tuition at age 17 and became professional upon moving to London in 1961, playing with several jazz bands in England and Europe before migrating to Adelaide in 1966. His versatility soon placed him in a range of stylistic settings ranging from the Campus Six (1966–76), Bottom of the Garden Goblins, Neville Dunn's big band, and in Bruce Gray's All Stars (1968–71). Coulson spent 1971 in Sydney with the Elizabethan Trust's National Training Orchestra, and back in Adelaide joined the Police Band in 1973. Also active as a studio musician and leading a trombone quartet. Took up bassoon in 1975. Led his Climax JB 1976–82, then rejoined Bruce Gray. With Peter Hooper's Royal Garden JB, 1984–85 and in 1986 was still with Bruce Gray, as well as being head of

music studies programmes at Prince Alfred College.

Cox, Barry Edward   b. 10/2/42, Perth, WA   dms
Began drums at age 15, and worked at dances and private functions. Worked with Bill Clowes on the ABC's *The Piano and Me*, spent a year with the Channel 7 band, played in first house band of the Perth Jazz Club/Society, and subsequently with other groups playing Society functions. Backed Danny Moss at the Festival of Perth and in the early '80s worked at the Brisbane Hotel with Clowes and Denise Dale. Cox has been one of Perth's most active freelance jazz drummers, and in 1985 he was also leading his own quartet, teaching, and running a music wholesale business.

Cox, Keith Thomas George   b. 6/2/26, Melbourne, Vic.   gtr/bs
Began on guitar in the '40s, moving to bass with lessons from Lou Silbereisen. Began sitting in with Bell at the Uptown Club and playing in practice sessions with a group of university students. Tony Newstead's group evolved from this, with Cox as a founder member. Cox joined Geoff Kitchen for two years in the early '50s, then went into Russ Jones's (vibes) band with Jack Varney and Ade Monsbourgh, until about 1956. Joined Kenn Jones's Powerhouse group, 1958, and the Melbourne Jazz Club house band, 1959. During the late '60s he freelanced, and spent six years in Smacka Fitzgibbon's restaurant. Following another period of casual work and a brief reunion of the Powerhouse band at the Victoria Hotel, Cox retired from music in the early '80s.

Coyle, Graham Francis   b. 10/8/32, Melbourne, Vic.   p.
In the late '40s played dances in his father's trio and by 1950 working in jazz settings, with Frank Traynor and Tony Newstead. To Shepparton as a surveyor, 1951–53, where his musical knowledge was extended by Rex Green. With Len Barnard's band, 1953–55, and with Alan Lee, 1957–59, an episode that enriched his style with more progressive ideas. Joined Kenn Jones's Powerhouse group, and became a founder member of the Melbourne

Jazz Club house band in 1958, working also with the latter under its Jazz Preachers title (Frank Traynor, ldr). In later years, has freelanced widely with virtually every traditional to mainstream group in Melbourne, frequently with Len Barnard's bands. In 1985 was with Swing Shift with Beverley Sheehan.

Crichton, Patrick Stuart (Pat)   b. 22/6/43, Palmerston Nth, NZ   tpt/arr./ldr
Began trumpet at age 14 and worked in the traditional style in Wellington NZ before moving to Sydney where he joined Ray Price in the mid '60s. Also played in jazz/rock settings, with the Daly-Wilson Big Band 1975–77 and session work. To Perth in 1979 and worked with the Will Upson Big Band. First head of jazz studies courses at the Mt Lawley College of Advanced Education and director of the Western Australia Youth Jazz Orchestra. Works with small groups often consisting of fellow teachers, Jim Cook, Mike Nelson, Bruno Pizzata.

Crow, Noel   b. 3/1/38, Somerville, Vic.   clt/ldr
Played in his first band, 1956, while living in Wangaratta, Vic. With Les Davis's (bjo) Out of Town Five at Nepean Hotel, Portsea, for holiday seasons, 1958–60. Very little activity in Melbourne or, from moving in 1968, in Sydney, until he began rehearsing a band in 1973. As Noel Crow's Jazzmen, the group opened the jazz policy at Red Ned's Wine Bar, Chatswood, in Sept., with Dave Ferrier (tpt), Jeff Hawes, Dave Robison (bjo), Rex Gazey, Rowan Wellsby (dms). AJC debut, Queanbeyan, 1973, and began a 12-month residency at the Woolpack Hotel, Parramatta. In the mid-'70s Dennis Tonge, Ken Sparkes, Terry Fowler replaced Robison, Wellsby, and Gazey respectively, followed by Col Best on electric bass replacing Fowler and Ted Sly on drums. Nancy Stuart performed with the band during this period, and the group played support for Dutch Swing College, June 1976. Ferrier and Sly were replaced by ex-New Zealander Kevin Keough and Geoff Allen, 1977. The band was now extremely busy, with casual work, a residency at Louis' at the Loo, and Red Ned's. Played its first jazz cruise on the *Minghua*, 1980, the Parkes Jazz Triduum, the Waratah Festival, and its first Sacramento Jazz Jubilee, 1982. Alumni from the late '70s incl. Dave Stevens and Verdon Morcom (p.), John Ryan (bs). The group made frequent AJC appearances and played the Manly Jazz Carnival, television and concerts. The Red Ned's residency finished in 1985, and in 1986 the band name was changed to Mr Crow and Phil Pryor (reeds) and Graham Cox (gtr) were added, with Keough, Hawes, Neville Byrnes (p.), Allen Riley (son of Perth-born pianist Jim) on bass, and Will Dower on drums. The band began residencies at Rock's Push and Liverpool RSL, with Will Dower directing the development of a 'show' presentation. In 1986 Noel Crow continued to lead the band which was scheduled for its third appearance at Sacramento and to present his weekly jazz information radio programme.

Cruickshanks, Robert William (Bob)   b. 24/8/19, Liverpool, NSW   reeds/vcl/comp.
Began noticing jazz in childhood, and in the '30s listened to swing programmes on radio and attended sessions of the Sydney Swing Music Club at the Ginger Jar, hearing musicians like Frank Coughlan, Harry Danslow, Jack Crotty, Dick Freeman, Reg Robinson (bs). Met Tony Newstead while serving in the RAAF in 1944, and they formed a quartet with Lee Gallagher on piano. While in Melbourne on leave, he came into contact with the burgeoning traditional scene of that period. Following discharge joined the PJJB, remaining (during its periods of existence) until 1948. Moved into the dance band area, with Harry Doyle on the Paradance circuit (1948–49), Cec Williams at the Albert Palais, Doyle again at the Surreyville (1952) and with Tut Coltman in suburban dance halls (1953). Concentrated on his public service career from the mid '50s, although continued to work in jazz, incl. as a member of the Black Opal JB in the early '60s. To Adelaide in 1965. Resumed freelance jazz activity, and began an association with Dave Dallwitz groups from 1972. Subsequently with Adelaide bands, incl. Blue Notoriety, formed by his daughter Penny Eames in 1976. Cruickshanks is a familiar face at AJCs and in 1986 continued to freelance in Adelaide.

Dallwitz David Frederick (Dave) b. 25/10/14, Freeling, SA   tbn./p./ldr/comp./arr
Following early exposure to jazz on records, tinkering on piano, and attempts to form bands, Dallwitz became involved in the Southern Jazz Group, with which his career was intertwined from 1945 until 1951, when he withdrew from the jazz scene. Continued to perform and compose in symphonic, chamber, and light revue contexts, playing cello and bassoon, while pursuing his career as a lecturer in art. Dallwitz began the second major stage of his jazz activity at the suggestion of Tas Brown (clt) who had worked in the SJG during its final period, and now returned to Adelaide from Gawler where he had maintained a practice in dentistry. With Brown, and other Adelaide musicians, Dallwitz resumed recording in 1972, the beginning of a prolific compositional and arranging corpus, much of it recorded by Australian musicians and by Earl Hines and Armand Hug in the US. He has been less visible in regular live performance, generally confining himself to leading various bands (the Hot Six, a big band, and a ragtime ensemble which reflects his most recent preoccupations), at concerts, festivals, and the AJC. He received the Order of Australia in 1986. For the most part his work as composer/arranger presents an ordered and elegant surface suggestive of Edwardian formality. He had been, however, founding chairman of the radical Contemporary Art Society in Adelaide, a reminder that there has always been beneath that surface a thrust towards more exploratory aesthetic gestures. The point is confirmed in the often disturbingly distorted reflections of traditional jazz forms in the Dallwitz masterwork, *The Ern Malley Suite*, which discloses many of the grotesque ironies of the episode that inspired it. Even in this instance however, and more unambiguously in his other suites, his musical idiom is consciously confined to pre-swing jazz styles, used as vehicles for the display of regional characteristics of Australian landscape and history. This convergence of style and substance has evoked from Clement Semmler the apt characterization of Dallwitz as a 'jazz Jindyworobak'.

Dave Dallwitz, 1986

Daly, Warren James   b. 22/8/43, Sydney, NSW   dms/ldr/comp./arr
Started drums 1958 and began working in pop groups 1959. Joined Gus Merzi at Romano's

(1964), Bill Barratt and Billy Weston (1965), and staff drummer at Channel 10 under Jack Grimsley (1966–67). Also worked in El Rocco. In 1967, to US to tour with the Kirby Stone Four then with the big bands of Si Zentner and Buddy de Franco. Returned to Australia 1968, and began to assemble a big band with Ed Wilson. With Don Burrows until 1970, leaving to concentrate on the Daly-Wilson Big Band. During the band's recess, 1971–73, Daly worked with Galapagos Duck at the Rocks Push. Resumed co-leadership of the Big Band from its resurrection in 1973, remaining until parting company with Wilson in 1983. During this period Daly had also played with the Toshiko Akiyoshi-Lew Tabackin big band on its Australian tour. In the mid-'70s he became the first Australian to be appointed to the Zildjian drum Hall of Fame. Daly formed a new big band under his own name in 1984 with sponsorship from the NSW State Lotteries Commission, and this has toured extensively

within Australia. He was MD for Channel 10's Olympic Games coverage in 1984, and in 1985, in addition to running his big band, was leading a quartet (Errol Buddle, Bob Gebert, Dieter Vogt, all of whom were in his big band) at the Rocks Push.

Daly-Wilson Big Band (Sydney, NSW)
Formed by Warren Daly and Ed Wilson from late 1968/early 1969, with arrangements provided by many of Sydney's most important session and jazz musicians incl. Paul McNamara, George Golla, Judy Bailey, Doug Foskett, Col Loughnan, George Brodbeck, Tommy Tycho, and visiting American drummer Louis Bellson, with whom Wilson had been working at the Hilton. The founding personnel during the long rehearsal period was Don Raverty, Kel Drady, Bernie Wilson, Dieter Vogt, Bob Barnard (tpts), Graeme Lyall, Foskett, Loughnan, Lee Hutchings, Bruce

The Daly-Wilson Big Band, 1974, with, from l. to r., back row: Larry Elam, Norm Harris, Don Raverty, Bob Bouffler, Warren Clarke (tpts), *middle row:* Mick Kenny (flugs), Ed Wilson, Herb Cannon (tbns), Peter Scott (bs tbn), *front row:* Paul Long, Doug Foskett, John Mitchell, Bob Pritchard (saxes), *rhythm:* Hugh Williams (gtr), Ray Alldridge (p.), John Helman (bs), Warren Daly (dms)

Johnston (reeds), Wilson, Brodbeck, John Costelloe, Merv Knott (tbns), Neville Whitehead (bs), Col Nolan, Ned Sutherland (gtr), Mark Bowden (perc.), and Daly. By the time of the band's public debut in August 1969 at the Stage Club, Whitehead and Johnston had been replaced by Ford Raye and Tony Buchanan respectively, and Lyall had gone to Melbourne. The band played concerts and worked in clubs and on the Miller hotel circuit, and made its first album in 1970. In January 1971 it toured with the Dudley Moore trio, which, according to reviews, it overshadowed. For economic reasons the band broke up in 1971, with a farewell concert on 8 December. In 1973, however, support from a tobacco company enabled it to be revived, with Daly and Wilson continuing to lead and Raverty resuming as band manager. In addition to local concerts, the band toured extensively, incl. overseas: NZ, SE Asia, Hong Kong, and to great acclaim in Russia, the Baltic states, and the US, in 1975. It won numerous awards, incl. the NSW Variety Star Award known as the 'Mo', in 1981 and 1982. The band made frequent TV and film appearances, and numerous LPs incl. several which achieved 'gold' status. In 1983 it played the opening of the new Regent Hotel, George St. In Sept. 1983 Daly and Wilson parted company and the band broke up, although each has gone on to form new orchestras. The Daly-Wilson Big Band was the most successful jazz-based big band in post-war Australia. It provided a performance workshop in which scores of young musicians served an apprenticeship well before the establishment of the jazz studies programme at the NSW Conservatorium. It also provided the setting in which many important performers both in jazz and in more commercial areas received increased exposure: singers Kerrie Biddell, Ricky May, Marcia Hines; sax players Col Loughnan, Tony Buchanan, Lee Hutchings; trumpeters Peter Cross, Mick Kenny, Paul Panichi; trombonists Herb Cannon, Bob Johnson, Dave Panichi. The great popularity of the Daly-Wilson Big Band made it a major force in the spread of jazz awareness in the '70s.

Davidson, James H. (Jim) b. 6/8/02, Sydney, NSW; d. 9/4/82, Bowral, NSW dms/ldr
Began on cornet, then euphonium, in local brass bands as a child. Switched to drums, c. 1919, playing in local dances and with a jazz quartet. Worked on the New Era Stage Presentations vaudeville circuit, 1922, later turning professional to join the pit band at the Lyric Theatre in George St with Jimmie Elkins, with whom he went into the Wentworth Hotel in 1927. Joined Jack Woods at the Ambassadors, 1928, until it was destroyed by fire in 1931. Formed a band (Gordon Rawlinson p., Frank McLaughlin reeds, Ray Tarrant tpt, Dud Cantrell tbn.) to play at Smith's Oriental in Her Majesty's Arcade, Pitt St, and later augmented this to a ten piece for a concert on 12 Aug. 1932. This led to his being booked by Jim Bendrodt for the Palais Royal from 5 May 1933 (the extra musicians being Jim Gussey tpt, Tom Stevenson tbn./gtr, Pete Cantrell and Chic Donovan reeds, Orm Wills bs). In addition to performing at the Palais, this band played society functions and recorded. The Palais season was interrupted by a season at Melbourne's St Kilda Palais (c. October 1933 to March 1934) where the band attracted a young audience but encountered conservative resistance to its more modern American-influenced style and rhythms. The band returned to Sydney to finish its Palais engagement in mid-1934. Davidson returned to Melbourne to take up a 16-month contract with the ABC to lead a broadcasting band as from 8 Jan. 1935, and this marked the end of his career as a drummer, and the beginning as full-time leader. Back in Sydney he resumed at the Palais Royal from 1 May 1936, then in March 1937 entered into a contract with the ABC in Sydney to lead a 25-piece dance band. With variety and vaudeville acts (incl. Peter Finch and Bob Dyer), this band broadcast and undertook a number of very influential national tours. From the outbreak of war it also played numerous services functions. Davidson's association with the ABC concluded in November 1940 and Jim Gussey took over the dance band. Davidson joined the AIF and throughout his service career he worked with concert parties in

Jim Davidson with his band, as identified, c. 1933

Africa, the Pacific islands, and throughout Australia. He set up the '50–50' (US and Australian) entertainment unit, and upon his discharge with the rank of lieutenant colonel in 1946 had become the officer in charge of all Australian army entertainment units. As a civilian he joined the Tivoli circuit in Sydney, but remained only briefly. In 1947 he joined Harry Wren's organization as MD, touring the eastern states with Evie Hayes, Tommy Trinder, and Will Mahoney. Davidson was an astute observer of the entertainment industry and it was clear to him that during his absence, the war had transformed the business. When the BBC in London invited him to become assistant head of Variety, he accepted, and arrived in London on 28 Jan. 1948 to begin a career that would see him influential in the post-war history of light entertainment in English broadcasting. This period of Davidson's career is not relevant here, but interested readers are refer-

red to his autobiography, *A Showman's Story* (Rigby, Adelaide, 1983) for further information. Davidson returned to Australia in 1964 and worked as a consultant for the ABC. He spent his retirement renovating old houses until his death from cancer.

Dawson, Charles Horsley (Horsley)   b. 7/2/32, Driffield, East Yorkshire, England; d. 2/1/87, Brisbane, Qld   bs

Took up bass at 17 and began working in dance bands. At 23 joined a jazz quartet in Scarborough, and worked with other groups in the district. In Brisbane in 1971, joined the Vintage group from 1974–79. With Ken Herron at the Melbourne, followed by 14 months with Mileham Hayes at the Cellar. In 1985 with David McCallum's City Stompers and working freelance until shortly before his death from cancer.

Dehn, Douglas (Doug)   b. 16/8/33, Melbourne, Vic.   tpt

Began playing in dance bands in 1949, and sitting in with the bands of Frank Johnson and the Barnards. First important engagement was a Melbourne TH Downbeat concert, followed by others in the '50s. In the traditional idiom he has worked with the Darktown Stompers, Dixie Rhythm, and later with Frank Traynor. In the meantime his experience broadened to include work with Brian Brown at Jazz Centre 44 and about a year and a half with Barry Duggan. Has also been involved with blues and rock-tinged groups like Original NO Rock Band and Dave Rankin Banned. With the Onyx JB, backing Clark Terry at the Victorian Jazz Club, 1974. In 1985, playing with Déjà Vu in Geelong, where he works as a school teacher.

Delamare, Roy McLeod   b. 27/10/13, Cottesloe, WA   p./tbn./vin

Began playing piano at three, first under his mother's instruction and from eight receiving more formal lessons. Switched to violin, taking lessons from age nine to 15, during which period he also took up trombone. He resumed piano in 1931, taking lessons from J.R. Nowotney, whose thorough instruction in harmony enabled Delamare to grasp the principles of improvisation at a time when few dance musicians in Perth had much understanding of the matter. He had first heard

Roy Delamare, 1985

jazz as played by Theo Walters's Knickerbockers at the Kit Kat Tea Rooms, Cottesloe in 1930, and he began working in dance bands in 1934, notably with that of Ron Jenkins (reeds). Following six months in Albany he went into the Cabarita with 'Skippy' Alexander, regarded as the most complete jazz musician of that time in Perth. Surviving acetates of broadcast material made by Delamare and Alexander, who died at 34 on Easter Thursday 1947, vindicate the high esteem in which both these men were held; as early as the late '30s Alexander was a fluent stylist in the Artie Shaw mould and Delamare was developing clearly towards the Art Tatum style which marked his mature work. The Cabarita residency lasted throughout most of the war, and in 1944 Delamare joined Viv Nylander at Cottesloe Palais and at the Embassy, where he continued under Bill Kirkham following Nylander's replacement. When Sam Sharp brought in a new band with Jim Beeson on piano in 1947, Delamare joined Merv Rowston, with whom he worked until 1963. This has been followed by freelance work in hotels and cabarets, incl. a period during which he replaced Jim Riley at the Royal Perth with Ron Jenkins while Riley was recuperating from injuries sustained in a car smash. Delamare has always held a day job, and since 1953 has run his own farm. In 1985 he was still freelancing, playing two or three engagements each week.

De Silva, Don   b. 31/1/37, London, England   bjo/gtr

Began on trumpet in a local band, then switched to banjo. Played cornet in a Royal Navy band for three years, then worked with various jazz bands in London, incl. more than three years with Micky Ashman. To Sydney, 1969, where he worked with most of the New Orleans style groups, though his original preferences had been in more modern areas. Worked for about five years in Nick Boston's band, for four years in Pam and Llew Hird's Sydney Stompers (also spent part of 1979 in Perth with the Hirds), and led the band at the Grand National Hotel, Paddington, for 12 years. In 1986, occasional freelancing.

Drinkwater, Ian John (Splash)   b. 10/11/28, Adelaide, SA   reeds/fl./arr/ldr

In 1938 J.E. Becker took Drinkwater into the Adelaide College of Music, where he learned clarinet from Les Mitchell. He took further lessons from Alf Holyoak and played in the Palladium and Embassy ballrooms 1948–52. In 1951 he formed his own progressive jazz group, Splash Drinkwater and his Four [later Five] Drops of Rhythm which was active on Bill Holyoak's concerts and for the swing picture nights on the Odeon cinema circuit. Members at various times incl. Brian Bowring (dms), Graham Schrader, John Bermingham, Roy Wooding (gtr), Billy Ross. During this period Drinkwater also worked with Frank Buller's dance band and a Latin American group in which he met Bryan Kelly who enlarged his knowledge of jazz. From 1959 Drinkwater worked in TV through to the 70s, and for a local music store during the day. In 1985 he was teaching and in his eleventh year with the band of the SA Police Force.

Duff, Edwin   b. 4/6/28, Dundee, Scotland   vcl/comp.

As a child Duff heard music in the local cinema where his mother worked, and showed an early aptitude as a singer. Migrated to Melbourne at age 10, and immediately won two radio talent quests. Worked with Ern Pettifer at Leggett's ballroom, singing jazz material in the small group features. During the war he sang in concert parties for the services and first heard jazz performed by Americans. Worked with Gren Gilmour, Billy Hyde, Bernie Duggan, at the Dugout with George Watson, and with the vocal group the Tunetwisters. Spent a period in Sydney in mid-'40s, and played the Trocadero and the Roosevelt. In Melbourne he worked with Jack Brokensha at the Plaza coffee lounge, 1947–48, then at the Galleone. Toured the eastern states with Brokensha until the latter's illness in 1949; Duff attracted a considerable following which incl. members of a fan club established in mid-1949. With Frank Marcy's band in the early '50s, then left for Canada, 1954, intending to join the Australian Jazz Quartet. The quartet, however, had moved to the US and established its reputation

as an instrumental group, so Duff took daytime employment, with casual jazz and cocktail engagements. Returned to Sydney in mid '60s, but found little opportunity for jazz performance. Apart from work with the Daly-Wilson band, c. 1971, most of his singing until the '80s was in TV. From the early '80s was regular guest with Georgina de Leon's Lucy Brown Quartet, and in 1986 was with Terry Rae's band at Vanity Fair Hotel. Duff has also composed, one of his songs being recorded by Mark Murphy, and has presented a recital/lecture on the history of jazz at schools. While earlier singers such as Des Tooley and Barbara James had worked with swing groups or presented current repertoire with a jazz feel, Duff was the first performer to establish a national reputation simply and explicitly as a jazz singer.

Duffy, Calvert John (Cal)   b. 1/10/49, Melbourne, Vic.   dms/wbd

First heard jazz on the ABC's *Music to Midnight*, and in 1966 he joined a practice band with Pat Miller, Sandro Donati, Paul Finnerty. Attended his first AJC that year, and in 1967 joined the Northside JB, moving into the Limehouse JB. Following its demise he worked casually, as well as with Lyn Thomas's ragtime group, and in association with Neville Stribling, with whom he appeared at the Sacramento Jazz Festival in 1982. In 1983 he formed Colonel Duffy's Privates (Ian Smith, Alf Hurst tbn., Dave Hetherington clt, Pip Avent tba, Tony Orr bjo). In 1984, briefly with Kid Sheik and the Bongo (Pat Miller, Sandro Donati, Conrad Joyce, Chris Summerville p.), and in June joined Chris Ludowyk's Society Syncopators, with whom he was still playing in 1986. Duffy is also an accomplished tap dancer, and as such, performed in a tribute to Bojangles in a production by Allan Leake's Jazz Repertory Co. in 1985.

Duggan, Barry (The Bat)   b. 8/12/42, Melbourne, Vic   reeds/fl.

Taught by his father (the whole family, incl. three sisters, were musicians; cousin Bernie was a prominent dance band p.). Began piano, violin, flute, at six, clarinet at 14, and alto (with coaching from

Frank Smith) at 16, when he turned professional. MD at Ritz Hotel St Kilda with Chuck Yates, Dave Tolley, Alan Turnbull. Led group at Opus 61 St Kilda and for town hall concerts, with Doug Dehn, Barry Veith (reeds), Steve Dunstan (bs), Brian Jones (dms), Ron Sedgeman (p.). To Sydney, 1963, working with Joe Lane and John Pochée. Back in Melbourne 1967, led David Martin (p.), Dunstan, George Neidorf (dms), in a post-bop experimental group. Late '70s, to Perth, working with Bill Gumbleton. Travelled to India, England, Europe, and the US, through the early '70s, working with jazz and rock groups incl. Max Merritt and the Meteors. To Sydney, 1976, and replaced Howie Smith in Jazz Co-op. Also worked with Serge Ermoll's Free Kata, the John Hoffman Big Band, led his own group (Bob Gebert, Stewie Speer, Ray Martin), and worked in studio and club settings. Performed with Sonny Stitt and Richie Cole, 1982. Moved to Adelaide, 1982, and since active as a teacher at SA CAE. Visited Sydney, 1985 and 1986 for Manly Jazz Carnival, Festival of Sydney, and concerts with Serge Ermoll. In 1986, resident in Adelaide and playing at Tooley's.

Dunbar, Ruth Kathleen, née Kelly (Kate, Sister Kate) b. 13/5/23, Manchester, England vcl
Arrived in Australia in the late '20s, took classical tuition at NSW Conservatorium with Cecily Atkins and, later, private lessons from Marianne Mathy. Gave up singing at the time of her marriage, 1946, but husband Eric introduced her to jazz. Attended rehearsals of the Paramount JB, 1952–53, sometimes sitting in. This came to little until 1955 when Pat Qua gave invaluable encouragement and advice. Dunbar sang at Cootamundra AJC, 1955. Joined an all-woman group, playing guitar, with Qua on piano, and played with Qua at the AJC in 1958, by which time she

had begun singing with Ray Price's group at Adams Hotel. In 1959 she joined the Paramount JB, now led by Ian Cuthbertson (cnt), playing in the rhythm section and gradually increasing her role as vocalist. Recorded on Pix with Ray Price, with whom she also played concerts in the early '60s. Performed with Graeme Bell for tours, concerts, television and radio (1963–64) and also sang with the Black Opals and Frank Traynor in Melbourne. In the late '60s she was still with Price, doing five nights a week at Hotel Australia. Played Waratah Festival, 1973, then retired from music. Resumed singing in the '80s, sitting in with different bands and becoming regular vocalist with Ted Sly's East Coast group at different residencies and for Sydney Opera House boardwalk concerts. In 1985 she became president of Sydney Jazz Club and editor of *Quarterly Rag*, (she had also edited the first series of *QR* for a period in the '60s) and founded the Kate Dunbar Presents record label, producing and funding records featuring Sydney jazz musicians. In 1986 she was still singing with Ted Sly's band. Throughout her career Dunbar has unwaveringly preserved the female blues style of Ma Rainey and Bessie Smith.

Dyer, Robert Warwick (Wocka) b. 6/6/28, Melbourne, Vic; d. 22/9/55, Nagambie, Vic
tbn./vcl/comp.
In 1946 Dyer began organizing practice sessions at his home, taking lessons from Roger Smith, and sitting in with the Bells at the Uptown Club. In May 1947 he joined Frank Johnson's Fabulous Dixielanders, and in 1952 won the *Music Maker* poll in the trombone category. Killed in a road accident, 1955. Although his career was brief and confined to one band, Dyer's exuberant and magnaminous personality left as significant a mark on Melbourne jazz as his driving, extroverted music.

E

Eames, Grahame   b. 29/3/43, Adelaide, SA   tpt
Began playing with the Hindmarsh Brass Band in
1954. Coached in jazz by Alex Frame whom
Eames met through the Wayville Institute functions. In 1961 he performed at the AJC with Mike
Rossiter, Kingsley Dignum, Peter Ubelhor, known
collectively as the South Town Jazz Four, which
provided the nucleus of the Adelaide All Stars.
Through the '60s with bands working the Tavern
venues, with Roger Hudson's Eumenthol Jazz
Jube groups and the Vencatachellum Jazz Peppers
which he co-led with Bruce Johnson. Following a
year in London (1969–70) he returned to Adelaide
and in 1972 joined Dave Dallwitz, working more
or less regularly with him ever since. Has also
worked in Penny Eames's Blue Notoriety, and
with the Avoca JB led by bjo/gtr player Tony
West. Although primarily a trumpet player in the
Armstrong tradition, Eames has also played occasional sax.

Eames, Penelope Jean (Penny)   b. 10/3/48,
Sydney, NSW   vcl/ldr
Her early interest in music was stimulated by
blowing sessions at home involving her father, Bob
Cruickshanks. She took music at school, and
formed a vocal trio which won a talent quest,
bringing a record contract with EMI and work on
the Miller circuit. Billie Holiday was a major influence in attracting her to jazz, and she began
singing casually in a jazz context after moving to
Adelaide in 1966. Joined Bruce Gray in 1971 and
Dave Dallwitz in 1972, recording with both bands,
notably on Dallwitz's *Ern Malley Suite* in 1975.
She worked with the Campus Six, and at the Melbourne AJC in 1974 performed with Clark Terry.
Broadened her range with her own band, Blue

Notoriety, formed by 1976 (with Cruickshanks,
Geoff Kluke, Bruce Hancock p., Jim Shaw dms
and husband Grahame Eames tpt). She has worked
on TV and in concert situations, and in 1977
toured Australia and SE Asia with Bob Barnard's
band. She has lectured in the SA CAE jazz studies
programme and in 1985 was working with the
band Body Heat.

Eclipse Alley Five (Sydney, NSW)

Formed in March 1969 to play in John Huie's
Wine Bar (later the Rocks Push), with Paul Furniss, John Colborne-Veel, Noel Foy (bjo), Dick
Gillespie (dms), Peter Gallen (bs), and later in the
year performed at a benefit concert for ex-licensee
of the Orient Hotel, Jack Button. The band name,
referring to a location in New Orleans, discloses
the group's initial stylistic affiliations. Began a
Saturday afternoon residency at the Vanity Fair
Hotel in Goulburn St, 20 June 1970; the enthusiasm of licensee Jim Hourigan inspired Furniss's
composition 'Hourigan Swing' which won the
original tunes competition at Dubbo AJC, 1970.
In 1971 the band replaced the Double Cross JB
which, earlier in the year, had begun Tuesday
nights at Vanity Fair (with Doc Willis ldr, Furniss,
Geoff Bull, Viv Carter, Foy, Danny Haggerty bs).
In the same year Carter replaced Gillespie in the
Eclipse Alley Five. When Bruce Johnson replaced
Colborne-Veel in 1972, the personnel remained
stable until 1983, with the exception of long-term
deputies as follows: for the Saturday gig, Peter
Buckland, then Allan English, replaced Furniss
who had a conflicting commitment with Graeme
Bell during periods in 1973–74; Graham 'Iggy'
Kellaway replaced Noel Foy from October 1974
to October 1976 while Foy was in New Guinea

The Eclipse Alley Five, Vanity Fair Hotel, c. 1972, with, from l. to r., Paul Furniss, Bruce Johnson, Noel Foy, Peter Gallen, Viv Carter

then England; Johnson was in England from mid-1975 to February 1976 and was replaced by various, incl. Roger Graham and Ian Barnes; Gallen visited New Orleans in 1977 and was replaced by Chris Qua. From 1978 Marty Mooney became a regular deputizer for Furniss on the Saturday gig. The band also worked other residencies, notably the Sydney Cove Tavern in the AMP Building, 1976–77, and the White Horse Hotel in Newtown for two separate terms, though with some variations in personnel and the addition of Nancy Stuart. In 1983 Noel Foy left, to be followed by a succession of replacements incl. Pat Wade (gtr), Paul Finnerty (bjo/gtr), Grahame Conlon, and finally from c. September, Peter Boothman. Gallen was replaced by Harry Harman in January 1983. Over its 17 years the band's style has inevitably evolved, partly by virtue of individual development of its long-standing members, and partly

through its occasional turnover in personnel, with Boothman's arrival having a particularly broadening effect. Departed founder members Gallen and Foy have continued to work primarily in New Orleans style settings, the former with Nick Boston on a long-term basis. The Eclipse Alley Five has also played numerous concerts for Sydney Jazz Club, whose unofficial headquarters was Vanity Fair during the '70s, though the band's development has estranged many members. The band also played the Tauranga Jazz Festival in New Zealand (1975) and for the Canberra Jazz Club, as well as numerous AJCs, and recordings. Other bands who have played the Vanity Fair include those led by Nick Boston, Peter Strohkorb, Inge le Lièvre (vcl) and Terry Rae. Although the Eclipse Alley Five's Tuesday night residency finished in March 1975, the Saturday afternoon sessions became a nationally known institution and the longest continuous jazz residency in Australia. On 31 May 1986 the band finished its residency at the Vanity Fair Hotel, which was scheduled for demolition, and transferred its base to the Crown Hotel on the same block.

Edgecombe, John   b. 17/2/25, d. 11/12/85, Sydney, NSW   gtr/bs/bjo/tba/tbn./comp./arr/vcl
Following early tinkering on ukelele, began sitting in at jam sessions at Palings music store, and took his first paid engagement in 1939 at the Amory reception venue. Freelance activity was interrupted by service in the AIF (1943–46) where he played tuba in a service band and also served with the Pioneers in Borneo and Moratai. Following discharge, Edgecombe became one of the most ubiquitous musicians in the night-club/concert/dance scene in Sydney. On the High Hat Dance Circuit with Bill Mannix (sax) then a founder member of Ralph Mallen's big band, and overlapping bands led by Bill Weston and Jack Grimsley. With Cyril Parker's Esplanade Orchestra (1946–47), the bop and swing group Harbour City Six (incl. Ron Falson, Don Burrows, Joe Singer) at the same period. In 1947 was a mainstay of jam sessions in the California Coffee Shop, played the State Theatre, the Celebrity Club, and was with Craig Crawford at Sammy Lee's. At Reg's Restaurant, 1948, and through the jazz concert era was active with (in addition to bands listed above) Bela Kanitz, Bob Gibson, Jack Allan, Les Welch. Married singer/film actress Ilma Adey in 1949. His hectic schedule continued through the '50s: with Abe Romain's State Show Band (1951–52), with Ron Gowans at the Roosevelt (1952–53), with Ralph Mallen's reformed big band (1954), with Graeme Bell for a TV series (1958), recording with Julian Lee (1958). Throughout, he was very active in studio and radio work, incl. with vocal groups. In 1954 he was, with Ron Gowans, one of the founders of the Australian Jazz Club which fostered progressive jazz styles. Played the Ling Nam (1960–61), and in the backing group for Dave Brubeck's tour (1960). Less visible in jazz settings during the '60s and '70s, Edgecombe was mostly involved in more commercial areas, incl. club and session work, though always continuing to freelance in jazz. In 1979 a gig with Noel Gilmour at the Redfern Courthouse Hotel led to his joining Dick Hughes's Famous Five (1980–85). At the time of his death he had been for many years with the Lucy Brown Quartet led by Georgina de Leon vcl (Frank Murray (dms), Ken Morrow (p.), and Charlie Munro, who predeceased Edgecombe by two days) and was with Merv Acheson's group. Edgecombe was one of the most prominent and active examples of the professional night-club/dance band musician who, in the '40s and '50s, also energetically fostered the progressive jazz movement.

John Edgecombe, Soup Plus Restaurant Sydney, 1981

Edwards, Desmond David (Desi, The Bear)
b. 30/11/26, Melbourne, Vic.   tpt
Taught himself on a borrowed cornet from 15, followed by experience in the Port Melbourne Brass Band. In the army at 18, serving in the Eastern Command (Sydney) and Southern Command (Melbourne) Bands. Most of his post-war work was in the dance band area with, *inter alia*, Mick

Walker and Denis Farrington, until his retirement from music in 1974 except for occasional engagements. Edwards was also intermittently visible in jazz settings, recording with Cy Watts for Jazzart and with other groups incl. those of Frank Traynor and Frank Gow. Edwards is an example of a musician whose talent was not fully compatible with current local tastes, so that he remained underrepresented on record and comparatively inactive in jazz performance. Private recordings reveal a fully formed and impressively accomplished jazz improviser in the big-toned tradition of Bunny Berigan, with unsurpassed drive and inventiveness.

Elhay, Sylvan (Schmoe)   b. 13/1/42, Cairo, Egypt   reeds/ldr

Took lessons from Ken Woolridge while at secondary school and attended Wayville Institute functions. Hearing the AJQ on its 1958 tour sharpened his interest in more progressive styles and he became an habitué of the Cellar. As an undergraduate, joined what became the University Jazz Group. As a foundation member of Bottom of the Garden Goblins in 1964 (with Jerry Wesley, Gordon Latta, Roger Swanson), he played at the Cellar, the Velvet Tavern, the Catacombs, the Somerset and Duke of York hotels. He took over as leader early in the lifetime of the band, which incl. at different times Roger Hudson, Don Bond (p.), Gordon Coulson, Geoff Kluke, Phil Cunneen (keyboards), Glen Bayliss, Phil Langford and Dean Birbeck (dms). Following the band's gradual demise by 1974, he worked in Onions, a semi-rock group. Since the late '70s he has worked with, *inter alia*, Ted Nettelbeck, Laurie Kennedy (dms) and Freddie Payne (tpt), and with Dave Colton in the Creole Room in 1978. Because there is no centre for the bop/post-bop style in Adelaide his work has been dispersed and varied—jazz performance, often with his group Schmoe & Co., incl. JAS concerts both in Adelaide and Sydney, a recording with Sue Barker, rock, and symphonic work on clarinet. Has written film scores and participated in jazz workshops as well as teaching in the jazz studies courses at the SA CAE since their incep-

Schmoe

tion in 1979. He succeeded Judy Bailey on the Music Board of the Australia Council in 1986. Schmoe's stature as a musician is impressive by any standards, but his decision to remain in Adelaide, where he has a university post as a computer scientist, has deprived him of the celebrity he would almost certainly have achieved in Sydney.

Ellis, David William (Dave)   b. 6/12/43, Perth, WA   bs/ldr

Learned ukelele from father from age three. Played home-made instruments throughout high school; bought his first bass at 17, taking lessons from John Mowson of the WA Symphony Orchestra and former Jack Brokensha sideman. Became active on the Perth jazz scene with numerous bands incl. the

Riverside Jazz Group, West Coast Dixielanders (1961–63), with more modern groups led by Bill Clowes (1962–69), Jim Beeson (1963–69), and with Bill Gumbleton's Soul Brothers (with Bob Cochran dms) from 1963–69. Also worked in the mainstream company of George Franklin and Horry King (1963–69) and on occasions with the WA Symphony Orchestra. To Sydney, 1969, working with Doug Foskett (1970–71), Warren Ford, and at various times with the Sydney Symphony Orchestra. With SCRA (Southern Contemporary Rock Assembly) with Peter Martin and Jim Kelly (gtrs), Ian Bloxsom (perc.) and Roger Sellers (dms) 1970–72. Began studio and session work (1972) and freelancing with, *inter alia*, Jeannie Lewis, Dave Levy, Margret Roadknight. With the Australian Elizabethan Theatre Trust Orchestra, 1973–74, with Peter Boothman at the Limerick Castle wine bar, 1974–75. Formed 50 Fingers, with Geoff Oakes, Ned Sutherland gtr, Ray Alldridge (later Tony Esterman) and John Swanton, 1978–81; the Sydney Jazztet with Steve Murphy (gtr), Dave Glyde (reeds), Willie Qua, c. 1979. Also worked in the late '70s with Jon Rose's Relative Band, with Serge Ermoll, and formed the group Flash, with Sutherland and Esterman. Mostly commercial work, early '80s, then formed Ozbop, 1984, with Esterman, Bloxsom, Warwick Alder and Peter Cross (tpts), Bob Bertles and Steve Giordano (reeds), Tom Ferris (gtr), James Greening (tbn.), Patricia O'Connor (vcl), playing regularly at Goulburn St Connection. Ellis has also toured with numerous visiting musicians, incl. Peggy Lee, Mike Nock, Herb Ellis/Barney Kessel, Phil Woods, Milt Jackson.

El Rocco (Sydney, NSW)
Opened as a coffee lounge in Brougham St, Kings Cross, in 1955. Manager Arthur James installed a TV for customers when the medium was still a novelty, and at the suggestion of Ralph Stock (dms) he hired a jazz group to play for the customers after the end of transmission on Sunday nights. The first band, beginning in October 1957, was the Ken Morrow (p./p. accordion) Quartet, with Stock, Wally Ledwidge (gtr), Jack Craber (bs).

Other musicians began sitting in, and in 1958 James extended the policy to Friday nights with a younger generation of modernists—Dave Levy, Bernie McGann, John Pochée, Dick Barnes (bs). In the meantime, Stock had taken over the Sunday nights, using musicians who had established their reputations in the post-war night-club scene, incl. Don Burrows, Nigel Rolphe (p.), Terry Wilkinson, and Dutch-born Freddie Logan (bs). Later, under different leaders, Frank Smith, Cyril Bevan, Cliff Barnett also played the Sunday nights. El Rocco became the centre for Sydney's progressive jazz activity throughout the '60s, providing older musicians with a chance to play unfettered jazz, introducing newcomers from within and beyond Australia, and ultimately, encouraging such fluid interchange that it is virtually impossible to reconstruct a detailed account of all musicians' and bands' movements. American Bob Gillette exercised his influence from 1958 through performances in El Rocco, and, at about the same time as Dave Levy, began introducing various experimental approaches incl. free form and poetry-and-jazz exercises. New Zealanders Mike Nock, Dave MacRae, Barry Woods, Andy Brown, Rick Laird (originally Irish), and Judy Bailey established their reputations in El Rocco. Nock (b. 27/19/40), was particularly influential during his brief residence in Australia. He arrived in Australia at 18, and was influenced by both Frank Smith, with whom he worked at The Embers in Melbourne, and by Gillette. Nock began playing at El Rocco in 1960, and led his own trio from about mid-year. With Colin Bailey dms (replaced after a few weeks by Chris Karan) and Freddie Logan, the group became the 3-Out, and soon had four nights at El Rocco. They became so successful that for a brief period towards the end of 1960 they were booked into the Primitif, Bayswater Road, a more commercial venue, and were scheduled to open in the renovated Pigalle in Oxford St, both clubs associated with the enterprises of entrepreneur Lee Gordon. The Pigalle failed to eventuate, and the atmosphere in the Primitif was not sympathetic to the uncompromising jazz approach of the group, who quickly returned to El Rocco where,

El Rocco, 1958, with, l. to r., Ralph Stock, Cliff Barnett, Frank Smith, and waitress Pam(?) James

throughout its lifetime, the undiluted centre of interest was the music. 3-Out played the International Jazz Festival (promoted by Gordon) in 1960, and toured New Zealand with Paula Langlands in 1961. In May 1961 the group left for Europe where, following its disbandment, the members dispersed to establish successful musical careers on their own. Nock settled in the US where he established a reputation at the top of the jazz profession and, incidentally, worked at times with ex-El Rocco colleague Rick Laird who himself achieved celebrity with John McLaughlin's Maha-

vishnu orchestra. In 1985 Nock returned to Australia and taught at the Qld Conservatorium before resettling in Sydney where he continues to be musically active.

El Rocco also nurtured the major modern jazz talents of the '60s and '70s, most of whom are recorded elsewhere in this volume. Its fame emerged from the underground through the '60s, with articles in the lay press, a TV feature focusing on manager Arthur James, and an enthusiastic review by Arnold Ross in the American journal *Down Beat*. James applied for a liquor license on several occasions, and was invited to undertake renovations and re-apply at about the time he left for an overseas visit in 1968. Upon his return, he became aware of changes taking place in Sydney entertainment, particularly the flourishing of the large sporting and RSL clubs which were providing well-paid work for musicians. He decided not to proceed with the projected extensions to the premises, and closed El Rocco in 1969. The room is now part of a restaurant.

For 12 years El Rocco maintained an uncompromising jazz policy which incubated the modern movement that has sustained the music in Sydney and elsewhere until the new generation which began emerging in the '80s. The club brought two generations of progressive musicians together (often under the magnanimous tutelage of Frank Smith), and in this way established continuity and encouraged stylistic mix and manifold permutations of talent. It hosted the earliest experiments that pushed beyond the bop movement into free jazz and other developments from the late '50s. It gave Australian musicians unprecedented confidence in their work, especially through the enthusiastic feedback from the host of visiting musicians, from Nancy Wilson, to Lou Rawls and Dizzy Gillespie, who jammed there. As El Rocco achieved a reputation with the lay public, it was also crucial in giving jazz a public image of aesthetic dignity. Its proximity to the ABC drew classical musicians on their way home from work, and this led to ABC and Musica Viva involvement in jazz concerts. It provided inspiration for similar venues in other cities, and also for the Wentworth Supper Club jazz room, which itself spread the public awareness of jazz. El Rocco was almost certainly the most important venue in the history of Australian jazz.

English, Allan (Spotty)   b. 15/1/40, Eugowra, NSW   reeds

Began clarinet as a child and in 1954 won a radio competition with a rendition of 'The Golden Wedding'. In the '60s worked in clubs, incl. with Johnny Golden at Balmain Leagues Club (1966), and with Jeff St John's rock/pop band. Also freelanced in jazz settings, led his own group at the Windsor Castle, 1966, and in one of the earliest jazz-wine bar residencies, in Darlinghurst. With Bill Haesler's band initiating the jazz policy at Soup Plus, 1974; replaced Paul Furniss for several periods in the Eclipse Alley Five at Vanity Fair. Led a band at Emma's bistro in the city, 1977, and in the '80s has been a regular member of the band playing at the East Sydney Hotel. Although generally reliant on club and other commercial work, English's intense, bop-based music has continued to be heard in jazz venues since the '60s.

Ermoll, Serge   b. 16/8/43, Shanghai, China
p./ldr/comp./arr

Born of Russian parents; his father was a musician and bandleader. Ermoll had been studying piano for five years when the family arrived in Sydney when he was nine. Because a piano was too expensive he moved to triumpet, and played it until he was 17, switching back to piano and travelling to Surfers Paradise where he worked with Stan Bourne, who broadened his repertoire. Returned to Sydney early '60s. A brief period with the NSW Conservatorium ended in disillusionment, and his most important lessons came from Chuck Yates, Billy Antmann, and Dave Levy. In 1963–64 leading a trio at El Rocco, then left for England, returning in early 1966. Drummer and mentor Len Young got Ermoll a trio gig at El Rocco with the addition of Neville Whitehead (bs), and during this period they jammed with many eminent visiting musicians. Ermoll was still playing at the El Rocco, with Jack Dougan (dms) and Bruce Cale, when it closed in 1969. In London 1970–74, working with major British names and at major venues such as Ronnie Scott's, as well as for the BBC with Chris Karan and Pete Morgan (bs), and en route to Australia played with Sonny Stitt in San Francisco. In the late '70s he performed and recorded uncompromising essays in free improvisation with his group Free Kata. The band used shifting personnel that incl. some of the most independent, but unpublicized musicians in the modern experimental area: Eddie Bronson (sax), Louis Burdett and Ross Rignold (dms), Graham Ruckley (bs). One album featured an improvised vocal recital by John Clare. Ermoll recorded a similar exercise with Jon Rose and Dave Ellis in 1978. In the '80s, his work has been generally more conventional, and he is equally at home within mainstream parameters. In 1982, with Barry Duggan and Speer he recorded with visiting Americans Herb Ellis, Ray Brown, Richie Cole. Since 1984, much of his energy has gone into promotion through his East Coast Productions, which have organized tours (Joachim Kuhn, 1984), the Paddington International Jazz Festival, and in January 1986, the East Coast Jazz Festival.

Throughout his career, Ermoll has been prolific as a composer, an arranger and performer, even though he remains comparatively unknown to the public. His work with Free Kata was of major importance. Working from a philosophical base connected with martial and meditative disciplines, he recorded extensively with uncompromising voices who have never achieved visibility in the rather conservative jazz establishment. In so doing, he helped to ensure the preservation of an important reservoir of experimental energy in Sydney.

Esterman, Anthony Bernard (Tony)   b. 19/4/39, London, England   p.
Began piano lessons aged eight and at 13 entered Guildhall School of Music on a scholarship, with bass as second instrument, on which he worked with traditional bands. At 19 while in National Service was posted in a regimental band, assigned clarinet and tenor. Arrived Sydney in 1960, but involved in little public playing until he formed a quartet for El Rocco, 1967. Esterman's importance since has been twofold. He replaced Dave Levy in 1975 as the regular pianist with The Last Straw, one of the major groups in the history of the modern movement in Sydney, performing again with the band when it was reformed for occasional work in 1983. Esterman has also had significant if more diffuse impact as one of those impressively gifted musicians who find it difficult to make a living out of jazz, but who are much in demand on a freelance basis for concerts, sessions and touring packages. He has backed Phil Woods and Richie

Cole, played occasional JAS concerts, and is found unexpectedly deputizing or playing short seasons at Sydney's main jazz venues such as The Basement, Soup Plus, or the Brasserie. He is heard to powerful effect on the album of the late Richard Ochalski's band, Straight Ahead.

Evans, Kenneth (Ken)   b. 26/10/26, Geelong, Vic.   tbn./tpt/french horn/tba/comp./arr
Began on clarinet in 1943, then switched to trombone. Attended first AJC in 1946 with some members of the Dixieland Ramblers, called for the occasion the Geelong Jazz Group (Evans tbn., Jack Connelly clt and/or bjo, Vern Dolheguy tpt, Ron Grimison clt, George Barby p.) Took up cornet when band instrumentation altered. Also joined Geelong Symphony Orchestra on horn. Became involved in Melbourne jazz scene and in 1949 joined Frank Johnson on trumpet. With John Sangster and Geoff Kitchen, 1953–54, during which time he wrote the *Park St Suite*, a classically flavoured jazz work. With Smacka Fitzgibbon, a founder member of Steamboat Stompers, which won 1951 Battle of the Bands and worked Leggett's ballroom for nine months. Also active in dance bands during this period, incl. with Jack Banston. Did not play 1958–75. Subsequently, various gigs including one year with Steve Waddell's Creole Bells, 1981–82, teaching brass, arranging for school orchestras, mainly Baroque and Renaissance. In 1984, composing under commission from the Music Board of the Australia Council, incl. a suite for wind quintet, *Family Album*, an extension of the *Park St Suite*.

Ken Evans (left) with John Sangster, 1978

F

Falson, Ronald Albert (Ron) b. 2/1/28, Sydney, NSW tpt/french horn/arr./comp.

Began playing piano by ear at 14, and learning trumpet from Arthur Leeman at 15 (later teachers were John Robertson and Harry Larsen). In a schoolboy group, incl. brother Christian (also tpt), Les Jones, Ray Horsnell, John Cerchi (saxes), Joe Singer (dms), playing local functions. Les Welch took over as leader, recruited Don Andrews, and in 1945 the band began two years of Red Cross shows for servicemen, for which Falson began arranging. Originally a dixieland player, the advent of bop altered his approach and in the Harbour City Six, making its broadcasting debut 15 May 1947, he was noted for its facility in the latest idiom. He played with this group and with Ralph Mallen (with whom he learned big band arranging) in the first Battle of the Bands, March 1948, and continued to be prominent in the jazz concert movement with various bands incl. those of Bill Cody, Bob Limb, Wally Norman, Les Welch, and his own Beboppers. Also active in the jazz-inflected night-club scene. Following a period (1948–49) in the Wentworth ballroom during which he took over leadership from Billy Walker, he worked at Romano's, the Celebrity Club, Prince's, and the Roosevelt, which he left in 1951. As well as freelance arranging (incl. for Don Burrows), spent two and a half years on Warren Gibson's Metronome Dance Circuit, returned to the Celebrity Club (1953–55), followed by André's, both with Burrows. In mid-1956 joined Bob Gibson and also began touring, often as featured soloist, with imported packages incl. Buddy Rich, Stan Kenton, Frank Sinatra, Lionel Hampton. Subsequently concentrated on session work, incl. with Jim Gussey, Jack Grimsley, on ABC. Associated with major TV studio bands, incl. Bobby Limb's for the live satellite telecast for the Montreal Expo '67. In 1984 involved in studio and club work, composing and arranging. Falson was a major contributor to the jazz concert phenomenon, and one of the first convincing bop trumpet players in Sydney.

Farbach, Rick b. 23/5/22, Koblenz, Germany gtr/p./arr./vcl/bs

Learned piano from age six, and played for British occupation forces after the war. Switched to guitar and worked in Munich while attending the conservatorium. Arrived in Adelaide in 1939, moving via Canberra to Sydney where he worked at Golds' with John Best (p.), taking over at Best's departure. Various cabaret/night-club engagements incl. the Roosevelt in 1953 when he formed a vocal group, the Harmonaires. To Surfers Paradise, Qld, in 1956. Took up bass in 1958 and began 10 years with Channel 7, then session and theatre work, with teaching, arranging and composing for film and TV. Retired to Tewantin in 1979, but remained interested in jazz and assisted in the establishment of the Sunshine Coast JAS.

Farrell, James Clifford (Duke) b. 22/9/25, Sydney, NSW bs/tba/cello/vcl

In 1942 took lessons on banjo and mandolin from ex-Palais Royal Californian Perc Watson, and at his suggestion took up bass. Met Merv Acheson who became an important influence. In 1943 worked with Les Welch, and in 1944 led his own group, The L for Leather Boys, at Kinneal Officers' Club. A founder member of the PJJB in 1945, left to work with, *inter alia*, Ralph Mallen and Bill Walker's Wentworth Hotel Orchestra, but rejoined the PJJB during its disastrous tour of NSW and Qld in 1948. In 1948 formed the Illawarra Jazz Gang, resident band at Ramsgate Baths, and active in jazz concerts. Married reed player Betty Smythe, 1950. Through the '50s, night-club and radio work, incl. with John Bamford. Played with the Paramount JB during early '60s, then moved into freelancing in clubs until 1974 when he retired from performing following a stroke. Has since been involved in teaching and promotion, and in 1985 was presenting jazz on community radio.

Featherstone, Geoffrey Benjamin (Benny)
b. 30/7/12, Brown's Creek near Kingston, Tas.;
d. 6/4/77, Melbourne, Vic.   tpt/tbn./reeds/
p./bs/tba/vcl/dms/ldr
Moved to Melbourne with his family as a child (his father was a doctor who became superintendent of the Caulfield Repatriation Hospital). In 1925 Featherstone heard Ray Tellier's band from America, and this inspired the ambition to play jazz.

Benny Featherstone's orchestra, Rex Cabaret, 1934, with unidentified tbn., Frank Lobb (bs), Mick Gardiner (tpt), Bill Dardis (p.), Marjorie Stedeford (vcl), unidentified dms, Benny Featherstone, Vin McCarthy (alt.), unidentified ten. sax

*Benny Featherstone,
probably c. 1930*

The Beachcombers, c. 1930, with Dave Pittendregh (bs sax.), Benny Featherstone (tpt), Gordon James (p.), Abe Napthine (alt.), Lindsay Sergeant (bjo), Dick Bentley (vln), Geoff Smith (dms), Don Binney (tbn.), Bill Van Cooth (ten.)

Virtually self-taught, he became equally impressive on a range of instruments, with emphasis on trumpet, trombone, drums. Played trombone in the Melbourne Grammar School Orchestra for a show *The Troubles of a Caliph*, 1926, and in 1927 was with The Footwarmers, another school band, for *The Patchwork Review* at His Majesty's. At 17, joined Joe Watson on drums at the Green Mill, and recorded with the group, moving with Watson to the Wentworth Hotel in 1930. Recorded with the Beachcombers on trumpet on Vocalian label for a

hospital charity in 1930. In 1932, with Maurice Guttridge's theatre and broadcasting band (dms), with Les Raphael on dms/tpt/vcl, and with Ern Pettifer at St Kilda Palais. He was by now being billed as 'Australia's Louis Armstrong'. With Raphael's band on a trade promotion cruise through SE Asia in 1933, and later that year went to England where he met Duke Ellington and Louis Armstrong and played at the Silver Slipper on dms/tpt/tbn./ten./p./vcl. Back in Melbourne in 1934 he wrote an article on his experiences in

England, demonstrating a highly perceptive grasp of current popular music. Joined Art Chapman's New Embassy Band, 1934, as percussionist and also doing a solo jazz act. The latter had him playing tbn./clt/alt./sop./bar./p./bs/tba/vcl/tin whistle, concluding with Armstrong style top Cs on the trumpet; contemporary reviews convey something of the frenetic energy of this performance as Featherstone jumped from instrument to instrument, singing 'I'm a Ding Dong Daddy from Dumas'. In 1934, led his first band, incl. Frank Coughlan and Dick Bentley (vcl), at the Rex Cabaret, and also Sydney's Manhattan cabaret. With Art Chapman at Wattle Path, 1935, and took over as leader of the 40 Club (formerly the Green Mill) orchestra in 1936; he possibly also visited the east again during this period and met Buck Clayton. In 1937 Featherstone was leading a quartet at the Junction coffee lounge, playing under Bob and later Ern Tough at Fawkner Park Kiosk (until at least 1940), broadcasting, and appeared as guest conductor of Harold Wray's Mayfair Band in Geelong, and played with Clarrie Gange's band (incl. Freddy Thomas) at Brunswick TH. A frequent participant in 3AW Swing Club sessions and in 1939 joined both Don Rankin at the Glaciarium and Mick Walker at Saul's, then in 1940 at Leonards. In 1939 led a group in Darrod's Basement, where the Bell brothers sat in. In 1940–41, with Bill O'Flynn in various venues and led bands at the Stage Door (later the Nut House), in Flinders St, and at the Palm Grove. From this point on, Featherstone's activities are difficult to document. Served in the Merchant Marine in 1942–43, and, according to some press reports, for longer or on another occasion. In 1943 he led his own Dixielanders (incl. Roger Bell) at the Manchester, but was in Mackay, Qld next year playing American service clubs. Visited the US, 1945, and sat in with Jimmie Lunceford. Apart from a very acute article on the local dance band scene for *Music Maker* in 1947, there is little evidence of Featherstone's post-war activities. Played a Downbeat concert, 1954, and sat in at various venues. He became a shipping clerk in 1958, retiring in 1975. His death in 1977 received no special notice in the daily press. Featherstone was, on contemporary accounts, the greatest natural jazz musician of his day in Australia. Graeme Bell accords him the status of legend and brackets Featherstone and Frank Coughlan as the most prominent Australian jazz musicians of the '30s. The few recordings of his work reveal a player of thrilling intensity with Roy Eldridge a clear inspiration on trumpet. Through the '20s and '30s, Featherstone was an awesome talent as well as a man of apparently irresistible charm and Renaissance magnanimity. He was also, however, possessed of a gypsy waywardness and immoderate appetites, which did not always assist his musical career. His natural musicianship carried him through the '30s, but was not enough to sustain him into an era requiring more formal musical disciplines. Furthermore, in the years following the war, the swing idiom, on which Featherstone's music was based, was receding. Featherstone was always forward looking, so the traditional path was not inviting to him. In the late '40s he was responsive to bop, but to his bitter frustration he was unable to assimilate it. This combination of musical conditions led him into comparative inactivity, which in turn is probably the main reason that this titan of the early years of Australian jazz has suffered such neglect, historiographical interest in jazz having concentrated mainly upon the post-war period. He remains one of the most fascinating and gifted musicians in the history of Australian jazz.

The foregoing has relied heavily upon articles, discography, and biography by Ernst Grossmann, published in *Jazzline*, March and June 1980 and June 1981, and on Featherstone's scrapbooks.

Fisher, Lewis Clark (Lew)   b. 5/8/27, Hobart, Tas.; d. 11/3/78, Adelaide, SA.   p./tpt/ldr
Came to Adelaide with his family at age 10, and began classical training on piano at 14. From 15 he was playing dances, and in 1943 was in the Swingphonics with Clare Bail. He worked in the Southern Jazz Group throughout its existence, but before it broke up had already begun to lead bands of his own. Following a brief spell on trumpet with

Bruce Gray's All Stars after the SJG broke up he formed a band, but within a short time reduced his musical activity for the sake of domestic commitments. In the mid-'50s he resumed a heavy schedule of playing, and for the rest of his career he worked in night-club/restaurant style venues—the Paprika, Lido, Wentworth Hotel, Ernest's Restaurant—developing until he had become one of the most respected modernists in Adelaide. He ceased playing when an aneurism left his right arm paralysed from 14 Feb. 1966, though continuing to teach until dying of a third aneurism on 11 Mar. 1978.

Flannery, Kenneth John (Ken)   b. 15/5/27, Sydney, NSW   tpt/ldr
Began trumpet in his teens and became a founder member of the Port Jackson JB, 1944. The contemporary documentation and the memories of those involved are somewhat confused, but it seems that Flannery was absent from the band while serving in a military entertainment unit (1945), then again

Ken Flannery, probably late 1940s

during late 1947 to early 1948 while he was in the US. Back with the band for the tour which was aborted in Brisbane (1948). Later in the year he re-formed the band with Bob Rowan (tbn.), Rex Kidney (clt), Kevin Ryder (p.), Ray McCormack (dms), Lyle Booth (gtr), and Wally Wickham (bs). In 1949 Jim Somerville replaced Ryder, and Flannery handed over leadership and left to join Billy Weston's big band at the Gaiety, returning shortly and playing with the PJJB until it disbanded in 1950. Flannery then freelanced in night-clubs and on jazz concerts, incl. a period with Frank Coughlan at Christy's (1951) and membership of the Ray Price Dixielanders through the early '50s. He resumed his association with the re-formed Port Jackson Band c. 1955–62, excepting for a 12 month period in 1957–58 when Bob Barnard replaced him. From the late '50s he was also active in TV orchestras incl. those of Les Welch and Tommy Tycho. Since the '60s Flannery has not been regularly visible in jazz settings. In addition to studio work he is with the NSW Police Military Band, with very occasional freelancing and deputizing, particularly in Dick Hughes's groups, and in the '80s, reunion concerts for the PJJB. Flannery's economical and elegant playing remains undiminished in its poise and freshness, and he is one of the most important and adaptable founders of the post-war jazz movement in Sydney.

Ford, Adrian de Brabander   b. 26/4/40, Sydney, NSW   mainly p./tbn./clt/comp./arr/ldr
At age eight began two years on piano. Began attending SJC functions, incl. the Saturday workshops. Took up trombone and joined the Jazz Pirates (incl. Rex Gazey tba) in about 1963. With Geoff Bull, 1965–67, first on trombone, then piano. Following a period with the York Gospel Singers and a rock group, the Big Apple Union, joined the Melbourne Yarra Yarra JB for its tour of England and Europe, 1969–71. Since returning to Sydney has been active in many bands—the Unity JB, Bill Haesler (incl. for the film *Sydney Jazz*), Nick Boston, the Charlestown Strutters and his own big band. These last two received sponsorship from private enterprise and recorded with considerable com-

mercial success. With the Yarra Yarra JB for its Aust. tour with Alvin Alcorn. Has recorded in many settings, incl. as a soloist. Has won the AJC original tunes competition on several occasions and wrote the music for the film *Between Wars*. In 1984 was playing freelance and teaching in the WEA jazz course.

Fortified Few (Canberra, Australian Capital Territory)

The Fortified Few was formed by Neil Steeper in 1966 with encouragement from traditional jazz enthusiasts Terry Pallet and Frank Peniquel, with Steeper, Terry Hillman (reeds), John Sharpe, Ken Tratt (bjo), Paul Herbert (tba), Ian Hill (dms). The band played a residency in the Federation Lounge of the new Hotel Dickson, until disbanding in 1968, having by then also had as members or frequent guests Roger Sherrington (bs), Tex Ihasz (bjo/gtr), Peter Voss (p.), Jim Cotter (cnt), Dick Jackson, the blind ex-Sydney reed player, and John Roberts. The band was reformed under joint leadership of Steeper and Sharpe in 1969, with Hillman, Voss, Ihasz, Hill, and Bill Murphy (bs). In addition to other residencies, they resumed at the Dickson in 1970. Steeper and Hillman were replaced during absences by Joe Sheard (tpt/reeds)

The Fortified Few at the Hotel Dickson, c. 1978, with Terry Wynn (clt), Tex Ihasz (bjo), Ian Hill (dms), Peter Voss (holding tba, though in fact the pianist), Peter Sheils (tpt), Bill Murphy (bs), John Sharpe (tbn.)

and Don Archer (reeds). In 1969 Ihasz left for the Northern Territory and during his six-month absence was replaced by Jim Styles (b. 30/9/42), ex-Melbourne tenor banjo player; Tony Thomas (tpt) replaced Sheard in the same year, to be replaced in turn by Peter Sheils (b. 26/2/37). Sheils had been a member of several important Victorian bands, incl. the Melbourne NOJB, the Southport JB, and a group led by Owen Yateman. Since 1979 he has been in retirement from music, concentrating on his career in law. Terry Wynne replaced Sheils during the latter's brief absence in 1974, stayed on and became the only reed player in the band (as well as its main arranger) after Hillman left in 1975; Wynne was followed by Greg Gibson (1978), whom he in turn replaced in 1981. Voss remained with the band (except for 1970–72 when Graham Coyle was on piano) until 1981 when Sterling Primmer took over, followed by Coyle (1981), and Vince Genova who was not replaced following his brief incumbency in 1982 until much later in the year by Primmer. Murphy retired owing to ill health in 1983, to be replaced by veteran Canberra bass player John Stear. Hill was replaced in the same year by Bob Everard, who commuted from Cooma over 100 km away. Sheils had meanwhile been replaced by Thomas in 1978. The Fortified Few became the most durable and the most famous band in Canberra's jazz history, with a national reputation spread through frequent AJC and other jazz festival appearances, recordings, and frequent performances for the Albury, Victorian, Storyville, and Sydney Jazz Clubs. They played important functions in Canberra, incl. for Government House, and backed numerous visitors incl. Merv Acheson, Nancy Stuart, and Ade Monsbourgh. In 1972 they played for a service conducted at Reid Methodist Church. The band held numerous residencies, incl. the Deakin Inn, the Statesman and Civic Hotels (all in 1969), the Shanty Steakhouse (1979), the Ainslie Hotel (1982). During the band's lifetime its main residency was the Federation Lounge at the Dickson, where it was playing up to the time of its disbandment in 1983, making it one of the longest continuous jazz residencies by one band in Australian jazz history.

Fowler, Terence John (Terry)   b. 17/7/40, England   tba

Learned from Harry Berry from 1957, with Jim Crint (a.k.a. Sid Smith) and the Swinging Sydneyside Jass Syncopators followed by Gary Dartnell's Washboard Wizards, then Bill Haesler's Washboard Band. Has also worked with Noel Crow, Neil Steeper's Raucous Arousal Brass Band and Peter Strohkorb's Early Morning Kings of Tempo. In the '60s played with Colonel Crint's Royal Band of Foot and Mouth Deserters, a comedy/jazz group which toured the UK and enjoyed celebrity in Australia. In 1985 with reformed Robbers Dog Trio, with Rod Lawliss clt, and Ken Tratt bjo. As a professional graphic designer, has done art work for LP covers and *Quarterly Rag*.

Frame, Alexander John (Alex)   b. 4/4/29, Newcastle, NSW   tpt/clt

Took up trumpet in Adelaide at 18 on advice from Dave Dallwitz and began practice sessions with John Pickering, Ralph Prisk (tpt/tba), and Ian McCarthy (clt). In 1949 with the addition of Geoff Ward (p.), Les Williams (bjo), and Doug Giles (dms), they formed the South City Seven, with which Frame played the first of many AJCs. Played two years in Bruce Gray's two trumpet band, overlapping with a period in the West Side JB led by Leon Atkinson (clt) and incl. Lew Fisher (also tpt) and Bryan Kelly (dms), until the band folded in 1954. A founder member of the

Alex Frame, with the Black Eagles, probably late 1950s. John Pickering (tbn.), Dick Frankel (clt), Alex Frame (cnt), Bob Wright (tba). Obscured are Col Shelley (p.), Glyn Walton (bjo), and unidentified dms

Black Eagles until its dissolution, after which he played in various bands associated with the Taverns. Has freelanced with most Adelaide traditional bands, on both trumpet and clarinet, which he took up in 1960. Although sparsely recorded, Frame has become nationally known mainly through the AJCs of which he was appointed a foundation trustee in 1961, serving continuously to the present. He has also won the AJC original tune competition on several occasions. With his hot attack, Frame was the pioneer in Adelaide of the strictly classic jazz style trumpet based on early Louis Armstrong, and as such, exercised an early influence on younger trumpeters, notably Grahame Eames and Bruce Johnson.

Frampton, Donald Roger (Roger)   b. 20/5/48, Portsmouth, UK   p./sopranino/alto/recorders/comp./ldr/arr

Started on tuba, then sax and piano at secondary school in Portstmouth, forming a quintet at 15. In 1967 the family migrated to Adelaide where Frampton played with Billy Ross and Dave Kemp (bs) at the Latin Quarter, and at The Cellar with Bob Jeffery, Ted Nettelbeck, Tony Gilbert. With Neville Dunn's big band at the 1968 Adelaide Festival of Arts. To Sydney in 1968 and worked at El Rocco with Barry Stewart (dms) and Bruce Cale, then Cliff Barnett (bs); also with the Dieter Vogt quartet. In 1972 he toured Europe with the experimental group Teletopa (David Ahern and Peter Evans). He formed a trio in 1972 with Phil Treloar and Jack Thorncraft. With the addition of Howie Smith this became Jazz Co-op, one of the major avant-garde groups of the '70s. Frampton began teaching in the jazz studies programme at the NSW Conservatorium in 1974 and was director of the programme in 1979. In addition to film and radio work, he has written a considerable corpus of work for live performance, winning a JAS award in 1977. He performed his *Double Entendre* with Serge Ermoll at the Opera House in 1981, and in May that year his *Five Reflections on Consciousness*, for saxophone sextet and percussion, was given its debut at the NSW Conservatorium. Has toured within Australia as a solo performer, and in

1978, through SE Asia with Galapagos Duck. Played the 1980 Adelaide Festival of Arts with Bruce Cale, and worked with Phil Treloar's Expansions in 1981. Other important concerts incl. duet performances with Mike Nock and Dave Liebman, and with the Conservatorium Symphony Orchestra in a performance of *Nightcries for Sopranino and Strings*, composed for Frampton by William Motzing. In 1985 his trio backed visiting sax player Lee Konitz. He has also appeared on the ABC TV series *The Burrows Collection*. In 1982 he formed Intersections with Treloar, Guy Strazullo (gtr) and Steve Elphick, with which he toured India for Musica Viva. Other musicians involved in the quartet have included Peter Boothman, Jeremy Sawkins (gtr), Lloyd Swanton, Gary Holgate (bs) and John Bartram (dms). Frampton coined the word 'comprehensivist' to describe his musical approach—being able to play with anyone. A musician of often intimidating energy and unpredictability, he continues to be a major focus of the experimental movement in Sydney.

Frank Johnson's Fabulous Dixielanders
(Melbourne, Vic.)

In 1945 Frank Johnson and Geoff Kitchen recruited Eric Washington (tbn.), Max Marginson (clt), Ken Thwaites (dms), and Geoff Bland, to rehearse at Bland's house. Their first engagement was for the Eureka Hot Jazz Society in the same year. In 1946 Kitchen moved from guitar to clarinet and Marginson left; Wes Brown and Doc Willis replaced Thwaites and Washington. In 1947 Wocka Dyer replaced Willis and Bill Tope joined on banjo. The band began a six-week trial for a Saturday night dance at Collingwood TH. Except for a few months in 1947 while Johnson, suffering from pleurisy, was replaced by Melbourne trumpet player Ron Webber, and during which time Kitchen, temporarily leader, brought in Jack Connelly on bass, this band, with a few changes, retained the Collingwood TH residency until what was in effect the break-up of the group. It made its first recordings in June 1948, and in March, May and July of 1949 won the Battle of the Bands

Frank Johnson's Fabulous Dixielanders, between 1952–55, with, l. to r., Frank Gow, Frank Johnson, Wes Brown, Ron Williamson, Nick Polites, Bill Tope, Wocka Dyer. Since Williamson seems to have replaced Jack Connelly (rather than occasionally deputizing) in about 1955, this seems the likely, though not definite, date of the photograph

events. Also in 1949 the group started at the Maison de Luxe, Elwood, Sunday afternoons, where Ken Evans from Geelong began sitting in and then joined, making it a two tpt band. It continued to record and also to be featured on *Australian Jazz Club of the Air* on 3UZ. In 1950 Bland was replaced by Ian Burns from Sydney. Following Burns's sudden death from poliomyelitis in Jan. 1951 Bland returned until August, when John Shaw came into the band. In the same year Geoff Kitchen left to form his own group and, following a brief interval during which Sydney musician John McCarthy played with the group, was replaced by Nick Polites. A proposed tour of England fell through, but Smacka Fitzgibbon, who would have replaced Tope for the trip, continued to deputize frequently during the latter's absences owing to professional commitments. (Fitzgibbon later became an important promoter through his restaurant, Smacka's Place, which featured local and imported jazz musicians, and he also achieved considerable celebrity in more commercial entertainment, particularly through singing the title song of the film *The Adventures of Barry McKenzie*, 1972. His children, Mark (p.) and Nichaud (vcl) have become active in jazz in the '80s.) In 1952 John Shaw was replaced by Frank Gow, and the personnel remained stable until 1955. This stability assisted in the development of a cohesive group sound, and in 1954 William A. Davis wrote that the band was now 'playing at its peak'. It was working in many Melbourne venues, as well as frequent out-of-town engagements. While returning from one of these, a ball at Nagambie, in the early hours of 22 Sept. 1955, members of the band were involved in a motor accident. In addition to minor injuries, Nick Polites suffered spinal damage, Wes Brown broken ribs, and Wocka Dyer was killed. Although Dyer was replaced by Frank Traynor, this virtually finished the band. There was occasional work through 1956 including Downbeat concerts in March and June, but by the end of the year the band had dispersed. A reconstituted version recorded in 1958, and there have been several reunions for single concerts. The Fabulous Dixielanders' significant era was from the beginning of the Collingwood TH engagement to 1955. Originally inspired by Graeme Bell's band, the Fabulous Dixielanders gradually came to rival and then to surpass it in the public estimation, until it became the most important embodiment of the traditional movement in Melbourne, if not in Australia.

Frankel, Richard Anstey (Dick)   b. 31/5/31, Waikerie, SA   clt/ldr
Began clarinet at 17 taking lessons from Ian Drinkwater and Bruce Gray. With the Ice Box Five at the 1951 AJC, and joined the Black Eagle JB in the mid '50s remaining to its demise as the St Vincent JB in 1961. Involved in the establishment of the first Tavern, for which he formed the first of his Jazz Disciples, a band name he has kept up to the present. Remained associated with the Taverns, but also branched out into the pub scene, beginning with the Somerset Hotel in 1967. With a succession of personnel that has included most traditional musicians in Adelaide, Dick Frankel's Jazz Disciples have become an institution, playing promotions, balls and support for visiting performers and maintaining continuity through various pub residencies incl. the Walkers Arms in the '70s, and the Sussex where he was still playing in 1985 after 10 years.

Dick Frankel, c. early 1980s

Franklin, George Thomas (Franko)   b. 11/3/22, Perth, WA   tpt

Began in the Youth Australia League brass band at age seven. In the RAAF 1940–46, and while initially posted locally played with Ron Moyle at The Embassy, receiving valued coaching from colleague Viv Nylander. Following a posting to the north-west, however, Franklin did no playing. After discharge returned to the Embassy under Bill Kirkham and conducted the YAL brass band. In Sydney 1948–51 where he played the Celebrity Club circuit and Charlie Coughlan's Railway Institute band, as well as attending jam sessions organized by Marsh and Gerry Goodwin. Returning to Perth he joined Sam Sharp and Jack Regan's Moderniques, in which contexts he established a reputation as Perth's foremost mainstream to West Coast jazz trumpeter through the '50s, being particularly visible at jazz concerts. Joined the army in 1958, playing in a service band, and during about the same period joined on tbn. a bop-based band led by 'Rocky' Thomas. Returned to dance band work in the early '60s, and formed his own trio which worked to the early '70s excepting for a lay-off while Franklin underwent surgery on his lip. He left the army in 1977, and has continued to take casual work since, incl. with local jazz groups.

Furniss, Paul Anthony   b. 27/11/44, Sydney, NSW   reeds/fl./vcl/comp./arr/ldr

Began on recorder and flute at age seven, moving to clarinet at 11. His teachers incl. Don Burrows and Burrows's own teacher, Victor McMahon. His jazz interest was strengthened when he saw a televized performance of *Black Nativity*. Began attending SJC workshops in his late teens, and playing with Bob Learmonth's Jazz Bandits. Joined Geoff Bull's Olympia JB in 1965. In 1969 began his nine-year association with Graeme Bell.

Paul Furniss, Sydney Cove Tavern, 1976

A founder member, then leader, of the Eclipse Alley Five since 1969. For three years participated in the newly established jazz studies programme at the NSW Conservatorium of Music, first as student then as tutor. By 1975 in addition to commitments with the Eclipse, Bell, and two conservatorium bands, was performing with Adrian Ford's Big Band, Saxafari (a sax quartet playing jazz and non-jazz material), and the newly formed Tom Baker San Francisco JB, with which he toured in the US (1978) and which he has led since its return to Australia. In addition to his Sydney-based work, he has toured frequently within Australia, incl. with visiting American shows *One Mo' Time* (1982) and Bill Dillard's Blues Serenaders (1983). His compositions include winners of AJC original tunes competitions.

In 1986, was in his seventh year with John Colborne-Veel Quartet, leading Eclipse Alley Five and San Francisco JB, the latter consisting of John Bates, Viv Carter, John Bartlett, Hans Karssemeyer, and bjo player Paul Baker (b. 19/3/46) who joined in 1978. Furniss is one of the most complete musicians working on the traditional scene. His instrumental and theoretical mastery and his broad knowledge of the whole jazz corpus make stylistic categorization largely irrelevant to his work.

G

Galapagos Duck (Sydney, NSW)

Around 1970 a group of jazz musicians began playing ski resorts in the Snowy Mountains. Tom Hare was one of the leading lights. Born 28 Jan. 1942 in England, he arrived in Victoria in the early '50s, began cornet c. 1958 and played country dances, later working on sax in a Melbourne rock group. To Sydney, early '60s, freelancing in traditional/mainstream groups, and went to England with Graeme Bell, 1967. Back in Sydney, he and the other ski lodge musicians Marty Mooney, the Qua brothers, and Des Windsor (p.) began rehearsing, then playing in Sydney, basing themselves in the newly opened Old Rocks Push (Oct. 1971) as Galapagos Duck. During this early period they used a number of other musicians to bring themselves up to strength, incl. John Helman (bs); Warren Daly, Laurie Bennett, Phil Treloar; Tony Buchanan (reeds), and Glynn Baker replacing Windsor. The Duck opened as the house band at The Basement in Aug. 1973, and this residency established their reputation. At that time the personnel was Mooney, Chris and Willie Qua, Doug Robson (p.), and Tom Hare leading. Every member doubled, to make a total of around 30 instruments which gave the band unusual textural variety for a five piece. This was enhanced by the eclecticism of its members, which produced a stylistic range overlapping the conventional jazz boundaries. Through subsequent changes of personnel these circumstances have continued to obtain. A sketch of the main outlines of those changes is as follows, each name followed by the year that member joined: Chris Qua replaced by John Conley (1981); Willie Qua replaced by Len Barnard (1977), Laurie Thompson (1980), Mal Morgan (ca. 1982), with Barnard returning

(1983). Marty Mooney was replaced by Greg Foster (tbn./hca) in 1977. The most unstable chair has been keyboard. Robson was replaced by Dave Levy (c. April 1974), followed by Paul McNamara (c. Dec. 1974). Roger Frampton succeeded McNamara, followed by Ray Alldridge (1977), although Frampton made the 1978 Asian tour with the band. Col Nolan was with the band in 1980, then, with Chris Qua and Len Barnard, left that year because Hare wished to alter its musical character, introducing more electronics, and two keyboard players, who became Bob Egger and Mick Jackman. The band began recording with Ebony Quill (1974), and subsequent recordings catch most stages in its development. The members wrote, and recorded, the music for the film of David Williamson's *The Removalists*; other albums continued to feature compositions by band members, and a recent LP *Endangered Species* is in aid of the World Wildlife Fund. The Duck has toured extensively within and beyond Australia, incl. Asia (embracing the first Jazz Yatra in Bombay) and Europe (1979), organized largely by their

Galapagos Duck at The Basement, 1976, with, l. to r., Paul McNamara, Chris Qua, Marty Mooney, Jim Piesse, Tom Hare

current agent Peter Brendlé. They have played with the Australian Ballet for John Butler's *Superman*, support at concerts by Dizzy Gillespie, Ray Charles, Nina Simone, John Mayall, and have played state balls for royalty. In 1986 they were still The Basement house band and playing one night a week at the Hotel Manly. Galapagos Duck (named from a phrase coined by Spike Milligan, a long-time fan of the group), is the best-known jazz-based band in Australia, with a broad repertoire ranging from bop to rock. Their stylistic accessibility has done much to introduce the general public to the idea of jazz in its post-traditional forms.

Garner, Percy Aloysius   b. 19/6/09, St Marys, NSW   reeds/tbn./p./bjo/uke/ldr

Played piano by ear as a teenager and began playing dances at age 14 in and around his home town, Winton, Qld. In 1926–27 played for silent movies with the Black Cat JB and took up alto in 1928. During the Depression he moved throughout Qld taking work where it was available, in Charters Towers, Innisfail, Townsville, and Gympie. Arrived in Brisbane in 1939 and played dances and symphony work while working as an instrument repairer. Following a year with Billy Romaine at Cloudland he formed his first Sweet 'n' Swing band in 1941. The ten piece broke down into a dixieland group to play occasional features, and was reported as the first band to present this style of jazz in City Hall at the first annual ABC staff ball in 1941. Garner took a number of young musicians under his tutelage, and gradually his Sweet 'n' Swing group came to be a youth band in which many of Brisbane's most important post-war musicians served their apprenticeship; they incl. Alan Arthy, Neil Wilkinson, Vern Thomson, Lloyd Adamson, and Gordon Benjamin (reeds) who later moved to Canberra where he is active in the jazz scene. Garner also worked with other dance bands at the Bellevue and Cocoanut Grove, where jazz-tinged material was encouraged by American audiences. The Sweet 'n' Swing units ceased operating around 1950 and Garner became a stalwart of the city hall concerts, mainly on baritone sax. Through the '60s he worked in radio, TV and on the night-club circuits. Retired from full-time performing in about 1970, though plays the occasional deputizing gig. In 1986 he was still running his own instrument repair shop.

Garood, Roger William   b. 20/8/37, London, England   reeds/fl./vcl

Studied piano from age four to 12. As a teenager he began frequenting Chris Barber's early gigs, and took up sax at 15. Garood spent 1955–58 in National Service in Germany playing in a military band which required him to take up other reeds. From his discharge to 1965 he worked throughout England and Europe with cabaret, show bands, dance bands, in pit orchestras, and occasionally finding jazz work from dixieland to the bop style to which he was increasingly attracted. After a brief retirement, farming in Cornwall, Garood joined cruise ships on the England–Australia runs through the late '60s, and settled in Perth in 1971. Following an interval of commercial playing, he began getting jazz engagements through the Perth Jazz Club/Society, and formed his own group with Barry Bruce, Paul Reynolds (bs) and Ian Daniel (dms). Active in session work, radio and TV, teaching, but remaining visible also in jazz contexts, incl. for the Festival of Perth.

Gaston, Edwin Porter (Ed)   b. 9/1/29, Granite Falls, Nth Carolina, USA   bs

Began clarinet at 13; first professional work at 14. Following graduation from Univ. of Nth Carolina drafted into army where he took up bass and met Wynton Kelly and Phineas Newborn. Following discharge toured with Hal McIntyre's big band and the Al Boletto sextet. Replaced Jack Lander in the touring Australian Jazz Quartet in 1957 and arrived with them in Australia in 1958. Following disbanding of AJQ in 1959, joined Bryce Rohde Quartet (with Colin Bailey, George Golla), working at El Rocco and touring with Dizzy Gillespie and Sarah Vaughan. Following its break-up in the US, Gaston worked out of the west coast, 1962–63, with, *inter alia*, June Christie, Shelly Manne, Russ Freeman, Jack Sheldon. Returned Australia

Oct. 1963, joining the first Don Burrows Quartet (John Sangster, George Golla). Engagements included El Rocco, six years at Wentworth Supper Club and, with Alan Turnbull, the Montreux and Newport Festivals in 1972. In 1974 toured Asia with the quartet (Laurie Thompson on dms). Played on Clark Terry's Sydney concerts, Jan. 1975. With Stephane Grappelli toured Australia/New Zealand, 1975, and played Hong Kong Arts Festival in 1976. From 1977–80 based in Los Angeles, working with leading performers incl. Barney Kessel, Terry Gibbs, Bill Holman, the Condoli brothers, Teddy Edwards. Since returning to Australia, active in session and concert work, incl. backing visiting musicians (Milt Jackson, Sonny Stitt, Richie Cole, Mark Murphy), and in jazz clinics.

Gaudion, Peter Graham   b. 19/10/47, Melbourne, Vic.   tpt./vcl/ldr
Began trumpet 1963, and played with a school jazz group. At 17 joined a NO style band which incl. Les Fithall, Andy Symes, John Kellock, and sat in with members of the traditional fraternity, incl. Ray Lewis (tbn.), Frank Stewart, and members of the Chicagoans, Peter Stagg (tpt), Steve Waddell (tbn.), Mike Longhurst (clt), Bob Gilbert (reeds), Don Standing (gtr/bjo), Geoff Thompson (tba) and Dave Daldy (dms). Formed the Kansas City Six, 1967, with Steve Miller (tbn.), using Longhurst, Len Cobbledick (gtr), Peter Grey, John Kent (dms), and later Ian and Barry (p.) Harrowfield, with Joanie Watts. Involved in the foundation of the Victorian Jazz Club and played at Smacka Fitzgibbon's restaurant with various musicians incl. Jim Loughnan, ex-Adelaide pianist Dave Eggleton, Keith Cox, Ron Williamson and Charlie Blott. In the '70s, with Frank Traynor who exercised a great influence. Left Traynor in 1978, and bought an interest in the Victoria Hotel where he presented his band, Blues Express, the stable personnel of which was Dick Miller, Mal Wilkinson, and Derek Capewell, Allan Browne and Vic Connor (p.), followed by Bob Sedergreen. Promoted other Australian and overseas performers incl. Phil Woods, Earl Hines, Barney Kessel, Herb Ellis,

Jimmy Witherspoon, with the last of whom the Blues Express made a highly praised record. Moved operations to the Beaconsfield Hotel, 1981, where he presented Sonny Stitt, Richie Cole, Kenny Ball, George Melly, Ruby Braff, Ralph Sutton. The Blues Express went into recess in 1984 and Gaudion joined Chris Ludowyk's Society Syncopators. In 1986 he was preparing to open a new club, Jazz Lane, in which the Blues Express would be revived.

Gebert, John Robert (Bobby or El Bobo)
b. 1/4/44, Adelaide, SA   p./comp.
In Adelaide studied classical piano with Thelma Dent and modern style with Wally Lund; six years of harmony and composition with London College of Music under Keith Barr. At 17 led his own trio in Melbourne at The Embers. Back in Adelaide, was influential in the modern scene in the early '60s, concentrated on the Cellar. Here he led his own quartet, worked with the Keith Barr/Bob Bertles sextet, and accompanied visiting US singer Helen Humes by whom he was much influenced. To Sydney 1966 where again he became a major figure in the modern movement growing out of El Rocco, where he worked with his own trio (Ron Carson, John Sangster) and with other groups

Bob Gebert, 1985

which incl. John Pochée, Bernie McGann, and Barr. Throughout the '70s active in session and theatre work. Worked with Renée Geyer, Jon English, played in the productions of *Hair* and *Jesus Christ Superstar*, toured with Jeannie Lewis and played on TV and film scores incl. that of *Crystal Voyager* which he composed. Spent 1980–81 studying in the US under Walter Davis Jr on a grant from the Australian Music Board. Although Gebert has not enjoyed great visibility as a jazz musician, the respect in which he is held within the jazz fraternity is reflected in Charlie Munro's assessment reported in 1971, 'Bobby Gebert is the most advanced piano player in this country', and John Sangster's description of him as being 'one of Australia's most brilliant composers'.

Geddes, Alan   b. 24/5/27, Lithgow, NSW   dms
Began drums, 1944, with Joe Singer an important influence. Established his reputation in the night-club and concert scene from the late '40s. At Ciros in Double Bay under Doug Burgess (1946), with Col Anderson (Frank Smith also in the band) at the Bondi Esplanade (1948), with Noel Gilmour at the Trocadero Restaurant in Katoomba (1948) and later at the Bobbin Inn (1951). Played with Keith Cerchi at Park View Restaurant and Gaby Rogers at Romano's (1949). In 1949 he also appeared in Gus Merzi's group at a Melbourne jazz concert (with Don Burrows and Jack Lander). Played Jack

Alan Geddes, 1979

Spooner's Hayden night-club (1950) and at the Gaiety and other concerts with Billy Weston's big band. Joined Warren Gibson's Metronome circuit from 1952, but continued also playing jazz concerts and, throughout the '50s was with the Port Jackson JB and Ray Price, and with Graeme Bell for periods after 1957. In 1960 spent four months in Melbourne with Frank Smith's band at The Embers, then back in Sydney played the opening of the Bird and Bottle jazz venue in Paddington with Keith Barr and John Sangster. With Graeme Bell's All Stars through 1962–63, freelanced, incl. for a tour of the Riverina with Barry Ward's band, then joined the Dick Hughes trio at French's Tavern, 1966–68. From the late '60s into the '70s Geddes worked in licensed clubs, incl. at the Paddington-Woollahra RSL under John McCarthy; at the same time, a member of American bass player Jack Lesberg's quartet (with Chris Taperell and Bob Barnard) incl. for a tour with Oscar Peterson. Founder member of Bob Barnard's jazz band (1974), then visited the US in 1975 and played at the Odessa Jazz Party in Texas. Founder member of Dick Hughes's Famous Five (1977–84); at the Courthouse Hotel in Redfern with Noel Gilmour (1979) and at the Jazzbah in Lewisham (1982). Led his own group (Don Heap, Gary Walford, Merv Acheson) at Bondi Icebergers (1982–86) and also worked regularly in the '80s with Tom Baker's Swing St Orchestra, Graeme Bell's All Stars, the Bruce Johnson Quartet, and led his own group, Red Bank, for a season at the Soup Plus in 1985. In 1986 he was a member of Roger Janes's band at the Unity Hall Hotel, Balmain. Geddes is one of the most unremittingly dedicated jazz drummers in Sydney, held in the highest respect and affection both as a person and as a musician. His ability to swing a group is unsurpassed, and although his style is swing/mainsteam based, he has inspired many younger drummers playing in bop and post-bop idioms.

Geelong (Vic.)
Geelong's size, its proximity to the state capital, and its importance as a port, make it in many respects analogous to Newcastle in NSW. It is,

however, much closer to the adjacent city, Melbourne. The distance of just over 70 km makes the two centres, by Australian standards, merely a lengthy commuter distance apart. It is perhaps this circumstance that has contributed to a more diverse and substantial jazz movement than in Newcastle. Geelong is close enough to Melbourne to attract, not only musicians who will be paid, but audiences from the larger city as well. Geelong musicians have also frequently worked on a regular basis in Melbourne, providing them with more stimulation and incentive than would be available if they were a little more remote from one of Australia's major jazz centres.

From the 1920s Geelong enjoyed the services of several dance bands, including the Moonlight Serenaders led during 1930–31 by teenage trumpet player Colin Brain, who continued leading bands throughout the '40s. In the later '30s, enjoying particularly the stimulus of swing, Frank Phelan's Harmonists and Ted McCoy's Revellers were mainstays of the dance circuit, centred mainly on the local Palais Royal and the Corio Club. During the war years, visits from Bob Gibson's Palm Grove Orchestra and from Frank Coughlan maintained enthusiasm for swing; Coughlan's appearance as guest conductor of McCoy's Revellers received enthusiastic responses in particular, which suggests the symbolic as well as the actual importance of visits from nationally respected figures. One of the more enduring swing bands was the Casino Band, led by Colin Brain through most of the '40s, and playing a four-year residency at the Hall of Honour in Yarra Street. The members included the best of Geelong's dance band musicians, and also for a brief period, Ken Evans.

Evans was also a member of Geelong's first traditional jazz group (not including those which, according to their own understanding, operated as such during the '20s). This was the Dixieland Ramblers (1943–c. 47), with Evans, founder Clem Harris (tpt) who was soon replaced by Vern Dolheguy, Ron Grimison and Gene Dolheguy (reeds), George Barby, Jack Connelly (bjo/gtr), Jack Costa (tba), Harry Michael (dms). The band played occasional local gigs, made a private recording, enjoyed

the visits of sit-in musicians incl. Stewie Speer and Doc Willis, and as the Geelong Jazz Group, some members performed at the 1st AJC. Late in its life, the band enjoyed the support of probably the first local jazz club, the Jazz Appreciation Society (1946–47), which also presented bands led by Costa, Evans, Ken Ingram, visiting Melbourne musicians including Manny Papas's group. During the decade following the war, Geelong enthusiasts profited by the stimulus of numerous working groups visiting from Melbourne: Graeme Bell's, Frank Johnson's and Len Barnard's bands between them gave the locals a feel for current Australian traditional at its best.

This inspiration in particular, the post-war interest in jazz, and the coastal resorts in the Geelong region, all contributed to a burgeoning of jazz groups in the '50s. Owen Yateman, Ron Moss, and Dick Tattam formed bands that worked with relative regularity. Yateman's Sand Dune Savages (which incl. Moss on dms and the impressive young cnt player Johnny 'Doc' Hughes) was particularly busy and popular. They recorded and broadcast, and played Downbeat concerts in Melbourne. The concert phenomenon spread directly to Geelong and Yateman led one of many groups at Battle of the Bands concerts held in local town halls and theatres. As in other capital cities, the interest in jazz signalled in such functions as the concerts also gave some support to more progressive styles. George Barby (b. 10/4/27) was an eclectic pianist who had been working various groups and as a soloist since the '40s. A regular performer at AJCs, he also played in groups outside of Geelong, and was at different times associated with Doc Willis, Tony Newstead, George Tack, Merv Acheson. His stylistic versatility made him one of the most respected musicians in Geelong. He died in a car accident on 4 Oct. 1971. Neil Butterworth's trio (with Graham Greenhalgh bs, Ron Moss dms) was one of the earliest in Geelong to play in the style of Oscar Peterson and Errol Garner. Butterworth led a trio from the mid-'50s to the late '60s (embracing a seven-year residency at the Victorian Hotel, Malop St) before he moved to Queensland in the late '70s. Dave Jeffreys (vibes) led the Uptown

Modern Jazz Quartet for the Uptown Club (founded in 1959), and demonstrated a firm commitment to bop and other progressive material. The quartet (with Doug McKenzie p., John Turner bs, Bill Ellis dms) was later augmented with Peter De Visser trombone, Tony Kulakauskas tenor, and Brian Fraser valve trombone; most of these musicians continued to play jazz into the '80s.

Geelong enjoyed the effects of the 'trad' boom, and, as was the case, elsewhere, the emphasis fell upon traditional styles. Jazz Club 78 (established in Sept. 1960), and Rivals (Oct. 1961) were two clubs which, along with casual dances at schools, colleges, and established surf and boat clubs, provided work for local groups in the early '60s. One of the most popular of these was the South Coast Stompers (1959–63), with Ian Burch (tbn.), David Syer (tpt), Peter Allen (bjo), Barry Hirst (p.), Ross Knights (clt) and Roger Bodey (dms). They enjoyed an enthusiastic teenage following, much like their colleagues in Melbourne, where they also made AJC appearances. The Barwon Valley Stompers also developed a cult teenage following. The nucleus of the band was made up of ex-students from Belmont High School, and regular members included Bruce Johnstone (tpt), Colin Garrett (clt), Bill Rogers (tbn.), Rod Green (p.), John Masterman (tba), Peter Haby (bjo), Dave Jeffery (dms). The group alternated with the South Coast Stompers in the Trades' Hall for the weekly Crescent City Jazz Dance, and later changed their name to the Crescent City JB. Although the late '60s jazz slump was felt in Geelong as elsewhere, the interest generated by the boom created a core of jazz enthusiasts who were able to maintain some continuity for the music. Owen Yateman led a band which, to a large extent, accommodated the rock idiom but without ever surrendering its base in jazz; his Big Fat Brarse operated throughout the '70s, and the name continued into the '80s. Des Camm (cnt) formed the Green Horse JB in 1965 and held down a two-year residency at the Old Grovedale Hotel, as well as playing casual functions and AJC and Geelong Jazz Festivals. His band included musicians who had appeared during the peak of the boom, and who have carried the flag into the '80s: Colin Elliott (clt), Chris Ludowyk, Peter Haby, Paul Ludowyk (bs), Ian Coots (dms). The Geelong Jazz Festivals (1965–71) were weekend affairs run by the local jazz club of that period. In format and character it was a miniature of the AJCs, and attracted visiting musicians mainly from Melbourne and Adelaide, as well as from provincial centres around Geelong. Their conclusion marked the beginning of the quietest period for local jazz in the post-war era.

A resurgence of interest was intimated in the formation of the Jazz City Five in 1978, with a residency at the Geelong Hotel. Later augmented (with appropriate name changes) to six and seven, the group finally took the name of its leader and founder, as the Des Camm JB, and as such continued to be active in 1985 both in pub residencies and casual gigs. A new Geelong Jazz Club was formed in December 1979, and organized the AJC held in Geelong in 1981, an event which produced the usual growth of local interest in the music. Geelong has also witnessed some upsurge in post-traditional jazz activity during the '80s. Barrie Edwards formed the mainstream Déjà Vu, which played a residency at the Sawyers' Arms Hotel (1981–82), and provided an opportunity for other musicians with progressive interests to sit in. He later formed the Buzz Edwards big band which played in restaurants and on concerts during 1983–84. Deakin University has also formed a big band which makes concert appearances.

Inevitably, the level of jazz activity has always been lower in Geelong than in Melbourne. But Melbourne is a near enough neighbour to exercise a stimulating influence, and to inspire the local enthusiasts and musicians with a feeling of solidarity as well as constituting a kind of exchange facility. Perhaps because of this, the Geelong scene manifests a resiliency not seen in other provincial centres.

In the preparation of this entry I have relied very heavily on research conducted by Des Camm (though my references to his contribution are based also on sources other than his very modest account of his own work). I wish gratefully to record my appreciation of the extensive photographic and written file, based on printed and oral sources, which he and his typist Pamela Kolb prepared for this enterprise.

Gibson, Gregory Edmund Campion (Greg)
b. 23/1/30, Stawell, Vic.   reeds/ldr
Took lessons on piano, 1938–41, played ukelele
and banjo ukelele, 1942–45, then took up clarinet
in June 1949. 1950–51, worked with Frank
Traynor, Max Collie, and Smacka Fitzgibbon. In
the RAAF, 1951, after which he joined the Cootamundra JB, remaining until moving to Canberra in
1957. Formed a quartet with Noel Bluett (accordion), John Hamon, Don Ross (dms), and at other
times Sterling Primmer and John Stear. As an
official with the diplomatic service, Gibson was
posted to Kuala Lumpur, 1959–61, where he
played for broadcasts in Alfonso Soliano's (vibes)
Rhythm for Six. Back in Canberra, 1961, played in
a quartet with Alan Pennay and recorded with the
Pix All Stars in 1962. Oct. 1962–66 posted at the
Australian Embassy in Brussels, during which
period he worked with local groups and with Ken
Colyer. Returned Canberra, 1967, and formed the
Capital JB (John Roberts, Derek Long tbn, Ray
Simpson gtr, Primmer, Stear, Terry Stephens bs);
then posted to Bangkok where he worked again
with Alfonso Soliano. Upon returning to Canberra, he formed Mood Indigo (with Graham Coyle,
John Stear, and Carl Witty followed by Colin
Horweeg, dms), with which band he played concerts locally, in Sydney and Melbourne, and a residency at the Park Royal. In Indonesia, 1976–78,
playing with the Indonesian All Stars led by Jack
(father of Indra) Lesmana. In Canberra, 1978–81,

*Greg Gibson's Jazz Australia UK Incorporated in Lon-
don in the early 1980s, with, l. to r., Greg Gibson, Chris
Karan, Denny Wright, Joy Yates, Dave MacRae, Barry
Dillon, Don Harper*

with the Fortified Few, then from March 1981 was
posted to the Australian High Commission in
London where he formed Jazz Australia UK In-
corporated with fellow Australasians Don Harper,
Dave MacRae, Barry Dillon (bs), Chris Karan, Joy
Yates (vcl); also worked with the bands of Bill
Brunskill and expatriate Australian Campbell Bur-
nap. Returned to Canberra, June 1985, and has
since freelanced. Gibson has recorded with the
Barnard's, and was guest musician at several of
Allan Leake's Jazz Parties in Melbourne.

Gibson, Robert Alan Franklin (Bob)   b. 23/5/12,
Perth, WA   reeds/vln/p./cnt./comp./arr/ldr
In Melbourne for three years as a child, then returned to Perth where he studied violin with Prof.
Joseph Nowotney, and became a founder member
of the Perth Symphony Orchestra at age 14.
Learned sax from Harold Moschetti while the latter was in Perth with Bert Howell's band. Began
playing in dance bands, graduating to Ron Moyle's
at the Embassy, 1932. Travelled to eastern states
with Moyle's musicians, 1936, and passed an audi-
tion for Ern Pettifer at Melbourne's St Kilda Palais.
Stayed on with Pettifer when Moyle returned to
Perth, remaining at the St Kilda Palais under subsequent leaders Moschetti, Jay Whidden, Roy Fox
with an interruption during the Americanadians'
residency during which he was with Bert Howell
at the Capitol and State Theatres. Assembled a
band to open the new Palm Grove (formerly Earl's
Court), 4 Nov. 1940, members incl. Fred Thomas,
Bob Storey (reeds), Alf Warne (p.), Keith Cerchi,
Don Rankin (reeds). During this residency, which
continued until late 1946 or early 1947, the orchestra also played for American and Australian servicemen functions, incl. a concert opposite the
Artie Shaw band, broadcast several radio series,
and gained the reputation as one of Australia's
leading swing bands. Following his departure from
the Palm Grove, Gibson freelanced as player and
arranger until leaving for England in 1948. There
he led his own big band, recorded, toured, and
broadcast. Returned to Australia, 1950, and settled
in Sydney where he worked for Radio 2UE. In
December 1950 he formed a 16-piece orchestra for
the Palais Royal, then led bands at the Surreyville

Bob Gibson's orchestra in the Surreyville dance hall, c. early 1950s, with, from l. to r., front row: Edwin Duff, Nola Lester, Bob Gibson, middle row: Jock McKenna, Errol Buddle, Charlie Munro, Johnny Green, Len Hailes, John Bamford, Jack Annersley, back row: Bob 'Beetles' Young, Chic Denny, Ken Brentnall, Bill Innes, Darrell May (partly out of frame), extreme rear on drums; Frank Marcy

ballroom and at the Bondi Esplanade and for the town hall jazz concerts in Sydney and in other cities, incl. Brisbane. Through the late '50s he was MD for the Ford Show on the Macquarie broadcasting network, and continued leading the band at the Surreyville, recording with the group in 1957 when it incl. Alan Nash, Eric 'Boof' Thomson, Charlie Munro, Eric Morrison, Jock McKenna, Clare Bail, Frank Marcy, George Thompson and Melbourne brass player Jack Grimsley (b. 1925), who had been intermittently with Gibson as musician/arranger since the Palm Grove days, and who continues to be active in studio work. During the '60s Gibson was primarily involved in TV, notably in *The Sound of Music* with Bob Limb, who had begun his association with Gib-

son at the Palm Grove. In 1972 Gibson became MD for Eastern Suburbs Leagues Club, leading a band that incl. at different times John Bamford, John Blevins (dms), Bob Barnard. He also joined Limb in a music production agency, in which he was still involved in 1986. In 1985 he participated in the revival of the town hall concerts, leading his own big band. In 1986 the durable and vigorous Gibson had resumed the violin and was rehearsing a floor show to take on the NSW club circuit.

Gilbert, Geoffrey (Geoff)   b. 2/10/32, London, England   bjo/vcl/wbd/ldr
First experience of jazz was Lu Watters records and the Graeme Bell band in London. Began

*Geoff Gilbert with the Harbour City Jazz Band, 1986:
standing, l. to r., Cliff Reese, Paul Andrews, Mick
Huggett, John Speight, Jim Fletcher; kneeling, Jim
Elliott; seated, Geoff Gilbert*

banjo, 1954, with lessons from Lonnie Donegan. To
Sydney (1956) and first band was led by Llew Hird
during the latter's brief sojourn in Sydney c. 1957–
58. Gilbert played with John Roberts, Fred Starkey (wbd) and Bob Learmonth (tbn.) at the Royal
George Hotel, 1958–59; joined Nat Oliver's NO
Jazzmen, 1960–62. In England, 1962–64: formed
the Gothic JB, and performed with visiting musicians Kid Thomas, Emmanuel Paul, Red Allen,
Alton Purnell. Returned to Australia and joined
Gary Dartnell's (tpt) Washboard Wizards. Formed
the Harbour City JB, evolving through 1966–67
from Dartnell's band and Ken Longman's Harbour City group. With Gilbert, the band incl.
Mike Hallam (ldr), Longman, Terry Fowler, Barry
Hallam and later Jim Fletcher (dms). Gilbert took
over as leader in 1970. Alumni incl. Bob Learmonth, Jean Williams (bs), Paul Simpson (reeds),
Darcy Wright, Jim Young (p.), Ken Crawford (p.),
Terry McCardell, Greg Foster, Vic Hatton (bs),
Dave Cross (reeds). Gilbert was also a long-
serving treasurer of the Sydney Jazz Club, and a
founder jazz broadcaster on 2MBS–FM in 1974.

In 1986 he was leading the Harbour City JB (Cliff
Reese, Jim Elliot tbn./reeds/p., John Ryan bs.
Paul Williams reeds, with Keith Cook and John
Speight alternating on p.), and in 1985 resumed
jazz broadcasting for 2MBS–FM.

Golla, George   b. 10/5/35, Chorzow, Poland
gtr/comp./arr
Took up reeds following his arrival in Australia in
1950. Began guitar in 1956 and played his first gig
in that year, virtually learning on the job, deputizing regularly for Bernice Lynch at the Bamboo
Restaurant, Pitt St, while Lynch fulfilled TV commitments. Joined Gus Merzi for four years in
1957, and the Bryce Rohde quartet for two years
prior to Rohde's departure for the US in 1960. In
1962 began a decade with Eric Jupp's studio
orchestra. Much of his work up to the mid '70s
was in radio and TV but his most celebrated
musical association has been with Don Burrows.
The two met in 1960 at Club 11 where Golla was
working with John Holman, and since then they
have worked together in groups of various sizes,
including as a duo, the possibilities of which have
been extended by the seventh bass string (B) of
Golla's Maton guitar. The names Burrows and
Golla have become inseparable in the Australian
public's awareness of jazz. Apart from local performances, they have toured extensively both

George Golla (right) with Don Burrows, early 1980s

within Australia and overseas, incl. the Montreal (1967) and Osaka (1970) Expos, the Newport and Montreux Festivals (1972) and Brazil (1977). Golla's compositions embrace a wide musical range, incl. a Concerto for Electric Guitar and Symphony Orchestra (1979). Has recorded extensively and written many books of guitar instruction. Very influential as a teacher, incl. six years at the Academy of the Guitar, four years at the Binna Burra Winter Arts Schools, and the jazz studies programme at the NSW Conservatorium of Music, where he still held an appointment in 1986.

Gould, Anthony James (Tony)   b. 2/2/40, Melbourne, Vic.   p./comp.
His multi-faceted career has incl. much studio and concert work. Played support for, *inter alia*, Sarah Vaughan, the LA Four, Jean-Luc Ponty, and accompanied Jeannie Lewis, Clark Terry, Ernestine Anderson and Mark Murphy in concert performances. In 1982, toured Asia for the Foreign Affairs Department with Keith Hounslow, together constituting McJad (Melbourne Contemporary Jazz Art Duo), which recorded two albums. Also recorded with John Sangster, and solo. Has composed in various genres for films, records, and concert presentation. A common thread in all his musical activities is the promotion of acoustic music. A professional musician with several degrees in music, Gould is involved in music education, lecturing in jazz studies at the Victorian College of the Arts, and is classical music critic for the *Sun-News Pictorial*.

Gow, Francis Kevin (Frank)   b. 29/12/30, Melbourne, Vic.   p.
Took piano lessons 1943–44. Joined Russ Spencer's Southland Rhythm, 1950, and with Frank Johnson's Fabulous Dixielanders, 1952–59. Formed his own group which incl. at various times Des Edwards, Alex Hutchinson, Bob Barnard. In 1964 he began playing commercial engagements in city hotels, using Peter Scott (bs), Stan Harris (dms) and Splinter Reeves. Full-time musician from 1966, since when most of his work has been as a soloist in piano bars, and with Frank Traynor.

Gowans, Ronald William Francis (Ron, Loud) b. 23/12/12, Sydney, NSW   reeds/fl/bassoon/arr/vcl
Began flute at eight, reeds at 16; introduced to jazz by Les Welch at 2KY Swing Club sessions in the mid '40s. Worked casually during late '40s, joining Eric Tann at Romano's in 1947. A stalwart of the jazz concert era both as leader and as sideman with, *inter alia*, Ralph Mallen and Wally Norman. At the same time, remained active in the night-club scene, incl. at Sammy Lee's (1950–51), Golds' (1951–52), the Bamboo Restaurant (1952), and at Segar's ballroom where he became leader in 1955. Joined Tommy Tycho's band for TV in 1956, and although he toured with numerous visitors (incl. Ella Fitzgerald, Buddy Rich, Artie Shaw, Gene Krupa), from the '60s he became less visible as a new entertainment era arrived. He moved to Qld where he worked in clubs, and in 1986 was leading a band at the Tweed Heads Bowling Club, and residing in Surfers Paradise.

Gray, Bruce Athol   b. 18/8/26, Adelaide, SA reeds/fl./ldr/arr
As a child studied violin for seven years, and played fife in a school band where J.E. Becker heard him and suggested that he take up clarinet. Gray became a foundation member of the Military Band of the Adelaide College of Music. In secondary school he joined with Bob Wright, Bill Munro and Colin Taylor to form a quartet. He established a reputation as a dixieland player working with Mal and Bob Badenoch in 1943 before joining Malcolm Bills, with whom he remained until the latter's band wound down 1946. A founder member of the Southern Jazz Group, through which he extended his reputation to the eastern states, Gray has always been a developing musician, and the desire to extend himself further was one of the causes of the dissolution of the SJG. In the meantime, he had made his debut as band leader at King's Ballroom on 6 Nov. 1949. Through the early '50s he continued to lead his All Star bands, recording in 1954 with Munro, Alex Frame, Bob Wright (tbn.), Geoff Ward (p.), John Malpas, Max Dickson (bs) and Jim McKenzie (dms). He also formed

Bruce Gray

a number of West Coast style groups, one of which played support during Dave Brubeck's first tour. Heavily involved in forms of studio work, incl. writing and leading his own All Stars in the ABC series *Evolution of Jazz*. Among the several big bands he has worked in are the ABC orchestra under Brian May, and he has also been active in TV studio bands. In the mid '80s he continues to lead his own group with life-long colleague Bill Munro, and recently joined the Dave Dallwitz Big Band. Hearing Gray's work today it is difficult to believe that he is one of the founders of the traditional jazz scene in Adelaide. He is still essentially a broad-based traditional/mainstream player, but the undiminished vigour and freshness of his work belies Gray's status as one of the founding fathers.

Green, Rex Withers (Hammerhead)   b. 18/5/22, Hobart, Tas   p.
Received classical piano tuition as a child, met Tom Pickering at age 16, and became a founder member of the Barrelhouse Four in c. 1938. Spent the war years in the services. Rejoined the Barrelhouse Four after the war, but moved to Melbourne in 1947 where he joined Tony Newstead until moving to Shepparton 1948–56, playing occasional solo work. In Geelong 1957–60, working in trios

and on occasions with Owen Yateman. Following his return to Melbourne, he played for the Melbourne Jazz Club in the early '60s, but has been musically inactive since, although maintaining an interest in jazz. Retired from his banking career in 1981.

Grey, Peter George   b. 3/3/37, Newport, Vic.   bs
Became interested in jazz in early teens and heard local groups through the Melbourne University Rhythm Club in mid '50s. Began bass in the early '60s, and joined the Glassy Alley JB, which incl. Richard Miller and Steve Waddell. Later played with Jeff Hawes's 431 Club Band and the Hot Sands JB. Joined a quartet which included Nick Polites and Ray Lewis; left Polites in 1979 and joined the Yarra Yarra JB, with whom he continued to work into the '80s.

Gumbleton, William Beresford (Bill)   b. 11/12/36, Perth   p./comp.
Following formal piano training as a child, Gumbleton became interested in jazz in the mid '50s listening to Terry Ingram, pianist with Jack Harrison, and started working out jazz approaches for himself, later taking lessons from Bill Clowes. In 1957 with the Modern Jazz Quintet (Frank Thomas tpt, Frank Smith bs, Tom Bone, Keith Carton gtr) doing Playhouse concerts and by 1958 was playing the concerts with bands led by Horry King, Theo Henderson, and his own. Most work however was of the non-jazz casual variety, with occasional jamming, until the early '60s. He began playing with various musicians incl. Bob Gillette,

Bill Gumbleton, 1978

Colin Scott, and Frank Smith at the Shiralee, and brought in his own trio with Bill Tattersall. In 1964 in Keith Stirling's group with Theo Henderson, Tony Ashford, Smith, Jim Cook, and Peter Hall (tbn.) at the Shiralee, moving into the Melpomene late in the year, then to the Hole in the Wall in June '65, with Tattersall coming in on drums. In 1967 the group was still there, with Alan Hale and Don Bibby replacing Tattersall and Smith, and when Stirling left in 1967 they continued as a quartet with Brian Bursey by now on bass. With the ending of the Hole in the Wall's jazz policy in 1969 there was little jazz activity until the foundation of the Perth Jazz Club/Society in 1973, for which Gumbleton played in the first house band with Colin Scott, Smith, Barry Cox, and leader Tony Ashford. Most of his subsequent jazz work has been for the society, and with the group Kaleidoscope which operated in the late '70s with Ray Walker, Bruno Pizzata, Guy Bart, Murray Wilkins, and Tony Ashford, breaking up when Ashford returned to England.

Gurr, Donald Cyril (Don)   b. 19/10/26, Launceston, Tas.   keyboards/reeds/comp./arr
Began piano tuition as a child and joined a school dance band in his teens. At 15 he joined Stan Huntington's dance band and during the war took up clarinet. Introduced to jazz primarily by Ted Herron whom Gurr joined in what became the Jazzmanians, with whom he recorded in 1946. To Hobart, 1951 to continue his training as a pharmacist, and began working with Don Denholm at Wrest Point, and sitting in with Tom Pickering at the 7HT Theatrette. Gurr, with a relatively progressive mainstream style, has not had regular opportunity for exposure in Tasmania, so that most of his work has been in revues and shows, and occasional functions such as a mainland tour with Roly Easton's big band to promote the opening of Wrest Point Casino in 1973. Gurr took over leadership, but the band dissolved for lack of work by 1980. He has been heard under jazz billing on local radio and TV.

Haesler, William John (Bill)   b. 20/4/31, Melbourne, Vic.   wbd/ldr
Led his own pick-up band at 1951 AJC on trombone, which he subsequently abandoned. Most of his playing post-dates his move to Sydney in 1966. With Robbers' Dogs (1968–70), and began leading his own groups in 1971 at the Royal George Hotel, Sydney, using as staple personnel John Roberts, Adrian Ford, Trevor Rippingale (reeds), Geoff Holden, and Eric Richards followed by Terry Fowler. His band also initiated the jazz policy at Soup Plus, 1975 and was the subject of a documentary film *Sydney Jazz*, 1976. He was subsequently assembled groups for special occasions incl. AJCs and jazz club functions, and joined Peter Strohkorb's (clt) JB in 1984. Haesler's primary importance to Australian jazz has been as administrator and writer. Began writing for *Australian Jazz Quarterly* 1951, and became editor, 1954–57. Also editor of *Matrix*, 1954–57. Wrote sleeve notes for numerous records, incl. the important Swaggie Jazz Collector Series 7″ EPs. Edited *Quarterly Rag*, 1976–77, and continues to contribute to various jazz publications incl. *Jazz*. Haesler was also co-founder and president of the Melbourne Jazz Club, president of Sydney Jazz Club for two terms, 1968 and 1978. A long-time committee member of AJCs: secretary (1952), vice-president (1960), president (1963), and programme co-ordinator (1970, 1975, 1980, 1983). He was the AJC's Victorian trustee, 1961–66, and has been its NSW trustee since 1979. Since 1982 he has presented a regular jazz programme on 2MBS–FM.

Hall, Eric   b. 25/10/26, Brisbane, Qld
reeds/fl./ldr
Began on fife at school, clarinet in 1938 and alto a few years later. In 1941 started playing in practice groups and joined Freddy Ailwood at Cocoanut Grove in 1943. Following RAAF service (1944–45), worked with Perc Garner's Sweet 'n' Swing band, with Jack Kenny at the Bellevue, and at Prince's Restaurant, 1946–47. To Sydney, 1949, to study at the conservatorium, while he worked with the Illawarra Jazz Gang, the Port Jackson JB, and with the bands of Wally Norman, Billy Weston, and Abe Romain. Returned to Brisbane 1952, played city hall concerts, then in the dance, pit band, and night-club scene and concert work supporting touring artists incl. Sammy Davis Jr., Dizzy Gillespie, Shirley Bassey. Since the resurgence of jazz in the late '70s has been a regular performer for the Brisbane Jazz Club and Qld JAS, and in the '80s has taught woodwinds with the state education department.

Hallam, Michael David (Mike)   b. 28/6/34, Nottingham, England   tpt/hca/gtr/ldr
Began in England, playing for 10 years with the Mercia JB in Nottingham. To Sydney, 1967, and co-founded the Harbour City JB. Early '70s, with Ray Price Qintet [*sic*], and with the Charlestown Strutters for a residency at the Charles Hotel, Chatswood. Left in 1977 and began forming his Hot Six which made its debut in 1978. Although remaining relatively stable in terms of personnel, there have been changes, and members have incl. Ken Crawford and Joe Allen (p.), John Fearnley (dms), Doc Willis, Norm Wyatt, Bill Benham, Barry Chew (bs), Brett Lockyer (reeds). The band has played long residencies at Red Ned's and the Marble Bar at Hilton Hotel, as well as monthly Sydney Opera House boardwalk concerts. Hallam's Louis Armstrong-influenced trumpet has been recorded on several occasions, with his own band and the Adrian Ford Big Band.

Hansson, Clare   b. 1/10/35, Brisbane, Qld   p.
Received classical training at school, and although interested in jazz she spent the early part of her career playing more commercial material. In 1974 she met and married Bernie Hansson (bs) from whom she learned much. Her introduction to playing in an uncompromising jazz setting came at the 1976 AJC, which spurred her to develop her improvisational skills. Following the sudden death of Bernie in 1979, she studied intensively, to be invited back to performance by Bob Barnard at the Melbourne Hotel. In the '80s she has been active with her own and other groups, incl. those of Rick Price, and in the Qld JAS of which she was foundation president. Has backed various visiting performers incl. Jimmy Witherspoon and Ernestine Anderson.

Harman, Henry Ernest (Harry)   b. 19/12/27, Melbourne, Vic.   bs/tba/bjo/ldr
Began on guitar 1946. Attended AJCs from 1948 and during visits to Sydney heard the PJJB and attended Sydney Swing Music Club functions. Settled in Sydney, 1949. In 1952 took up tuba and began rehearsing what became the Paramount JB which made its debut at the opening of the Sydney Jazz Club which Harman founded as a venue for the band, in August 1953. The personnel of the Paramount JB was drawn largely from the Bridgeview JB (formerly Bob Learmonth's JB) and consisted of Harman (bs), Trevor Pepper (tpt), Dan Hardie (clt), Peter Towson (p.), Don Hardie (bjo), Bob Leggett (dms) and Learmonth (tbn.) although he was replaced on opening night by Tony Howarth. The Paramount JB continued in existence as the Sydney Jazz Club's house band into the '60s, and subsequent personnel incl. Jack Parkes, Jimmy Roach (p.), Duke Farrell, Gordon Hastie (dms), Peter Neubauer (clt) and Ian Cuthbertson (tpt) who took over as leader until leaving the band in 1963 and retiring from music. Harman began playing with the Port Jackson JB in 1956 and dropped out of the Paramount JB c.1958. Left the Port Jackson to return to the Paramount on banjo, 1960, then joined Graeme Bell's All Stars, 1962, until moving to Newcastle in 1976, and

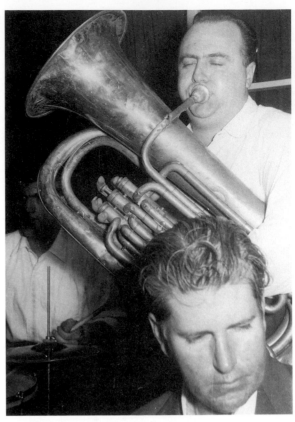

Harry Harman (tba), Sydney Jazz Club Function at the Real Estate Institute, 1954; Bob Leggett on drums, with Merv Acheson in the foreground

ceased musical activity. Returned to Sydney, 1981, and joined the Eclipse Alley Five in 1983. With the formation of the Sydney Jazz Club and of the Paramount JB, Harman made two of the most important contributions to the post-war history of Sydney jazz. He continued to be fully active musically in 1986 as a member of the Eclipse Alley Five and with John Fearnley's Pacific Coast Jazzmen.

Harrison, Kenneth Herbert (Kenny)   b. 6/3/33, Southampton, England   dms/ldr
Took lessons from Bosh Lacy-Moody of the Edmundo Ross orchestra; first gigs with Albert Harrison (sax), 1950. Called up in 1954 and played in RAF band and its jazz unit. Following discharge, 1957, worked in Southampton with a trio that backed visiting guests incl. Kathy Stobart, Tubby Hayes, Ronnie Scott, Joe Harriott, George

Chisholm. To London, late '50s, working with Nat Gonella, Owen Bryce, and Doug Richford. Following a period as an acrobatic comedian, formed a modern quartet in 1962. To Australia, 1973, joined Doc Willis then Pam and Llew Hird's Sydney Stompers (1975–76). With Graeme Bell (1977–82), then in 1983 his own band resident at Manly Pacific Hotel. With the formation of Compass, 1985 (with Jim Somerville, Paul Williams sax, Stan Kenton bs) he has resumed the mainstream/bop style with which he began his career.

Hatton, Richard Traquair (Dick)   b. 27/7/26, Perth, WA   tbn./p./alto sax
Heard jazz as a teenager and began playing in 1946. Founder member of Perth's pioneering West Side Jazz Group, followed by the West Coast Dixielanders until the band's dissolution. Subsequently freelanced and played with various short-lived groups, and attended the 1960 AJC in Melbourne and the 1979 in Fremantle. Has played in Europe during visits, and some broadcasts in Perth. In 1985 was in a mainstream group, Fine and Dandy, led by his son Philip, a six-piece band, Sunnyside Up, and the 17-piece Magic Dream Band.

Hawes, Jeffrey Douglas (Jeff)   b. 13/2/37, Melbourne, Vic.   tbn./perc./ldr/vcl
After brief period on clarinet, began trombone at 18. Introduced to jazz by Nevill Sherburn, with coaching from Frank Traynor. With the Saints JB (1955–56), which won the band battle at the Rochester Easter Jazz Festival. Also with Port Phillip JB and the band at the 431 Club (c.1956–61), of which he took over leadership. Also worked occasionally with the Yarra Yarra JB from 1959. Retired from music, 1964. To Sydney, 1969, and gradually resumed playing in the early '70s, becoming a founder member of Noel Crow's Jazzmen, with which, under its new name, Mr Crow, he was still working in 1986.

Hawthorne, Michael Frank (Mike)   b. 7/5/40, London, England   tbn./vcl
Arrived in Brisbane in 1951 and became interested in the dixieland features at city hall concerts. In 1960 began sitting in with the Varsity Five. Played in the Moreton Bay JB with Rod Byrne (p.), Ashley Keating (bjo), Alan Whitworth (dms), and formed the Bay City NO JB with Marty Mooney. In 1964 Nick Boston arrived from UK and Hawthorne switched to trombone as founder member of Boston's Colonial JB. In 1966 the group folded when Boston and Mooney moved to Sydney, and Hawthorne later worked with the Pacific Jazzmen. Since 1973, apart from some work with Ken Herron (tba) in a band called Chile Beans, his main contribution has been with the Vintage Jazz and Blues Band.

Hayes, Mileham Geoffrey   b. 27/11/40, Brisbane, Qld   clt/bjo/ldr
Became interested in jazz at 14 and in 1957 started a jazz club at school. Attended jazz functions at La Boheme, where Sid Bromley encouraged his interest. Bought a clarinet, but shortly switched to banjo after meeting Lachie Thomson, an association which led to the formation of the Varsity Five with which Hayes remained, with brief interruptions for study, throughout its lifetime. From 1964 to 1974, apart from AJCs, he retired from music to concentrate on his career as a doctor. Upon returning from a period in Edinburgh in 1974, he lobbied the ABC regarding the lack of broadcast jazz, and was given his own programme, *Stomp Off, Let's Go*. President of the 1976 AJC in Brisbane, and during the mid to late '70s his jazz activities incl. functions at the Melbourne Hotel (from whence he presented a TV series, *Dr Jazz*), journalism, the establishment of the Cellar Club, and taking a band to the Commonwealth Games in Edmonton in 1978, during which he also performed at Eddie Condon's in New York. From 1978–82, he promoted a series of jazz festivals which gave Brisbane audiences exposure to top American and Australian musicians on an unprecedented scale. He established the Queensland Jazz Club (1980–83), and organized the first national conference on jazz. Bought the old Pelican Tavern 1983, and reopened it as the jazz restaurant Sweet Patootie, which, as well as featuring local and imported bands, has been the main venue for his own group, Dr Jazz.

Healy, John Michael   b. 21/10/45, Melbourne, Vic. bs

Began with various traditional groups around Melbourne, incl. the Jericho JB. To England, 1970 where he joined Max Collie's Rhythm Aces, with whom he toured. Returned to Melbourne, 1972, and worked with the Yarra Yarra JB before settling in Perth, 1974. Has since become one of the most visible bass players on the traditional scene, working with most bands of that style in Perth. President of the 34th AJC, and in 1985 was working with the Lazy River JB and Pam and Llew Hird's Perth City Stompers.

Heap, Don   b. 25/12/38, Melbourne, Vic. bs

Following retirement from professional cycling because of knee injuries, took up bass while living in New Zealand, 1960–65. Returned to Melbourne and worked with various groups incl. Frank Turville's. To Sydney, 1968, to work with Geoff Bull; returned to Melbourne, 1973, to join the Yarra Yarra JB, led by Heap's brother-in-law Maurie Garbutt. To Sydney to join Tom Baker's San Francisco JB and, in 1977 as a founder member of Dick Hughes's Famous Five. Also worked with groups led by Paul Furniss and by Geoff Bull. Joined Roger Janes's band at Unity Hall Hotel, Balmain, 1979. In 1986 he was still with Janes and with Alan Geddes's band at Bondi Icebergers' Club. Heap is remarkable in being probably the only professional bass player in Australian jazz who plays without amplification. A player of great power and immaculate intonation.

Henderson, Robert James (Bob)   b. 2/11/41, Sydney, NSW   tpt/vcl

Began cornet, 1951, playing in Glen Innes Municipal Brass Band. Moved to Singleton, played with town band. Formed a dance band, 1957. In Newcastle, played with Harbourside Six, 1962–68. Through the '70s, played commercial work—clubs, television, cabaret, theatre with occasional jazz work; also attended Newcastle Conservatorium and played in its brass ensemble. Joined Graeme Bell, 1979, and since then has freelanced and worked with Freddie Wilson at the Cat and

The Jazzmanians, c. 1975, with, l. to r., Kevin Robinson (bs), Bill Browne, Ted Herron, Bruce Gourlay, Max Gourlay

Fiddle, Balmain. In 1986 living in Newcastle and commuting to Sydney, still with Graeme Bell and joined John Fearnley's Pacific Coast Jazzmen.

Herron, Edwin Martin (Ted)   b. 16/6/23, Queenstown, Tas.; d. 21/1/86, Launceston, Tas. tpt/gtr/bs

His family moved to Launceston in 1934, and Herron started on harmonica, then ukelele, moving to guitar on which he began working with Bill Massey's dance band as a teenager. At 18 became interested in jazz and took what limited opportunities were available locally for private jamming. Conscripted in 1941 and posted to the mainland and New Guinea; guided by Ron Wills's reviews in *Tempo*, instructed his mother to build up his record collection in his absence. Took up trumpet while in service and heard a broader range of music than was available in Launceston. Contact with the Bells, Cy Watts, and Frank Johnson deepened his knowledge of the traditional jazz idiom. After the war he resumed work in dance bands, and played in a group called the Moderns with Don Gurr, Bill Browne, Jack Madden (sax) and leader Sid Cane (p.), as well as other dance/cabaret work with Jack Duffy, later led by Jack McBain who emphasized a more progressive style of swing. Also playing in private jam sessions with Gurr, Browne, McBain, Bill Sutcliffe and Cane, and, with some changes, this band became the Jazzmanians which recorded material released on Ampersand in 1946. The Jazzmanians were to become the rallying point for traditional/swing style jazz in Launceston up to the mid '80s, and the core of the band was Herron and Bill Browne, probably the most important drummer in Launceston since the war and especially valuable as an energetic promoter in a town where jazz had little commercial appeal. Most of Herron's playing was, by force of circumstance, with dance bands, usually with musicians who incl. current members of the Jazzmanians. Among alumni were Ted's younger brother Ken Herron. The band was frequently in recess for lack of work and during Herron's absence in the early '60s, in Hobart where he played mainly commercial work. Returned to Launceston, c. 1964, and worked in a commercial group with Browne and the Gourlay brothers Max (clt/vln) and Bruce (p.); this was the closest thing to jazz that was heard in the town until the late '70s. The foundation of the JAS, and in 1977 the AJC in Hobart, generated renewed jazz interest and the Jazzmanians gradually regrouped around Herron, Browne, and the Gourlays, who remained the nucleus of the band until Herron's death. The band was able to work more regularly, incl. a two-year residency at Rose's. The multi-instrumentality of its members gave it considerable range, from its usual dixieland/swing approach to a group within the group which used Herron on guitar and Max Gourlay on violin. Gourlay is an original clarinet player, a lyrical violinist, and a writer of song lyrics with wit and grace, and would certainly have achieved greater recognition on the mainland. Hot Strings, the group 'within' the Jazzmanians, played sensitive interpretations in the Hot Club of France style. In a centre where there was little public awareness of jazz until the late '70s, Herron's Jazzmanians and the musicians involved in the band have been vital to the music.

Herron, Kenneth Arthur (Ken)   b. 24/7/34, Queenstown Tas.; d. 20/5/81, Sydney, NSW   tbn./tpt/tba
Began at 12 in a brass band. Heard the jazz records being ordered by brother Ted and subsequently played occasionally with the Jazzmanians. Lived in Melbourne in the '50s, working on the jazz scene incl. with John Tucker, and at the 431 Club with Frank Johnson. Moved to Sydney where he recorded with the Pix All Stars, 1962, and joined Graeme Bell for the Chevron Hotel's Oasis Room residency which began April 1964. In addition to long though intermittent association with Bell, Herron was very active as a session musician. To Expo '70, Japan, with Don Burrows. To Brisbane, 1976, as entertainment manager at Melbourne Hotel, where he also led the band for a regular TV programme. Rejoined Bell in Sydney, 1978, for the Salute to Satchmo tour, and freelanced. Was with Bell when he died following several heart attacks.

Hird, Llewelyn John (Llew)   b. 23/1/29, Birkenhead, England   tbn./bjo/vcl/ldr
Hird, Pamela Rose, née Parrot   b. 14/6/36, Portsmouth, England   tpt
Llew immigrated to Brisbane, 1952, and began trombone, then moved to Melbourne where he joined the Southern City Seven which incl. Frank Turville, Allan Leake, Willie Watt. Formed his New Orleans JB, 1957, which later became the

The Hirds with the Don De Silva Band, c. 1973: Pam Hird (tpt), Don De Silva (bjo), Llew Hird (tbn.), Don Heap (bs), Alan Geddes (dms)

Melbourne New Orleans JB. In the meantime Pam arrived in Melbourne in 1948 and began attending Llew's gigs, which inspired her to take up trumpet. Her first work was in Llew's band deputizing for Turville. Following their marriage, Llew and Pam's careers have been coeval. Following the birth of son Karl (29/11/59), the family moved back to England, 1960, and formed the Llew Hird JB which enjoyed success in England and Europe during the trad boom. Returning overland to Australia in 1964, the Hirds were stranded by lack of funds in Egypt, and developed a floor show which led to a six-month booking in Alexandria, after which they took it on tour through Libya and Greece. From 1965–67, the Hirds alternated between Sydney and Perth, working mostly in clubs. From 1967–68 they took a jazz-based show on tour in the Far East, (Dennis Tonge bjo/gtr, Viv Carter, Laurie Gooding reeds, and Ian Lawson keyboard), in the course of which Pam was awarded an honorary Purple Heart after the band came under fire in Vietnam. The Hirds continued to play club work in Australia and in Asia, until Don de Silva invited them to join his jazz band in Sydney in 1973. Formed their Sydney Stompers, 1974, with son Karl on reeds, Peter Gallen (bs), Kenny Harrison (dms) and Bill Brunskill (bjo), shortly to be replaced by de Silva. Except for breaks while they toured to Japan and a period in Perth, the Stompers operated until the Hirds returned to settle in Perth where they formed the Perth City Stompers, with which they were still working in 1985. Since 1977 they have travelled often to New Orleans where they are members of the Society Brass Band of New Orleans, and have been awarded the Keys of the City.

Holden, Geoffrey Norman Francis Daniel (Geoff)   b. 1/10/40, Sydney, NSW   bjo/gtr
Began hearing jazz on radio and at Sydney Jazz Club functions in the '50s. Began on washboard, 1958, sitting in at the Royal George Hotel, 1959 and joining the band (incl. Jeff Hawes, Geoff Bull, Peter Neubauer, Dick Edser bs) on standard banjo, 1960. Replaced Harry Harman in the Paramount JB, 1962, then joined Geoff Bull's Olym-

pia JB with which he continued an association through the '60s and '70s. Also worked with other groups, replacing Geoff Gilbert in Nat Oliver's band, 1962, for its residencies at the Ling Nam and Suzie Wong restaurants, and playing at the Orient Hotel with Adrian Ford, Dick Gillespie (dms), Edser, and Neubauer followed by Paul Furniss. Spent 1971 and 1974 in New Orleans and Europe, at other times continuing the intermittent association with Bull. Also with Roger Graham's (tpt) band, 1972, and with Pam and Llew Hird at the Limerick Castle, 1978. Retired from music, 1982, then resumed playing in 1984, first with Nick Boston then in 1985 with Bull. Holden has been a well-informed mainstay of New Orleans-style jazz since the early '60s, and its most durable exponent on banjo.

Holyoak, Alf   b. 26/12/13, d. 27/3/85, Adelaide, SA   reeds/ldr/arr
In 1929 elder brother Bill organized lessons for Alf with ex-Melbourne reed player Frank McMahon, who introduced his pupil to jazz. Holyoak began playing off-season dances at the Palais Royale in late 1932. He built his reputation with Ron Wallace, broadcasting with Hedley Smith, and in the pit band at the Majestic; when he entered the RAAF in 1941 his departure left a serious gap. Served in No. 1 Mobile Entertainment Unit under F/O Harry Dearth, extending his reputation interstate and receiving numerous offers of post-war work incl. from the ABC in Melbourne. Following medical discharge, 1945, returned to Adelaide and found a growing interest in jazz, partly as a consequence of his brother's vigorous proselytizing. Alf formed his own groups and taught. His students and protégés incl. Maurie Le Doeuff, and most of the progressive young musicians who emerged in Adelaide during the '40s and took their influence to Melbourne and Sydney: Laurie Parr, Bob Limb, Bob Young, Clare Bail, Bryce Rohde, Errol Buddle. Holyoak's bands were a mainstay of the swing/jazz concerts of the '40s and '50s in Adelaide, and in 1946 he directed for broadcast a group with the audacious instrumention of piano (Wally Lund) and four saxes—himself, Le Doeuff, Buddle

Alf Holyoak Sextet, c. 1949, with, l. to r., Bryce Rohde, John 'Slick' Osborne, Alf Holyoak, John 'Jazza' Hall, Jim Hogan, Milton Hunter

and Bail. In later years jazz opportunities diminished and he worked more in 'society' music, leading the South Australian Hotel band playing for Government House and mayoral balls. He continued to teach, and several generations of musicians passed through his open-minded tutelage. As a musician, his work complemented the entrepreneurial jazz work of his brother; at his death he was a revered father figure, whose influence had indirectly touched many significant areas of Australian jazz.

Hooper, Peter   b. 12/2/59, Adelaide, SA
bjo/gtr/bs/ldr
Started banjo at age 11. With Bruce Gray in 1975, Gordon Coulson's Climax JB (1978–82), Pam and Llew Hird in the late '70s, and with Tommy

Richardson (dms) at the Union Hotel, 1979–83. Joined Richardson at The Patch in about 1980, and later took over as leader. In July 1983 formed the Royal Garden JB which was still resident at the Brittania Hotel in 1986. Hooper has played with Dave Dallwitz's Euphonic Sounds, incl. the band's recordings. He has become an energetic force not only as musician but in forming bands and organizing work. In 1985 formed the BBC Trio (BBC: bjo—Hooper, bs—James Clark, clt—Andrew Firth, doubling on alt.) for ABC–FM broadcasts.

Hounslow, Keith   b. 19/9/28, Perth, WA   tpt
Began playing in Perth late '40s, and joined West Side Jazz Group in 1947. Visited Melbourne for the 1947 AJC, then in 1949 toured as baggage boy with Rex Stewart, a major influence on Houns-

low's musicial attitudes. The tour concluded in Melbourne, from whence Hounslow departed, with Doc Willis, for Adelaide and Perth. In Adelaide, briefly with the Southern Jazz group, with whom he recorded. In Melbourne by December 1951, freelancing over the next three years, with Doc Willis, John Sangster, Frank Traynor, Charlie Blott, and with Frank Coughlan at the Trocadero. Associated with Brian Brown, at Jazz Centre 44, 1955–60. Following a period with Alan Lee, Hounslow embarked upon what was to become an award-winning career in independent film and TV production, and from 1962 went into semi-retirement from jazz. Returned to music through casual gigs throughout the early '70s, and met Tony Gould when he recorded with the Datsun Dixielanders in 1975. The two formed the duo McJad, presenting melodic spontaneous compositions, recording and touring in Australia and Asia. In the meantime Hounslow joined Frank Traynor from 1978–83, after which he moved to Sydney where he freelances. Self-taught and with no formal theoretical knowledge, Hounslow is a melodic player who has instinctively drawn upon the broadest possible range of influences. Throughout his career has been involved in some of the most important events in the post-war history of Australian jazz.

Hudson, Roger Massey   b. 9/1/34, Richmond, Vic.; d. 8/7/86, Melbourne, Vic.   p.
Began playing piano in Adelaide, where he had lived since early infancy. First bands were John Pickering's Cross Road JB, the Adelaide University Four, the West Side JB, and his own Baron Hudson's Eumenthol Jazz Jubes. Co-founded Adelaide Jazz Society in 1956, remaining secretary until 1961. Secretary, publicity officer, and president of the Adelaide AJCs in 1957, 1961, and 1968 respectively. From 1962–70, a partner in various jazz venues including the Taverns, and playing mostly with Dick Frankel's Jazz Disciples. In Melbourne, with Frank Traynor's Jazz Preachers, 1972–80; Kenn Jones's Powerhouse Band, 1980–83, and later led his own group, Hot Lips, formerly Jazz Lips. An executive with World Record Club, and

co-ordinated its Jazz Unlimited programme, from 1958 until the organization moved to Sydney in 1983. At the time of his death, was associated with Allan Leake in jazz enterpreneurial and performance activity.

Hughes, Richard (Dick)   b. 8/7/31, Melbourne, Vic.   p./vcl/ldr, occasional dms/comp
With Will McIntyre an important early inspiration, Hughes played casually with Melbourne traditionalists and was president of the Melbourne University Rhythm Club (1950–51). In England (1952–55), he worked mainly with Cy Laurie and conducted ABC interviews with visitors incl. Sidney Bechet, Billie Holiday, and Mary Lou Williams from whom he took lessons. From 1955 settled in Sydney, and with PJJB and Ray Price groups into the '60s. Began leading groups with a quartet at Macquarie Hotel, 1962, then residencies at Windsor Castle Hotel, Adams Hotel, French's Tavern, into the '70s. Much of this work was without bass or drums, leading a perceptive writer at the time to predict accurately that this would lead to an unusually strong left hand in Hughes's Fats Waller/Jelly Roll Morton-influenced style. Played support for and toured as baggage boy with Eddie Condon's group in 1964. Toured with Alvin Ailey's American Dance Theatre (1965) and attended Newport Jazz Festival (1972) as guest

Dick Hughes, 1977, with Alan Stott (tba), Playhouse Theatre, Hobart, during the 32nd Australian Jazz Convention

Dick Hughes's Famous Five, Soup Plus Restaurant, 1982, with, l. to r., John Edgecombe, Merv Acheson, Alan Geddes, Bruce Johnson, Dick Hughes

journalist of the US government. First solo pianist at Sydney Opera House (1973) and recorded the first solo piano jazz album in Australia. Led a trio at the Journalists' Club from early '70s to 1983 (various drummers incl. Ron Bowen, Gordon Dennis, Lindsay Arnold; Bruce Johnson, tpt). Formed his Famous Five (Chris Taperell reeds, Bruce Johnson, Alan Geddes, Don Heap) in 1977 as a recording group and began residency with the band at Soup Plus. With Johnson the remaining original, other members have been: dms: Bryan Kelly; reeds: Clare Bail, Errol Buddle, Ian Wallace, Marty Mooney, Merv Acheson, John McCarthy; bs: Richard Ochalski, Mal Rees, John Edgecombe, Cliff Barnett. The Famous Five were still resident at Soup Plus in 1986. In addition the band has recorded, broadcast, and played numerous concerts incl. as support for Chris Barber (1978), Kenny Ball (1982), Stan Getz (1980). Hughes has continued also as a soloist incl. as support for George Melly, 1986, and in revivals of the PJJB.

His exuberant personality and piano style have attracted considerable publicity for jazz, as have his activities as a jazz journalist on several daily papers, and as the author of Australia's first jazz autobiography, *Daddy's Practising Again*. In 1986 he was working on a second volume of memoirs.

Hyde, William Garnet (Billy)   b. 12/11/18, d. 27/12/76, Melbourne, Vic.   dms/vibes/ldr
Began during at age 11. First work with a local suburban dance band, the Night Owls, then joined Fred Hocking at the Casino in Brunswick. Left to join Charles Rainsford on the Hoyts theatre circuit, and later worked in concert bands under Glen Williams, with Bob Gibson at the Palm Grove, in radio, and as a symphony orchestra tympanist. From the late '30s Hyde also established a reputation in swing-based jazz circles, leading groups for 3AW Swing Club functions and in St Kilda coffee lounges. Joined Wally Nash at Tate's Teahouse in 1942, and continued extensive cabaret/dance work through the '40s and '50s, incl. as bandleader at the Maison de Luxe, Elwood. Also active in the jazz concert scene throughout the early '50s. In 1949 he began working with the ABC, becoming staff drummer from c. 1950. From 1956 he was primarily involved in television orchestras, incl. as staff drummer at Channel 2, 1956–62. Retired from playing in 1970 and ran a very successful drum studio and shop which was taken over following his death by his son Gary, also a drummer. Hyde was one of the most admired drummers of his time, and exercised considerable influence both as a performer and teacher.

I

Isaacs, Mark   b. 22/6/58, London, England
p./comp.
Father (Saul) and uncle (Ike) are well-known guitarists. Mark began at six, jamming with father from nine. Immigrated to Australia; conservatorium training in Sydney under Peter Sculthorpe. Winner of the JAS composition competition (1977), joined Kerrie Biddell's Compared to What (1978). Studied in New York (1979), Jerusalem (1980). With Errol Buddle (1981–82); writing for TV and theatre, and recorded an album of his own compositions (1981). Won the Don Banks Memorial Fellowship for further study in the US, 1984.

J

Jacques, Judy   b. 17/5/44, Melbourne, Vic.   vcl
In 1960 she was singing at a local dance with a pop band when the Yarra Yarra JB replaced the band. Jacques stayed on, sitting in with and finally joining the Yarras with whom she recorded in 1962. During her period with the band, 1960–63, she became something of a cult figure for the young audiences who followed the jazz and folk scene. Joined the Gospel Four in 1963, then during the next decade, moved into TV and club work. In 1975 she began collaborating with the poet Eric Beach and performing at the Pram Factory, La Mama, and Clifton Hill Music Centre. In 1977 she moved into experimental voice/instrument areas with Barry Veith, leading to a tour under a grant from the Australia Council in 1979. Reunited with the Yarras at Bell's Hotel, 1981, and worked at Athol's Abbey with Veith, Sandro Donati (to whom she is married), Mark Fitzgibbon, Gary Costello and Larry Keane in the band Blues by Five. Formed a group with Ken Schroder, Peter Jones, Matt Kirsch, Jeremy Allsop and Graham Morgan, 1984, also working with Bob Sedergreen's Blues on the Boil and with the Brian Brown Ensemble. In 1986 she was teaching contemporary singing at the Victorian College of the Arts. Jacques is one of the very few singers to emerge from the jazz/folk boom of the early '60s and define a new phase in her musical career in Australia.

Judy Jacques, 1985

James, Kenneth Ian (Ken)   b. 4/1/44, Sydney, NSW   reeds/fl./comp./arr/ldr
After freelancing in various musical settings during the early '70s, James joined Judy Bailey (1974–77) and The Last Straw (1975–78). Led his own jazz/funk group, playing concerts in WA, 1978, and a group called News which also broadcast for Perth ABC, 1979. Formed his Reunion, 1982, with John Pochée, Vince Genova (p.), Col Brown (bs) and sometimes adding Carlinhos Goncalves (perc.). This band broadcast on 2MBS, supported visiting performers incl. Eddie Daniels, Mike Nock, Herb Ellis/Ray Brown, Monty Alexander, and played the Festival of Sydney, 1982, 1983. James is primarily a studio and club musician, but regularly reappears with significant groups on the contemporary scene. Has performed with Herbie Mann and Stephane Grappelli, and has also contributed substantially to jazz education at clinics, and JAS workshops.

Roger Janes, 1987

Janes, Roger Colin   b. 19/5/42, Bristol, England   tbn./vcl/ldr

Began trombone at 19 and played in a trio at the Victoria Hotel, Colac, 1963, and his first AJC in the same year. Moved to Melbourne, worked with Penfield JB at Jazz Junction and with Eddie Robbins (clt). Joined Yarra Yarra JB, 1965; moved to Sydney, then England, 1967. Toured Europe with Graeme Bell, 1968 and rejoined the Yarras in England, 1969. Returned with the band to Melbourne where he remained with them until moving to Sydney to join Geoff Bull, 1976. Formed his own group, 1978, with which he shortly began a residency at Unity Hall Hotel, Balmain. The band recorded in 1983 (Janes, Marty Mooney, Don Heap, Lynn Wallis, Gary Walford p.). Janes was still leading the group at Unity Hall in 1985, with Mal McGillivray and Paul Martin having replaced Wallis and Mooney. Beginning in the New Orleans style, Janes's spontaneous eclecticism has made this one of the freshest sounding bands in Sydney. He is also a graphic artist.

Jazz Centre 44 (Melbourne, Vic.)

Started by Horst Liepolt in 1957, probably April, as Club 44 on Sunday afternoons, opening with Graham Morgan Quintet (with Peter Martin sax, Keith Hounslow, Brian Rangott gtr, Ron Terry bs). Jazz Centre 44 was distinctive in presenting concerts rather than dances, and in promoting modern as well as traditional jazz. Its main base was the Katherina,* St Kilda, but at other times it operated from the Lido, St Kilda, and Albert Park LSC. In the first few months several house bands were used, until the installation of the Brian Brown Quintet in July. Brown became the most regular leader, and also played as sideman in Keith Hounslow's quintet at various times. There were few if any post-traditional musicians in Melbourne who did not at some time play at Jazz Centre 44. Graeme Lyall, Barry Buckley, David Martin, Derek Capewell, Ted Vining, Keith Stirling, Alan Turnbull, all worked at the centre, usually under the leadership of Alan Lee, Hounslow, or Brown. Jazz Centre 44 also promoted traditional jazz as from July 1957, presenting the Melbourne NOJB, the Barnards, Max Collie, and the young Yarra Yarra JB. The centre also saw experimentation with 'poetry and jazz' functions, beginning October 1960 and involving poets Adrian Rawlins and Bob Cumming with Mookie Herman's band. During its heyday, c. 1958–60, Jazz Centre 44 was the most fertile jazz venue in Melbourne, with its willingness to sponsor experimentation and new talent, and in so doing, it incubated a generation of modern jazz musicians who would dominate the scene into the '80s. In November 1960 Horst Liepolt moved to Sydney. Jazz Centre 44 continued to operate, but without its former vitality and adventurousness. The proportion of traditional jazz gradually increased in response to the traditional boom until, by July 1963 the fare was wholly traditional with bands like the Yarra Yarra JB and John Hawes's group. It has not been possible to document the centre's closure precisely, but in 1966 it ceases to be mentioned in any of the usual printed sources. German-born Liepolt went on to make a mark in Sydney. He began producing records in 1975 established the 44 label (named

---

*This has been variously spelled: Katherina, Catherina, Catharina, etc. a confusion exacerbated by the existence of another cabaret with the same name also in St Kilda at the other end of Acland St. The Jazz Centre 44 Katherina was at the seafront end of Acland St, and is the same building referred to by this name elsewhere in this volume.

for Jazz Centre 44, which in turn commemorated the year he discovered jazz). Some of the most innovative Sydney bands of the late '70s were documented by 44 records. Liepolt also produced the regular magazine *Jazz Down Under*, was involved in the establishment of The Basement, acted for a period as manager of Galapagos Duck, and organized numerous jazz concerts incl. for the Festival of Sydney. He moved to New York, 1981, where he quickly established himself as an entrepreneur, notably at the jazz clubs Sweet Basil's and Lush Life, and through concerts and festivals incl. the Village Jazz Festival.

Jeffery, Robert (Bob)   b. 28/12/41, Adelaide, SA   reeds/fl./bs/vcl

Began alto at age nine and first broadcast on radio the following year. Although his primary interest was in jazz, there was little opportunity in this area, and from the late '50s he worked in rock bands incl. the Penny Rockets and the Hi Marks, and toured with Johnny O'Keefe in the early '60s. The establishment of the Cellar in 1961, however, created a platform for a modern jazz movement in which Jeffery was a central figure. Also worked in Neville Dunn's Big Band. The end of the jazz boom in the early '60s saw Jeffery working in commercial areas, incl. TV orchestras and pop groups. More extensive jazz activity has been resumed since the late '70s, with radio and television broadcasts, a lengthy residency with Alan Hewitt (p.), Bill Ross, Dave Siedel (bs) at the Gateway Hotel, North Terrace, concerts for the JAS backing Milt Jackson, Georgie Fame, and on bass with Ruby Braff and Ralph Sutton. In 1986 Jeffery was working casually and with Bruce Gray.

Jenner, Andrew Lathan (Andy)   b. 12/7/43, New Brighton, Cheshire, England   reeds, with some keyboard/fl./gtr/vcl

Active in England during the '60s with various groups also involving John Braben, Jenner arrived in Brisbane around 1970. Played casually until Vic Sanderson and Braben arrived in 1973, when Jenner recruited them as founding members of what became the Vintage Jazz and Blues Band. While

Andy Jenner at the Twelfth Night Theatre Club, 1978, with John Braben (tpt) obscured

still working with them he also established the Caxton St JB in 1977, leaving in 1979. Left the Vintage band, 1981, to freelance. Has also set up a studio in which he composes and records in a broad range of musical styles. In addition to his authoritative reed playing, Jenner's main contribution to jazz in Brisbane lies in his having founded its two most important traditional-oriented groups of the last 15 years.

Jennings, Herbert Vincent (Herb)   b. 21/7/42, Melbourne, Vic.   tbn./sax./vcl/comp./arr/ldr

Became involved in collecting, research and organization from late '50s. Started trombone seriously in 1969 and played that year at the Ballarat AJC of which he was secretary. 1970–73, with the Golden City Six at Golden City Hotel with Bob Pattie cnt, Barry Currie clt, Eric Goon p., Alf Armstrong or Dick Barnes bs, Ron Rosser dms, Bruce Walsh hca/vcl. In 1976, formed New Golden City Six, becoming Golden City Seven in 1979. In 1985 the personnel was: Jennings, Pattie, Rosser or Gary Richardson reeds, Bob Franklin p., Graeme Day bjo, Alan Houghton tba, Ben Systermans dms. In 1977–78 with Tour de Force, a quintet incorporating some pop music in its repertoire, for which he took up tenor. Jennings and the Golden City

Herb Jennings with the Golden City Seven, at the Ballarat Begonia Festival 1983; Ron Rosser (clt), Bob Pattie (cnt), Herb Jennings (tbn.), Ben Systermans (dms), Bob Franklin (p.), Graeme Day (bjo), Alan Houghton (tba)

groups have been at the forefront of jazz in the Riverina since 1970. The band regularly works throughout the rural area around Ballarat, and has made TV appearances. Its several recordings include material from the AJC's original tunes competitions (which Jennings has himself won on several occasions), with financial assistance from the Australia Council. Brother of Mal, an ex-teacher, now farmer, he currently lives at Ross Creek near Ballarat.

Jennings, Malcolm Thomas (Mal)   b. 22/12/44, Melbourne, Vic.   tpt/vcl/some arr
After about five years in the brass band in rural Beaufort, began playing for school dances in a quartet, then joined the Neil Barrett (p.) Trio playing various local functions. At this time, also studying classical piano, and won a South St competition in Ballarat as solo pianist in 1961. Played with other Ballarat groups, and developed a dedication to jazz while with the Hot Potatoes JB. Following its dissolution, formed his own Ballarat Jazz Messengers, which played AJCs, local festivals, Ballarat Jazz Club functions, and recorded. For three or four years through the mid-'60s the Messengers were based at Gallery 321, a coffee lounge which Jennings ran with his brother Herb. Following the slump in jazz in the late '60s, Mal spent more time freelancing locally and in Melbourne, incl. with the bands of Owen Yateman, Frank Traynor, Allan Leake, Barry Young, Steve Waddell, and the St Arnaud and Grange-Burn JBs.

Although most of Jenning's career has been rurally based, his fiery style, owing much to Louis Armstrong and Bob Barnard, has become nationally known and admired through appearances at AJCs, particularly with Lachie Thomson's New Whispering Gold Orchestra. Since 1980 Jennings has been living in Brisbane, Qld, where he teaches sheetmetal and welding at a TAFE College, and freelances on an amateur basis.

Jesse, Graham   b. 19/10/55, Sydney, NSW   reeds/fl./arr/comp.
Began on recorder at 10, then alto at 12. Began jazz studies programme at the NSW Conservatorium, and playing casual gigs, in 1973. Spent some years in the Daly-Wilson Big Band, and in c. 1977 joined Compared to What, the backing band for Kerrie Biddell. Spent a year studying in New York (1980–81) and back in Sydney led his own small groups for seasons at The Basement and Jenny's Wine Bar, later Paco's. Has been performing in duo with Mike Bartolomei since 1984, incl. for the Sydney Improvized Music Association. Also with the John Hoffman Big Band, Julian Lee, and freelancing.

Johnson, Francis Walter (Frank)   b. 22/5/27, Melbourne, Vic.   tpt/vcl/comp./ldr
Began cornet, 1944, with the ambition of working in a swing orchestra. Hearing Graeme Bell's band in 1945 inspired him and Geoff Kitchen to form a band in that style: Frank Johnson's Fabulous

Frank Johnson at the Playhouse Theatre, Hobart during the 32nd Australian Jazz Convention, 1977: l. to r., Dick Hughes, Frank Johnson, Alan Stott, Bill Miller, Graham Spedding

Dixielanders was one of the most celebrated bands in Melbourne from roughly 1947 to 1956. After the break-up of this band Johnson formed several groups, including a quartet which played to crowds at the 431 Club. In 1959 he began a two-year period with a five-piece dance band at the Federal Hotel. Returned to jazz in mid-1961, and since that time has been involved in occasional concerts and casual work. In the '80s residing in Ipswich, Qld, and playing around Brisbane.

Johnson's most important work as a musician was as leader of his Fabulous Dixielanders. In addition, however, he was an active presence at the earliest AJCs, and has continued this association. He has also written on jazz in various journals, and has played a significant role in the promotion of jazz concerts, such as the Downbeat series, with Bob Clemens, in Melbourne.

Jones, Kenneth Albert (Kenn)   b. 18/6/33, Melbourne, Vic.   reeds
Began alto at 16; at about 19 joined a quartet which incl. Fred Parkes and Graham Coyle, playing 50/50 dances. Received a baritone sax for 21st birthday, which altered his direction in jazz. Played with various bands, incl. Max Collie's and Len Barnard's. In 1958 formed his Powerhouse Crew (also known as the Mainstream Crew), playing at the Powerhouse Rowing Club Sunday nights, primarily with Jones, Parkes, Coyle, Harry Price, Bob Barnard, Alan Lee, Stewie Speer, Keith Cox (bs), Gaynor Bunning (vcl). This highly successful residency finished in 1966. 1969–75, with Storyville Jazzmen. Withdrew from music and moved to Bairnsdale 1976, returning to Melbourne in 1979 and joined Storyville All Stars until mid-1980. Formed Jazz Powerhouse, which had a two-year residency at the Victoria Hotel, Albert Park, and played Sacramento Jazz Jubilee in 1982. Also formed Three and Easy in 1980 with guitarists Pierre Jaquinot and Vince Hopkins, working the Melbourne Hilton through 1981. In 1984 was leading Jazz Powerhouse, with Coyle, John Hawes tpt/vcl, John Turner dms/vcl, Ivan Videky bs.

Karssemeyer, Hans   b. 1/7/39, Leidschendam, Holland   p.

Began piano at age 6, became attracted to jazz at 13. Played in local jazz groups and Leyden University JB on banjo, guitar, bass, trombone, moving to piano in 1957. Joined the merchant marine, and settled in Melbourne, 1957. Joined Max Collie's band until Collie left for England, 1962, and Karssemeyer then joined the Melbourne Dixieland JB. In Brisbane during 1969, playing with Tony Ashby and Mike Hawthorne, then moved to Sydney. Spent five years with Nick Boston's Colonial JB, and joined Tom Baker's San Francisco JB in 1975, continuing with the band to 1986 under Paul Furniss's leadership. In the meantime has toured with Ray Price, 1976–77 and worked casually with the Abbey and Harbour City JBs, and groups led by 'Kipper' Kearsley at the Manly Pacific Hotel and John Colborne-Veel at the Purple Grape.

Kelly, Bryan   b. 24/4/27, Adelaide, SA   dms

Began drums, 1945, playing in a band led by Len Perkins (clt). To Brisbane (1947–48), where he was musically inactive. Back in Adelaide, worked with Lew Fisher, then in the early '50s at the Hotel South Australia with Carlos Brunos (a.k.a. Ian Brown) then Ian Drinkwater. Also with Bruce Gray at the Dollar Club. To Surfers Paradise, Qld (1956), playing with Stan Walker, then to Brisbane (1958), where he led a quartet (Gordon Benjamin or Jack Thomson, Darcy Kelly, Maurie Dowden), and deputized in the Varsity Five. To Sydney (1962), playing freelance incl. with Allan English at the Orient Hotel (1965). From the mid-'60s he booked and often worked with the bands for the Windsor Castle Hotel, Paddington. With Mike Hallam's Charlestown Strutters (1976–77) and his

Hot Six (1978). With a quartet (Jack Wiard, Dennis Tonge, Richard Ochalski) at the Paragon Hotel (1977) and with Jack Allan's trio at Wollongong (1978). A member of the Abbey JB (1979–82). In 1986, in addition to freelancing, Kelly was with Dick Hughes's Famous Five. Kelly has also played at AJCs, incl. with Bud Freeman (1975), Sam Price (1983), and in Dave Dallwitz groups.

King, Horace (Horry)   b. 7/1/25, Perth, WA   reeds

From the time he joined his first band playing cabaret work in 1952, King has been in heavy demand in a wide range of settings and styles from traditional to the more progressive developments of swing. He was recruited for Jim Beeson's trio and by Sam Sharp for the latter's grandiose Concerts in Jazz concert in 1952. King was one of the busiest musicians on the jazz concert stage during the '50s, playing with a variety of bands incl. those of Sam Sharp, Frank Thomas, later Bill Gumbleton, and with groups of his own that included at various times George Franklin, Harry Rettig (p.), Ralph Filmer (bs) and Guy Bart (dms). When Bob Gibson assembled local musicians for a special All Star concert in 1953, King was one of them, and he worked again for Gibson during a visit to Sydney in 1962. He was also among the top-ranking dance band musicians at the Embassy ballroom, and continued to work in pubs and cabarets throughout his career, particularly following the decline of the jazz concerts. With the resurgence of jazz activity in the '70s, he began working for jazz clubs, leading groups for Perth Jazz Club/Society functions. He has worked with most pre-contemporary style jazz bands in Perth including the Lazy River JB, Dixie Kidd's groups, and the Gentlemen of Jazz for the Jazz Club of WA functions at the Highway Hotel, Claremont. He has also been featured at the WA Jazz Festivals at York with different groups including Sweet Substitute with ex-Melbourne vocalist Elvie Simmons in 1985.

Kitchen, Geoffrey William (Geoff)   b. 23/10/29, d. 26/1/86, Melbourne, Vic.   reeds/ldr/comp./arr Began on guitar, with lessons at Harry Mawson's academy, then took up clarinet, 1944. Founder member of Frank Johnson's Fabulous Dixielanders, until 1951. In the meantime played at 1st AJC, 1946, and with Tony Newstead's band when it deputized for Graeme Bell at the Uptown Club. Played Leggett's ballroom, 1951–52; led a quintet in the feature film *Night Club* (1952), recorded with 'Splinter' Reeves (1952), and led a group (Geoff

Bland or Ivan Hutchinson p., Keith Cox, Bud Baker, Max Wally dms) at Mentone TH dances (1952–54). Freelanced (1955–57) then moved to Sydney to join Bob Gibson on the Ford radio show. Returned to Melbourne (1959), to work for GTV–9 as a casual arranger until being appointed staff arranger in the early '60s. Left GTV–9 (1965), freelanced then joined Media sound jingle and music production company, remaining until 1976, except for a freelancing period (1972–73). Formed Jazz Foundations (1976), combining a jazz frontline with a pop rhythm section. This group recorded compositions by Ken Evans and Kitchen on a Music Board grant (1977). From the late '70s Kitchen played at Potter's Cottage restaurant, Warrandyte, with his own and Dick Tattam's groups, continuing to work as a studio arranger. In 1983 he

Geoff Kitchen at The Embers, 1961; l. to r., Geoff Kitchen, Chet Clark, Barbara Virgil, Alan Geddes, unidentified Canadian(?) pianist, Ivan Videky, Billy Weston, Frank Smith (leader of The Embers group), Bobby Tucker, Paddy Fitzallen, Alex Hutchinson

operated a music production company. During the '80s he worked as a duo with Bruce Clarke, taught privately, and continued to serve as president of the Music Arrangers' Guild of Australia, a position he had held from the early '70s, until his death from cancer.

Kluke, Geoffrey William (Geoff)   b. 1947, Adelaide, SA   bs

Began playing early '70s, working around Adelaide with the more modern musicians. Regular work with Sue Barker, with whom he recorded. Much session work, commercials, studio bands, pit work, and with visiting musicians from Sydney incl. Paul McNamara, Alan Turnbull, Don Burrows. Also worked with US visitors incl. Herb Ellis, Barney Kessel, and Blossom Dearie, during whose tour he suffered serious kidney illness which led to a six-month break from music. Following recovery, joined Glenn Henrich quartet. 1980, to northern NSW for regular club work. Also working in Brisbane with Clare Hansson trio and with visiting artists incl. Richie Cole, the New York Jazz Giants, Ruby Braff, Mark Murphy. 1982 with the Australian Jazz All Stars in Brisbane. In 1983 to Victorian College of the Arts teaching jazz studies with Brian Brown. Currently with Brown's quartet.

Koch, Norman Clarence (Norm)   b. 3/9/29, Adelaide, SA   bjo/vcl

Began in 1946 with lessons from Monty Rossiter, with whom he performed as a duo act. Koch's first jazz work, in 1950, was with the Jubilee Jazz Group which incl. later colleagues Rod Porter and Col Shelley. By 1953 he was working with groups led by Richie Gunn (incl. Porter and Ted Nettelbeck), Roger Hudson, and Leon Atkinson's West Side JB. His association with Hudson was maintained intermittently throughout the '50s. In 1960 he began playing with the University Jazz Group which ultimately became the Campus Six with Ernie Alderslade (tbn.), Bill Munro, Porter, Bob Lott (bs), Jerry Wesley-Smith, and Kent Fuller (dms). Koch remained with them for a further two years, and intermittently thereafter. In 1962 he began a 20 year association with Dick Frankel's Jazz Disciples, with whom he played in the Hindley St. Velvet Tavern, in which he was an investor. He joined Bruce Gray in 1965, and recorded with him in the early '70s, as well as broadcasting with him. During the same period he recorded with Dave Dallwitz bands as well as making an album of his own. Koch has been a ubiquitous presence in primarily traditional jazz, and has also been involved in jazz education through the ABC. Work for Barossa Valley tourist excursions led to a highly successful tour of SE Asia in 1982 under state government auspices. In various bands Koch has played support to, and with, a number of visitors, incl. the Dutch Swing College, Kenny Ball, and the World's Greatest JB. In the mid '80s he has been active with the Adelaide Stompers and has also formed his own group.

L

Lane, Keith Joseph (Joe, Bebop) b. 21/3/27, Sydney, NSW vcl/dms/lrd/comp.
Attended 2KY Swing Club sessions during Second World War, and learned drums from Mark Bowden. Began night-club work in the post-war period, incl. Sammy Lee's with Ian Gunter, and the Golden Key; played jazz concerts and in 1949 was with Roy Maling at Actors' Equity Club. To Melbourne c. 1953–56, where he played Downbeat concerts and worked with George Wallace Sr. Back in Sydney, fronted his own band, the Dee Jays, which Johnny O'Keefe took over. With Ken Morrow (with Jack Craber bs and Wally Ledwidge gtr) at the Arabian coffee lounge in Kings Cross, and worked in Joe Taylor's Corinthian Room where he jammed with many visiting Americans incl. members of Stan Kenton's and Lionel Hampton's groups. Returned to Melbourne, 1958, playing at the Downbeat and Monmartre clubs, and the *Cool for Cats* TV show. He was an important proselytizer for bop during this period, especially among the younger modernists incl. Bernie McGann, John Pochée, Andy Brown, Graeme Lyall, Dave MacRae. He also worked and jammed with, *inter alia*, Chuck Yates, Freddie Wilson, and Brian Rangott (gtr) at this time. Led a trio, and played the Blue Note, St Kilda (1961) and the New Blues, Ripponlea for a brief period in 1962. Returned to Sydney, 1963, and began a club contract in Wollongong, 1964, which was terminated when injuries sustained in a road accident 11 May necessitated a layoff. Played Katoomba jazz festival, 1965. During the subsequent decade there were few opportunities for a jazz singer, especially one so uncompromising as Lane. There is little evidence of activity during this period, for about five years of which he was in New Zealand. From the late '70s he has been increasingly visible. Presented shows at The Basement with his Tailwaggers (1978), and a 'history of jazz' which incl. Peter Boothman and Murray Hill. (Hill is representative of those Sydney jazz musicians who are impressively talented, particularly in the bop idiom, but who have received little recognition partly because of the size of the city's jazz population and partly because of a temperament not disposed to seek publicity.) With drummer Dennis Sutherland (who also wrote the arrangements), Lane formed Killer Joe, a big band with which he has played concerts, incl. for JAS and SIMA, and at The Basement. He also co-leads with Sutherland a small group called the Jazz Cats which has played residencies at the Criterion Hotel. Lane's raw and intense vocal delivery and his unpredictable repertoire and treatment of material have gained him the status of a minor legend in Sydney, a cult figure in the jazz underground.

Joe Lane, 1986

Lazy River Jazz Band (Perth, WA)

This traditional-based band was formed in 1980, playing its first engagement at the opening of the Matilda room at the Nookemburra Hotel, with Bob Barnard as guest. Foundation personnel were Vince Holmes (gtr/bjo/ldr), Bob Anderson, Lew Smith, Lloyd Napier, John Healy, Keith Whittle, Bob Robinson (tbn.). The personnel has remained stable except for a succession of reed players incl. Jim Cook, Ross Nicholson, and Horry King, and the loss of a pianist for reasons of economy. The band's longest residency has been at the Castle Hotel, North Beach, where it began in 1982. In Oct. 1983 it toured the north-west with American trombone player George Masso, and in 1985 played for the Kalgoorlie Jazz Society.

Leake, Allan Bruce   b. 16/8/35, Melbourne, Vic.   dms/ldr

Began drums, 1954, and shortly afterwards joined Frank Turville's Southern City Seven, taking over as leader after Turville left for England. In the early '60s he formed his Jazzmakers, the alumni incl. Tony Newstead, Derek Reynolds, Denis Ball, Peter Pretty (clt), and John Murray. The Jazz-makers lasted until 1968, when Leake formed the first of his Storyville groups, the Jazzmen, playing its first engagement in October, and operating in close conjunction with the Victorian Jazz Club. The band personnel stabilized in early 1969 with Leake, Tony Newstead, John Murray, Denis Ball, Kenn Jones, Dave Campbell, John Brown, Fred Stephenson (bs). In 1970 Leake founded the Story-ville Club, which ran for more than 12 years. In the same year, Dick Tattam (who had played initially) rejoined, Ian Harrowfield replaced Ball and Joanie Watts (vcl) worked with the band. When Ian Walkear replaced Harrowfield in 1972, the personnel remain stable for the next three years, and recorded in 1974 with Penny Eames and Dutch Tilders, the latter a major figure in the blues field in Melbourne. In 1975 the name was changed to the Storyville All Stars, and the personnel altered to Newstead, John Adams, Ross Anderson (bs), with Leake, Murray and Walkear. Bob Whetstone, then Derek Reynolds, succeeded Newstead in

1976, when Ian Harrowfield also augmented the front line for several months. Tony Orr (bjo/gtr) was added in 1977, Beverley Sheehan (vcl) in 1978, and Kenn Jones in 1978. Changes in 1982 left the band as: Michael Poore (tbn.), Walkear, Adams, Leake, Ian Orr, Graham Coyle, Graham Taylor (bs), Sheehan, and Maurie Dann (bjo). Dann died in 1984.

The Storyville groups attempted to bridge a gap between traditional and mainstream which had existed in Melbourne since the '50s. They also addressed the question of 'packaging' as well as music. For all these reasons they incurred some suspicion among purists, but have been significant in broadening the traditional movement from within.

With his wife Jean, Leake has also had considerable impact as a jazz entrepreneur. He was instrumental in the establishment of the Jazznote record label, and organized the annual Jazz Party for five years, beginning in 1976—a weekend mini-festival bringing together paying and invited guests from all over Australia. In 1979 he became full-time musician and founded Perdido Musical Enterprises, he reopened the Storyville Club in 1983 and 1985 was producing jazz shows through his Melbourne Jazz Repertory Company. In 1986 he was still leading a Storyville band, as well as Swing Shift with Beverley Sheehan, formed in 1982.

The Storyville All-Stars, 1979, with, l. to r., front row: *Kenn Jones, Ian Walkear, Beverley Sheehan, Derek Reynolds, John Murray,* back row: *John Adams, Ross Anderson, Allan Leake, Tony Orr*

Le Doeuff, Maurice Alan (Maurie)   b. 27/8/18,
Adelaide, SA   reeds/lrd/arr
His interest in music kindled by movie musicals
and record recitals at the Semaphore Kiosk, he met
like-minded youths while taking tap-dancing lessons, and they formed the Six Rhythm Masters in
1938, playing dances. Sitting in at the Bondezvous
with Frank Cleves's band led to work there following the outbreak of war. Coached in sound
production by Alf Holyoak, his growing reputation led to casual work and in 1940 he was with
Mark Ollington's Swing Club Band. Le Doeuff
was also active in the Adelaide Jazz Lovers' Society, of which he became founder treasurer in 1941,
when he also joined the Glenelg Palais band and
worked under Jack Barter at the Astoria which,
with the Bondezvous, was favoured by US servicemen for its swing orientation. Inspired by the
dynamics of the American Red Arrows, the 32nd
Division Special Services band, Le Doeuff and his
colleagues spent much instructive time in their
company. Called up in 1942, he served in entertainment units with, *inter alia*, Charlie Munro and
Lou Campara. Following his discharge, he rejoined Harry Boake-Smith's Palladium Band and
later formed jazz groups for broadcast and concert
performance. With a brief interval in Melbourne,
when he played with many of the bop pioneers, Le
Doeuff spent 1948–49 in Sydney working with

*Maurie Le Doeuff (alt.) leading the Palais Royal Dance
Band, Adelaide, c. mid-1950s*

Ralph Mallen and with Wally Norman at the
Roosevelt. Returning to Adelaide he took over the
Palais Royale orchestra from Colin Bergerson,
until 1957. 1959–70 was chiefly involved with the
Channel 9 orchestras and worked at Spurden's
Music Shop, but also formed jazz groups for radio
and backed visiting performers like Mel Tormé and
took the occasional gig such as at the Toll Gate in
1964. From 1970 he moved into education, finally
becoming a lecturer at the Adelaide College of
Advanced Education. Retired in 1983. Although
primarily a dance band and studio musician, Le
Doeuff created opportunities to play modern styles
of jazz in a city which has rarely been able to
sustain them. The broadcasting groups which he
directed in the '60s with, *inter alia*, Gary Haines,
Bill Munro, Alan Slater, John Bermingham, Don
Knispel, Ron Ackfield, produced some of the most
convincing West Coast style jazz ever played in
Adelaide.

Lee, Alan Whiteley   b. 29/7/36, Melbourne,
Vic.   vibes/hand perc./ldr
Took piano lessons as a child, switched to ukelele,
then guitar. Attracted to vibes by Lionel Hampton
on the radio in 1950, he had to wait until 1956 for
an available second-hand set. Formed a band in
National Service, with himself on p., vibes, gtr.
Following his discharge he took his first gig at the
Hobohemian Club in Middle Park LSC rooms

Maurie Le Doeuff, c. 1957

Alan Lee, 1977

with John Allen, Len Barnard, Brian Brown, Frank May and Ross Fusedale. At Jazz Centre 44 he played briefly with Graham Morgan's Jazz Disciples in 1957, then with a MJQ inspired group (Barnard, May, Allen) alternating with Brian Brown on Sundays. 1959–63 he worked with various groups incl. those of Frank Gow and Kenn Jones, and led his own quartet generally with Ted Vining, Derek Capewell, and Graham Coyle, succeeded by Grant Jones. The quartet played at Jazz Centre 44 and concerts, such as for the CAS at Argus Gallery, Nov. 1960, and the 25th anniversary of the Melbourne University Rhythm Club in June 1962. They aroused controversy when, having undertaken to present a Jelly Roll Morton set at the 1960 AJC, they instead played a more modern bracket. Lee moved to Sydney in 1963 where he played in various stylistic contexts: El Rocco, Mocambo, Windsor Castle (John McCarthy, Dick Hughes, Bryan Kelly), French's Tavern (David Martin, Phil Treloar, Jeannie Lewis), and was involved with Dick Lowe in an attempt to establish a jazz coffee lounge, Café des Artists, at Cronulla. Back in Melbourne to take a job in Thomas's record store, he formed a band of flexible instrumentation and personnel which, after his departure to play for *Hair* in 1970, became the group called Plant. In the early '70s he teamed with Tony Gould, and became involved with Sunday concerts at the Arts Centre Gallery, presenting

both jazz and classical with Ron Anderson, Murray Wall (bs), Ted Vining, Ron Sandilands (dms), et al. He developed a large percussion group inspired by Brazilian music, which performed, sometimes unannounced, at the gallery, and also interstate. In the late '70s Lee suffered health problems and, following a brief period in Sydney, it was not until 1982 that he resumed regular work. In the '80s as a duo with Jex Saarehlaht he has been represented on Anteater cassette. Lee has also been important as a journalist, notably for *Music Maker*, in which he published interviews with jazz musicians. Throughout his career he has been a charged point of contact between disparate styles, a flexible catalyst of unpredictable musical events, a true improviser in its largest sense.

Lee, Julian   b. 11/11/23, NZ   p./tpt/reeds/arr/comp.

Began on piano at six. Later moved to Auckland to study at the NZ Foundation for the Blind. Returned Dunedin and played for radio and in dance band work. Back in Auckland he became bandmaster at the Foundation for the Blind, and studied music theory at the university. In 1947 became leader (on a rotation system) of Auckland Radio Dance Band, and began freelance arranging. In 1952 joined Derek Heine's band at Radio Theatre and began leading his own bop group for concert work. Returned Dunedin, 1953, led local radio band and opened his own recording studio. Married singer Elaine Richardson, who became his musical amanuensis. To Sydney, October 1956, joining John Cerchi's group working for Wally Norman's Rex Hotel chain. MD at the Ling Nam, then following a break, resumed at the Ling Nam with Dave Rutledge as leader. Also with Jim Gussey's radio band as arranger/pianist, and recorded under his own name. Became arranger for Tommy Tycho, 1960, when he also met George Shearing, who suggested a visit to the US. Lee moved to America, 1963, and worked as arranger for Shearing, Gerry Mulligan, Chet Baker, James Moody, Stan Kenton, Frank Sinatra, Peggy Lee. Played at Shelly Manne's, and worked as a producer for Capitol records. Returned Sydney, 1968; played El

Rocco with Don Burrows, freelanced in session work as performer and arr. To US, 1972, then New Zealand, 1974. Back in Sydney, 1978, since when he has continued to work extensively in studio settings, incl. musical co-ordinator for Channel 10 and as arranger for the ABC. Since 1985, he has been a resident pianist at the Don Burrows Supper Club in the Regent Hotel, leading a trio with Craig Scott and David Jones. Lee continues to work as a sound engineer, record producer, and arranger. His musical credentials are of the highest professional order, an extraordinary tribute to the highly developed ear and memory of a musician who has been blind from birth.

Charlie Lees, c. 1937

Lees, Charles Anton (Charlie)   b. 1913, Sydney, NSW   gtr/bjo/vibes/dms/arr

Lees spent his childhood travelling the world with his parents, who settled with him in NZ when he was 15. Left school in 1930 for professional music. In 1935 arrived as a stowaway in Sydney, and joined Frank Coughlan, playing the opening of the Trocadero, 3 April 1936, on the first Gibson Super 400 imported into Australia. Joined Stan Bourne at the Ginger Jar and in 1937 played the first electric 'Spanish' guitar, a Rickenbacker, in Australia. In 1939 with Craig Crawford's band, with Bob Lyon then Dick Freeman at the Trocadero, later joined Bert Howell at the Prince Edward Theatre. Imported Australia's first vibraharp, and organized

jam sessions under the name Swing Socials. With Roy Royston and Max Daley he started a music school which merged with Nicholson's in 1941, and in the meantime was organizing the music for the Cinesound film company. He regularly won the annual *Tempo* award for guitar. In 1942 he entered the AIF, serving in a commando unit, followed by a period involving rumour-shrouded activities at sea. In 1945 he settled in Qld, playing in Gold Cast cabarets, incl. the Currumbin Playroom and at Southport, into the '60s. Apart from a brief episode in Sydney playing banjo duets with Ron Carson at the Texas Tavern, he remained in Qld, settling in Townsville. Stopped full-time playing in the '70s but continued to teach until disappearing at sea while sailing his trawler to Cooktown in March 1981. Lees was virtually the father of Australian jazz guitar; Coughlan described him as 'the most outstanding guitar stylist in the antipodes' (a personal reference dated 18 August 1936) and fellow guitarist Bruce Clarke wrote that Lees was 'among the three or four world class guitarists that Australia has produced' (*Journal of Australian Music and Musicians*).

Levis, Neil Frederick   b. 17/11/35, Hobart, Tas.   p.

His first jazz work was with a traditional group, the Wildcats, in the mid-'50s although he was drawn to George Shearing. Subsequently he worked in Peter Webster's quintet, but throughout the '60s and '70s conditions in Hobart provided little opportunity to perform in more progressive styles, with the main exception of ABC work organized by Jack Smith in a trio that incl. Alan Park (dms). The general pattern during this period is representative of the activities of other post-traditional (swing) musicians in Tasmania, like Terry Root (bs), Neil Heather (gtr), and Brian Martin: cabaret work with jazz smuggled in in a form unobjectionable to management and patrons. Most of Levis's improvisational urge has expressed itself in pop groups from the late '70s, but since the advent of the JAS and the Contemporary Jazz Society there have also been occasional opportunities for explicit jazz work. In 1982–83 he led a trio (Steve Martin bs, Phil Stubbes dms) which played a residency at the West Side Motor Inn and performed for the JAS in Launceston.

Levy, David Michael (Dave)   b. 13/4/36, Hobart, Tas.   p./comp./arr/ldr/and occasional vibes/ marimbah

Family moved to Sydney in 1938; Levy began classical studies in 1947 and developed a jazz interest in the early '50s, leading to conflict with his classical teacher. Active in Sydney jazz from mid '50s, at the Mocambo and El Rocco. Studied with Igor Hmelnitzky in 1961, and spent 1964–67 in Europe, and England where he toured in 1965 with Jimmy Witherspoon. Back in Sydney, began teaching in 1969 and later became involved with other jazz education programmes: lecturing to secondary school teachers from 1978; jazz workshop director for the JAS of NSW from 1980; joined the jazz studies faculty of the NSW Conservatorium of Music in 1982. 1970–76, played with Ray Price's school concert/lecture programme. Has worked in commercial areas, touring with the Four Kinsmen (1981) and MD for club performer Daryl Somers (1979–82). Has played with visiting American jazz musicians incl. members of bands ranging from the MJQ to Blood, Sweat, and Tears. There is hardly a Sydney group with whom he has not performed, from traditional-oriented (Mike Hallam, Bob Barnard, Dick Hughes, Harbour City), to bands representing the leading edge of current modernism. Founding pianist with The Last Straw until retiring temporarily from music in 1975, and in the early '80s with Ken James's Reunion; also led groups of his own such as Phingus (modern) and Hot Phingus (traditional). Levy's compositions incl. work performed by the Australian Ballet, and, collaboratively, music for the film of David Williamson's *The Removalists* during the 1972–75 period he spent with Galapagos Duck. Some of his compositions have the rare Australian distinction of having become part of the repertoire of other bands. With his catholic receptivity to influences, including many Australian musicians, Graeme Bell, Dick Hughes, to Frank Smith, Mike Nock, Levy's approach is as amphibious as it is intense. He demonstrates the exceptional gift of being able to adapt to any idiom, yet remains fully creative and distinctive. He is a major representative of the generation of modernists who came to the fore in the late '50s.

Limb, Robert (Bobby)   b. 10/10/24, Adelaide, SA   ten./ldr

Following fife and banjo-mandolin, Limb began sax at about nine, with a six year scholarship at the Adelaide College of Music, taught by Les Mitchell. In 1942 he made his jazz debut at a jam session for the Adelaide Jazz Lovers' society. Joined Maurie Le Doeuff's band with Bob 'Beetles' Young, and they broadcast for the ABC. After just over four years with Harry Boake-Smith's Palladium band, Limb and Young moved to Melbourne in February 1947 to join Bob Gibson. With other progressive Adelaide and local musicians, they became involved in the brief but energetic bop movement in Melbourne in the late '40s. Limb formed his own septet with Don Banks, Charlie Blott, Joe Washington (gtr/arr), John Foster (bs) Ken Brentnall, Jack Brokensha, and played New Theatre concerts for the Modern Music Society of Victoria. When Gibson left for London in 1948, Limb formed a band for Admiralty House, refurbished as Ciro's night-club, and remained resident there until 1950 when, with Brentnall and John Bamford, he went to Sydney to play at Sammy Lee's. In 1951 he took a band into the Colony Club, at the same time working on the jazz concert scene and moving into radio and other commercial areas. In the UK 1953–57, he consolidated a career in comedy and since returning to Australia, worked in TV variety, then concentrated on the production, promotion, and administrative areas of entertainment.

Lockyer, Brian Claude (Brett)   b. 25/10/34, Perth, WA   reeds

Started clarinet in 1951 and used an 18-month convalescence after a motor cycle accident for intensive practice. Joined Don Thomson's Riverside Jazz Group, 1954, coming with the band to Sydney, 1962, where he settled, playing with the group

Bobby Limb at Ciro's, Melbourne, 1959: l. to r., Fred O'Kane (dms), Bob Limb (reeds), Bob 'Beetles' Young (p.),
Tommy Trinder, Ken Brentnall (tpt), Joan Clarke (vcl), Neville Maddison (reeds), John Foster (bs), John Bamford
(tbn.)

until it disbanded under the name King Fisher All Stars. With Ray Price Qintet [*sic*], 1964–67, and retired from music, 1969–77. Founder member of Mike Hallam's Hot Six, 1977, and still with the band in 1986.

Longman, Ken   b. 21/7/31, London, England   tbn.

To Australia, 1948. Jimmy Bell (dms) of Riverside JB encouraged him to take up an instrument, and Longman practised privately with other beginners Ian Cuthbertson (cnt), Peter Neubauer (clt), Jim Roach (p.), Willie Gilder (bjo), Arthur Croydon (dms). Joined Ross St Ramblers with Eric Curry (tpt), Peter Towson (p.), Doug Lampard (bjo) and sometimes Neubauer. Formed Sophia St Stompers, 1956, with Mike Ward (cnt), Eric Richards (tba), Bob Paul (bjo), Graham Spedding. With Spedding he joined the Southern Cross JB formed by Terry

McCardell (tpt) and Mick Fowler (dms), with Alan Lynch (p./p.accordion) at the request of the communist Eureka Youth League to participate in the sixth World Youth Festival in Moscow. The band played in China and Russia in the course of this venture, in concerts, on television, and for recording, from 24 June to 4 Sept. 1957. The first Australian jazz band to perform in these countries, it was received with great enthusiasm. An account of this tour written by Terry McCardell for *Quarterly Rag* was reprinted as a booklet, *Jazz Speaks All Languages* by the AJC Trust Fund, 1985. McCardell went on to concentrate on a variety act, though played occasional jazz work. Lynch continued in jazz and dance music and at the time of his death on his 42nd birthday in 1971 was with Merv Acheson. Mick Fowler continued working in jazz on drums/banjo as often as his occupation as a sailor permitted. A strong proponent of radical causes, he entered Sydney mythology as one of the most determined opponents of the commercial re-development of the inner-city residential areas. He died in 1979. Longman became a founder member of the Black Opal JB (1958), incl. Eric Richards, Ken Le Lievre (bjo), and Jim Young or Helen O'Sullivan (p.) This became the second house band of the Sydney Jazz Club. In 1964 Longman formed the Harbour City Jazzmakers with Pat Qua, Rupert Buckland (bjo), Brian Hutchinson (dms), Dennis Noone (tpt), Rex Gazey (tba). With changing personnel incl. Noel Foy (bjo), Len Smith (dms) this band held a long residency at the Brooklyn Hotel. The group broke up, then re-formed with Mike Hallam, Dennis Stevens (clt), Geoff Gilbert, Barry Hallam (dms), Rex Gazey/Terry Fowler (tba) and Longman, resuming at the Brooklyn in the early '70s. Longman also worked during the '70s with John Hubbard's (tpt) band, with Ian McCann (tpt). In 1986 he was with the Abbey JB which he joined in 1974, and which has been his primary jazz setting throughout that period.

Loughhead, Ronald Arnold (Ron)   b. 7/11/29, Gawler, SA   p./arr/comp.
Took lessons on piano from mother as a child, but did not study seriously until after moving to Adelaide in 1941. At 16 formed a 10-piece band incl. Errol Buddle, John Bamford, John Foster, Laurie Parr, for the Amateur Hour radio competition. Played the Cairo Club (1946). While holidaying in Sydney, 1947, wrote a bop composition for Ralph Mallen's band; while returning through Melbourne accepted a job with Ken Lester at Sammy Lee's Storklub. Joined Jack Brokensha, 1948, playing at the Galleone, and composing and arranging both for Brokensha and for Freddy Thomas's concert bands. While in Sydney with Brokensha also wrote for Bill McColl concerts. Left Brokensha in 1950 to study in Melbourne, where he played at Claridge's. While at Ciro's, 1952, with 'Splinter' Reeves and leading his own band, Loughhead won a CBS composer's competition with 'The Thrill of Your Kiss'. Continued activity in night-club settings was followed in 1955

by an Australian tour with English singer David Hughes. Also writing for Winifred Atwell. Moved to England, 1955, to work with Hughes, and began an illustrious career in studio settings: staff arranger for Decca records, and for BBC TV incl. *Make Mine Music* with Wally Stott's orchestra. Played the Dorchester Hotel, at Quaglino's with fellow expatriate Don Harper, and London Palladium. Also played restaurant work in Sweden. Returned Sydney, 1962, and became active as a studio arranger for Channel 9 (incl. Graham Kennedy shows) and the ABC with Eric Cook. Since 1976 played solo restaurant work in Adelaide, Melbourne, Sydney. In 1986 resident in Sydney, working with Judy Robertson (vcl) with whom he has recently recorded.

Loughnan, Colin John (Col)   b. 26/10/42, Sydney, NSW   reeds/fl./arr
Studied with Graeme Lyall, Victor McMahon, Neville Thomas, and later during visits to the US, with Joe Allard, Victor Morosco, Eddie Daniels, George Trembley. Began playing professionally, 1967. Joined Col Nolan's Soul Syndicate (1968–69), Daly-Wilson Big Band (1969–72) and staff arranger on Channel 9, 1970–72. Toured Canada with Sandy Scott and studied in US, 1972; based in London playing session work and with English group Kala, 1972–73. In Australia joined rock group Ayers Rock, 1974. Toured, then resident, in US until returning to Australia, 1977. Resumed session work, incl. MD for Marcia Hines and for Johnny Farnham-Debbie Byrne ABC-TV series. Toured Asia with Judy Bailey, 1978; in 1979 formed quintet Endeavour, incl. Melbourne guitarist Steve Murphy (b. 21/8/54) who had moved to Sydney in 1975 and studied at the NSW Conservatorium under Roger Frampton. With Judy Bailey and John Hoffman's Big Band, 1984. Loughnan has continued in session and concert work, with Sammy Davis Jr, Johnny Mathis, Debbie Reynolds, Peter Allen. Also with visiting jazz performers incl. Toshiko Akiyoshi-Lew Tabackin, Don Rader, Joachim Kuhn. As arr/MD has won awards incl. best arrangement Award, Yamaha Song Contest, Tokyo, 1983. MD for Aus-

tralian entries in Pacific Song Contest, 1979 and 1980. Since 1984 has taught in jazz studies course at NSW Conservatorium and in workshops. In 1986 was playing with the big band Supermarket.

Loughnan, James Anthony (Jim)   b. 30/12/28, Melbourne, Vic.   reeds
Became interested in jazz at age 17 through records of Benny Goodman and Artie Shaw, taking lessons from dance band musician Bob Storey. He began attending Frank Johnson's Collingwood TH dances where he was inspired by the work of Geoff Kitchen. In 1949 joined Abby Ross's Dixielanders, which incl. Ron Williamson. In the early '50s he played with Smacka Fitzgibbon's band then joined Nevill Sherburn in 1953. Becoming interested in Sidney Bechet; took up soprano sax (1956). Following a period of commercial work, he joined the Southern City JB in 1961. Most of his subsequent career has been on a casual and freelance basis. Loughnan has regularly appeared as both sideman and leader at Melbourne AJCs as well as in numerous short-lived pub residencies, and in 1985 his work continued to follow that pattern. He was the first musician in the traditional to mainstream area in Melbourne to become associated with the soprano sax.

Ludowyk, Christopher John (Chris)   b. 6/3/44, Sri Lanka   tbn./bs/ldr/arr
By 1959, playing private sessions in a trio. The family moved to Geelong, Vic., 1962, where Ludowyk played with the Crescent City JB, 1964–65, the Green Horse JB, 1965–66, then the Baton Rouge JB until moving to Melbourne in 1968. He had also begun playing with what became the New Whispering Gold Orchestra at the Geelong Jazz Festival and the AJC. In Melbourne Ludowyk joined the New Harlem JB, becoming leader in 1979. In 1980 he formed the Prahran Market JB with Ian Smith, Dave Hetherington, Tony Orr, Pip Avent, Cal Duffy. To England, Dec. 1982, working with Tony Ashwell and Cuff Billett. To Melbourne, 1984, and formed the Society Syncopators to play at the Abbey Jazz Cellar. Following the Cellar's demolition in Dec. 1984 the band started at the Emerald Hotel, South Melbourne. The band has played the WA Jazz Festival, concerts in Tasmania and Canberra, and appeared in the film *Death of a Soldier*. In 1986 the band's line-up was Peter Gaudion, Dick Miller, Tony Orr, Pip Avent, Cal Duffy, with Ludowyk, tbn. and ldr.

Lyall, Graeme William   b. 25/1/42, Melbourne, Vic.   reeds/fl./comp./arr/ldr
Began as a brass player at age 10, switching to reeds in 1958, studying with Frank Smith. To Sydney, 1961, and studied at the conservatorium under Raymond Hanson, Victor McMahon, Donald Westlake. Joined the TCN–9 Orchestra (1963–71) and while in Sydney led his own group which played at El Rocco, was a founder member of the Daly-Wilson Big Band during its initial rehearsal period, and travelled to Japan in 1970 as a member of Don Burrows's Septet. Returned to Melbourne, 1971, and joined a music production company. With the ABC Show Band (1974–77), and appointed MD for GTV–9 in 1977. Left GTV–9 in 1980 to work in educational music projects. Returned to GTV–9, 1981, and formed the first Balwyn Youth Concert Band, with a second in 1983. MD for the Royal Gala Concert 1983, and for the opening of the Sydney Entertainment Centre in the same year. In 1984, MD, conductor, and arranger for Olivia Newton-John's Australian segment in the Olympic Gala Concert emanating from ABC television in the US. In 1984, also teaching in the jazz studies programme, Victorian College of the Arts. Retired, May 1986, and moved to the Benalla district (Vic.) to work with young children in school and community programmes. Although Lyall has been primarily a studio musician for most of his career, his prodigious talent established for him a jazz reputation which is scarcely equalled, and which is also a tribute to his mentor Frank Smith. His work can be heard to effect on John Sangster's *Lord of the Rings* LPs.

Macbeth, Neil (Chauncey)   b. 3/9/37, Sydney,
NSW   dms/wbd
Began drums at Sydney Univ., 1957–58 and joined
the original Black Opal JB mid 1958, remaining
during the band's Sydney Jazz Club residency. To
Melbourne, 1960, joining Frank Traynor's bands
the Preachers and the Melbourne Jazz Club house
band. Macbeth remained with Traynor until returning to Sydney, 1964, where he freelanced on
the traditional scene—Geoff Bull, Nick Boston,
Black Opal JB—and spent two years in the early
'70s at the Stage Door with Alan Murray (tpt). Has
continued to freelance on an occasional basis while
pursuing his career as a clinical psychologist.

Mallen, Ralph   b. 27/1/26, d. 14/12/56, Sydney,
NSW   tbn./ldr
While in the services during the war Mallen also
played suburban dances. Upon discharge in 1946,
took lessons at the NSW Conservatorium under
Harry Larsen. In 1946 Mallen joined Ike Holbor-
ough's band, later taking over as leader and, with
advice from Wally Norman and extensive changes
in personnel, built it into the first of his big bands.
In 1947 Mallen's band was playing concerts for Bill
McColl at Macquarie Auditorium, at Bondi Esplanade and on Sundays at Gaiety ballroom. In addition to others as indexed in this volume, the band
incl. Joe Singer (dms), Keith Cerchi (reeds), Ron
Hogan (bs/gtr), Dick McNally (tpt), and later
Norm Wyatt (tbn), Marsh Goodwin (tpt), Ray
Horsnell (reeds). These, with their colleagues,
would become important sidemen under Mallen
and others during the jazz concert era. Mallen became less visible through 1949, dropping out of
concerts and the Gaiety though still playing at the
Esplanade. Following comparative inactivity he
led a band at Newtown Ballroom and for engagements in Wollongong in 1953, and in March returned to the jazz concert stage to much acclaim,
both as leader and with a group led by Norman.
He was again resident at the Gaiety with a big
band incl. Frank Smith, John Edgecombe, Noel
Gilmour, Alan Nash, in 1954 when the building
was destroyed by fire, taking with it the band
library and much equipment. Although Mallen
played briefly at the Sheridan (formerly Sammy
Lee's) in mid-1954, this blow led him virtually to
retire from music. He established a trucking company, in which enterprise he was killed jumping
from a runaway lorry.

Mallen's band was the major pioneer in Sydney
of the Stan Kenton/Woody Herman inspired progressive style. It also provided an apprenticeship for
a post-war generation of young musicians—Terry
Wilkinson was only 15 when he joined as a founder member—and established jazz reputations for
musicians who would otherwise have remained in
the relative anonymity of night-clubs. Mallen's
group was the standard-bearer and -setter of the
earliest jazz concert big bands.

Malpas, John Ray   b. 28/3/21, Adelaide,
SA   bjo/gtr
His first significant jazz activity was with the
Southern Jazz Group, 1945–50. With the Bruce
Gray All Stars, for the first half of the '50s. Fol-
lowing casual work he joined the Campus Six
through the '60s and into the early '70s, when he
also resumed his association with Dave Dallwitz
through the latter's Euphonic Sounds and the
Jubilee Stompers, the latter formed in 1985. Since
the mid-'70s he has worked regularly in two bands
led by Ron Flack: the Jazz Quartet and the Unity
JB. In 1984 Malpas played and recorded with Ade
Monsbourgh's Adelaide Connection.

Ralph Mallen's band at the Parisienne, Campsie, probably late 1946; l. to r., Terry Wilkinson (p.), Ralph Mallen (tbn.), Billy Mannix (ten.), Norm Wyatt or Les Nelson (tbn.), Ron Hogan (gtr), Andy Mackintosh (clt), Don Osborn (dms), Dick McNally (tpt), Ron Gowans (ten., seated), John Cerchi (ten., standing), Dick Tuckey (bs), Bob Anderson (alt); the partially obscured musician between Gowans and Anderson is possibly Ron Mannix

Martin, David   b. 30/7/39, Melbourne, Vic. p./ldr/arr

Received classical tuition in childhood then jazz study with Ron Rosenberg. With Brian Brown Quintet at Jazz Centre 44 for three years, recording with the band in 1958. In England, 1961–65. Settled in Sydney c. 1966 and worked with many groups on semi-permanent and freelance basis incl. Energy and Friends, Joyce Hurley (vcl), with the Australian Ballet for *Caravan*. Worked at The Basement with his own and other groups and played concerts interstate. In the late '70s he assembled a big band which he led for three years, playing the Festival of Sydney, 1978 and 1979. Also formed a quintet, 1978, which broadcast for 2MBS and played for JAS, augmented by Norma Stoneman vcl (later married to Martin). In the US, early 1979, then returned to Sydney where his quintet played

support for Count Basie. Formed a new quintet, 1980 (Alan Turnbull, Lloyd Swanton, Dale Barlow, James Morrison), playing the Festival of Sydney, The Basement, Paradise Jazz Cellar, Qantas International Jazz Festival, Manly Jazz Carnival, and support for Jimmy Witherspoon. In 1982, assembled a band with singers Sandie White, Trude Aspeling, Doug Williams, for a concert tribute to Duke Ellington. Since 1984 has been less visible as a performer, but has taught and acted as MD for the museum concerts and for a period at the Sydney Hilton.

Martin, Paul Charles   b. 17/8/37, London, England   reeds

Began clarinet, 1956, attending music lessons with brother Mick (tpt). Moved to Sydney, 1957, and played with other expatriates Ken Longman, Eric

Richards, sitting in sessions at Foresters' Hall, Croydon. Founder member, Black Opal JB, 1958. To Melbourne, 1961 freelancing with Sny Chambers, Cliff Tierney (tpt), Max Collie, Allan Leake, John Adams. Took over Max Collie's band, 1962, changing the name to Paul Martin's Jazzmakers. Joined the Hot Sands JB, mid '60s, then the Yarra Yarra JB, 1966, when he also took up alto. Remained in Melbourne after the Yarras' departure for England, 1969, and played in a rock group. Joined the Red Onions, 1972, then Frank Traynor, and rejoined the Yarras upon their return, 1975, remaining to 1981. Moved to Sydney and joined Roger Janes at Unity Hall and Alan Geddes at Bondi Icebergers. Toured with Frank Traynor, 1984. Recorded under his own name, 1985. With Janes and Geddes until moving to Perth, 1986.

Martin, Raymond John (Ray)   b. 12/8/42, Colac, Vic.   bs

Born of musical parents. Moved to Melbourne, 1960, where he was much impressed by Oscar Peterson at The Embers. In Port Moresby, New Guinea, c. 1961–67, where he took up bass and worked with other Australians Allan Murray (tpt) and Cec Regan (tbn.). In London, 1967–68, playing mainly pub work. Returned Melbourne, 1969, and worked in the modern scene with, *inter alia*, Brian Brown, Tony Gould, Alan Lee, Ted Vining, Bob Sedergreen. To Sydney, November 1986. Joined Harbour City JB, 1977, but most of his work has been with more modern musicians incl. Bernie McGann, David Martin, Bob Gebert, Vince Genova, Alan Turnbull, John Pochée, Barry Duggan, incl. concerts, broadcasts, recording, and in regular jazz venues incl. Soup Plus. Martin is a ubiquitous freelancer, and has also presented jazz programmes on 2MBS–FM.

Matthews, Helen   b. 22/11/43, Guildford, WA   vcl

Started her jazz career with J.T. and the Jazzmen, leading to work with ABC TV. Left J.T., 1972, and worked in various settings incl. jazz, such as with Judy Bailey at the Festival of Perth in 1974. Subsequently joined Barry Bruce's Chicago JB and through the early '80s worked with A Slight Diversion—Ray Walker, Brian Bursey, Gary Ridge (dms)—at the Highway Hotel, Claremont. Has guested with Don Burrows, George Golla, Stephane Grappelli, with the WA Youth Jazz Orchestra, the band of the jazz studies faculty at Mt Lawley College, and the WA Symphony Orchestra. The brief sample of her work on a recording with Uwe Stengel's Manteca indicates why she is regarded, with June Smith, as one of Perth's leading jazz-based vocalists.

McCarthy, John Grant   b. 6/1/30, Sydney, NSW   reeds

Began clarinet at age 15. Joined Riverside JB, 1947, with which he played town hall concerts. Very briefly with Frank Johnson's Fabulous Dixielanders, 1950, then returned to Riverside band until c. 1953, having replaced Dick Jackson in the Port Jackson JB in 1952 on an intermittent basis. Joined PJJB full time in 1955, and also a member of the Ray Price Quartet from 1958. Left both bands, 1962, and joined Dick Hughes's quartet, then replaced Laurie Gooding in Graeme Bell's band in the mid-'60s. Played with, then led, the band at Paddington-Woollahra RSL for a very long period; although this was not a jazz gig, many of Sydney's leading jazz musicians passed through the band.

John McCarthy, the Seymour Centre, Sydney, for the Jazz Action Society, 1976

The RSL residency finished in 1978, by which time McCarthy had joined Bob Barnard's JB at its foundation in 1974. In 1978 he led his own group at Rocks Push. Replaced Merv Acheson in Dick Hughes's Famous Five at Soup Plus in 1985. In 1986 was still with Hughes and Barnard, and working extensively on a freelance basis. McCarthy has also played session work and recorded extensively, including on John Sangster's *Lord of the Rings* project, where his spare but intense clarinet style is heard to great effect, notably on the track 'Vale Theoden'. McCarthy is in continuous demand as a sideman, and for many years worked in Nancy Stuart's groups. His career, spanning virtually the whole post-war era in Sydney jazz, has established him as one of the most versatile yet distinctive, and admired, musicians in the traditional to mainstream area.

Bernie McGann, c. 1986

McConechy, Howard Norman (Joe)
b. 22/8/35, Melbourne, Vic.   bs/tba
Began on trombone in 1952, switched to bass in 1953. Joined William Flynn's dance orchestra in 1956 and toured with Graeme Bell (1957) followed by TV studio work. Worked with Frank Johnson in England (1962) and with Teddy Preston's quintet on the ABC in the '60s. With Frank Traynor's Jazz Preachers, 1974–79 and joined Allan Leake's Storyville group in 1982. McConechy is a professional musician with considerable studio experience and has backed many visiting musicians incl. the Mills Brothers, Dill Jones (p.) and Teddy Wilson, as well as being on the famous *Naked Dance* album with Len Barnard's band.

McGann, Bernard Francis (Bernie)   b. 22/6/37, Sydney, NSW   alto sax/comp.
Exposed from an early age to jazz through his father, a drummer, for whom McGann occasionally deputized. Took up alto in 1955, with lessons from Kevin O'Connell, followed by harmony instruction from pianist John Newton. By the late '50s he was sitting in with Dave Levy and John Pochée at the Mocambo in Newtown. In May 1962 he played a two-week season with the American Ballet in Melbourne, with Barry McKimm (tpt) and recent arrivals from New Zealand, Barry Woods (dms) and Dave MacRae (p.) and English bassist Mike Ross. Returned to Melbourne in 1964 for a residency at the Fat Black Pussycat with a band called the Heads (Pochée, MacRae, Andy Brown bs) for five nights a week. Returning to Sydney in 1965 he worked casually in rock-oriented bands and played jazz at the El Rocco. Following its closure in 1969, McGann found little opportunity for work and finally took employment as a postman. Although he has never enjoyed the same level of musical employment of many lesser, though more entrepreneurial musicians, the upsurge in modern jazz interest through the late '70s provided new opportunities. He played some concerts at the Kirk Gallery with Kindred Spirit (Chuck Yates, Phil Treloar, Ron Philpott, Ned Sutherland gtr). From 1975–78 he played with The Last Straw at The Basement and, in 1977, at the adjacent Pinball Wizz where he also worked with Wendy Saddington and his own trio. From 1979–80 he led a group at the Glebe restaurant, Morgan's Feedwell, and in Jan. 1980 played support for the Art Ensemble of Chicago. Through the '80s, while opportunities for long residencies continue to elude him, there has been a growing awareness of his importance. He played support for Lester

Bowie in 1981, Freddie Hubbard in 1982, and recorded with Sonny Stitt during the latter's 1981 tour. A reconstituted Last Straw played Jenny's, 1982–83, and, with Lloyd Swanton and Bob Bertles replacing Jack Thorncraft and Ken James, for SIMA in 1985. McGann also played a lengthy season in 1985 and 1986 at the Soup Plus. In 1983 he visited the US on a performance fellowship from the Music Board of the Australia Council; in 1984 recorded for Anteater with the Ted Vining Trio; in 1986 Dewey Redman performed with McGann and his trio, again for SIMA.

Initially influenced by Paul Desmond, McGann later drew upon the music of players like Ornette Coleman and Albert Ayler, arriving at a highly personal style which has led him to be regarded as one of the most significant and original players and composers in Australian jazz.

McGillivray, Malcolm (Mal)   b. 29/8/39, Melbourne, Vic.   dms/wbd
Began drums in 1960, working casually until invited to join the Melbourne NOJB in England. Returned to Melbourne in 1963, freelancing and spending 18 months with Frank Traynor. To Sydney in 1972 and joined Geoff Bull, then Nick Boston. A founder member of Tom Baker's San Francisco JB, with which he toured in the US and recorded. Following a layoff in 1982–83, has resumed playing on a freelance basis.

McIntyre, William Landale (Will, Willie The Lion)   b. 24/5/19, Benalla, Vic.   p./vcl
Began piano 1930; to Melbourne 1937 and formed a trio with Laurie Cowan (alt) and Wes Brown. 1933–41, increasingly involved in record and jam sessions. 1942–46, with AIF in Queensland, New Guinea, New Britain. In Brisbane 1943, played with Maurice Goode Orchestra at Dr Carver Club for black US servicemen. 1947–51 with Tony Newstead's South Side Gang. 1952–54, with the Portsea Trio. From 1955 played with various pick-up groups including the Baron's Footwarmers led by Lou Silbereisen, Jim Loughnan quartet and trio, Don Roberts's Trio, George Tack's Jazzmen. Reduced jazz activity since 1960, though still plays occasionally.

McNamara, Paul John   b. 1/2/45 Sydney, NSW   p./comp./ldr/arr
Following formal classical training from an early age, McNamara's interests turned towards jazz, with lessons from Chuck Yates a vital turning point. Working with Doug Foskett's band for three years at the Wentworth Hotel extended his association with the local modern scene. With Galapagos Duck, 1974–76, and also playing duo work with Phil Treloar and Alan Turnbull. With Bruce Cale, 1978, and played a short season with Dick Montz's big band in 1979. He joined Bob Bertles's Moontrane, with which he recorded some of his own compositions, in 1979, and in 1980 (and again in 1983) worked with Laurie Bennett at Soup Plus, also performing with John Hoffman's big band. In 1982, on an Australia Council grant, he studied in New York with Barry Harris, Hal Galper and Jim McNeeley. In general, he has had an extremely active career as a performer, frequently on a freelance basis, and also leading his own groups at major venues like The Basement, the Regent and Jenny's. Also involved in more commercial work, incl. a period as MD for singer Judy Stone in 1974. He has played opposite and with a host of eminent visiting Americans: Buddy Rich (1973), Ella Fitzgerald (1978), Dave Liebman (1980), Milt Jackson (1981), Woody Shaw (1981), Mark Murphy (1983), and has toured extensively with, *inter alia*, the Daly-Wilson band (1979), for

Will McIntyre

whom he had earlier arranged, Joe Henderson (1981), and Don Burrows, incl. the Festival of Asia Arts in Hong Kong (1983). McNamara is also prolific as a composer, much of his work having been recorded, incl. by himself. Very active in jazz education, he has produced a piano course on cassette and has written a booklet analysing a 12-tone concept of improvisation. Began teaching at the NSW Conservatorium in 1979, and also contributes to jazz clinics in Sydney and elsewhere in Australia. He delivered a paper on jazz education at the 1984 annual conference of the ASME. McNamara is a continuous and energetic presence on the Australian jazz scene; although increasingly occupied with administrative functions (he serves on the advisory board to the NSW state jazz coordinator), he still maintains a heavy and stylistically wide-ranging performing schedule.

Melbourne

As in Sydney, the magnitude of jazz activity in Melbourne requires its own book. Melbourne has been a crucial centre of Australian jazz, in particular its more traditional styles. The contribution made by Melbourne to attitudes regarding the music has been sketched in some of the major essays in this volume. More detailed aspects of the subject are incorporated in the relevant shorter entries on individual musicians and bands. The purpose of this essay is to deal with matters which fall somewhere between the general and the particular, and which have been distinctive to or unusually important in Melbourne.

The city experienced the first wave during the '20s in much the same way as, though at a lower level than, Sydney. As elsewhere, those characteristics of the history of its jazz that were peculiar to Melbourne began to emerge most clearly during subsequent activity. The onset of the swing movement found its focus in Melbourne in the 3AW Swing Club, inaugurated in August 1936 under the presidency of Colin Keon-Cohen. With programmes of record recitals and discussion groups, its membership mushroomed to more than 300 within a few weeks, and over 500 by the end of the year. It was clear that there was an enthusiastic public for some alternative to the rather bland dance band fare hitherto available. The club was soon able to present live music, in which Bob Tough and Benny Featherstone were linchpins, as they were also at the most important public venue for jazz or small group swing at that time, the Fawkner Park Kiosk on Sunday afternoons. In 1937 the beginnings of a parallel movement with a more finely articulated jazz philosophy were signalled in the foundation of the Melbourne University Rhythm Club, of which Ade Monsbourgh was a co-founder.

Swing, and other forms of jazz-based popular music, quickly filtered through the entertainment industry. Most dance bands absorbed the idiom to a greater or lesser extent, with the most important leaders during this early period being Jim Davidson and Frank Coughlan, the latter dividing his time over the years between Sydney and Melbourne. The late '30s also saw the growth in importance of cafés and coffee lounges as workshops for jazz, particularly in the St Kilda area where musicians including Featherstone, Billy Hyde and Stan Bourne introduced a strong swing inflexion into the commercial fare. Radio was also a major medium through which newcomers to the music were initiated, aided by the beginning during the war of significant reissue programmes of early jazz records. The arrival of the Mills Brothers in 1939 was greeted with near-hysteria, and in December 1940 Melbourne had its first 'band battle', a Battle of Swing bands at the Trocadero Palais involving Frank Coughlan's Trocadero Band, Mickey Walker's Radio Band, and Ern Tough's Fawkner Swing Band featuring Neville Maddison.

During the war, the different strands of jazz in Melbourne became increasingly identifiable, though musicians and the general public for the most part moved freely between them. The convergence of the Bell brothers, Pixie Roberts, and Ade Monsbourgh, laid the foundations of the traditional jazz revival, although the initial spirit was not exclusively revivalist. The establishment of *Jazz Notes* in 1941 and the release of the first Ampersand records in 1945 manifested and further reinforced a focus on jazz as distinct from other forms of modern music which, as early as this, was distinctive to Melbourne. In this, Bill Miller was

Bob Gibson's orchestra, Palm Grove, Melbourne, 1940; l. to r., saxes: *Vin McCarthy, Bob Storey, Lester Young, Bob Gibson (clt), Bernice Lynch, Mae Knuckey (vcls),* brass: *Harold Broadbent (tbn.), Bob Trembath, Fred Thomas, Billy Weston (tpts), Keith Cerche (dms), Alf Warne (p.), Gordon Peake (bs)*

the crucial figure. At the same time, swing continued to be the staple fare at the major dance band and cabaret venues, augmented during the war by the Palm Grove (1940) where Bob Gibson opened, and the Dugout, for allied servicemen. Mark Solomon led the band at the opening in May 1942, but shortly was replaced by drummer George Watson. The coffee lounges burgeoned with the increased demand for entertainment, and continued to provide venues for dixieland and small band swing, as well as incubating experiments which would lead to the brief but energetic bop movement in Melbourne. The Plaza, the Junction, the Galleone (a.k.a. Galleon), Saul's, the Swing In, the Manchester, all provided opportunities for musicians to play less diluted jazz programmes, from traditional to increasingly progressive.

In the post-war explosion of jazz activity the coffee lounges continued to play a major role, aided by night-clubs like Ciro's and the Storklub which frequently played host to after-hours jam sessions, where the core of progressive jazz stylists grew, with musicians incl. Jack Brokensha, Bob Limb, 'Splinter' Reeves, Orm Stewart, Doug Beck, Ken Brentnall, Don Harper, Ted Preston, and in the late '40s young newcomers Bruce Clarke and Ron Rosenberg. The focus for traditional jazz activity began at the Uptown Club with the Bell band, but spread under the leadership of Frank Johnson, Tony Newstead, and Ken Owen, with Ade Monsbourgh and Kelly Smith also leading groups, notably at the jazz concerts. In Melbourne some of the earliest of these were at the New Theatre, where the progressive fraternity presented its first function in 1947. Suburban town halls, like the Brunswick, as well as Melbourne TH, became concert venues, and the Downbeat series organized primarily by Bob Clemens was a continuous thread through the concert era. These functions brought virtually all the jazz strands

1946

SUNDAY, JULY 7th

8 p.m. at EUREKA CLUB
104 Queensberry St., Nth. Melb.

ONE NIGHT OF

HOT JAZZ

FEATURING MELBOURNE'S GREATEST GATHERING OF JAZZ MUSICIANS

* ★ Graeme Bell's Famous Dixielanders
* ★ Adrian Monsborough's Backroom Boys
* ★ Frankie Johnson's Hot Five
* ★ Don Banks' Jump Quartette

| WILL McINTYRE, BLUES SHOUTER BOOGIE WOOGIE PIANIST | HARRY BAKER NEW ORLEANS PIANO MAN |

They will tell the Story of Jazz in Music with Compere Tony Underhill (late A.B.C. Hobart)

PRESENTED BY THE EUREKA LEAGUE   Phone : FJ 2947

Poster for a Eureka Youth League jazz concert, Melbourne, July 1946

together, including some of the larger swing and Stan Kenton inspired bands, notably under the leadership of Fred Thomas.

From 1950, music in Melbourne entered a slump. The coal strike of 1949 was followed in Melbourne by a long public transport strike in 1950, and these were among the more obvious blows dealt to the entertainment industry. Naturally, a minority music like jazz suffered most, and through the first half of the decade the centre of gravity of the music gradually moved to Sydney. Smaller gigs began to close and bands to break up. Tony Newstead's venture at the Katherina was short-lived, the Copacabana night-club closed, Geoff Kitchen was unable to get his band any work and it never got past the rehearsal stages. By the end of 1951 only a handful of traditional groups, incl. those of Frank Johnson and Max Collie, was in regular work. Coffee lounges resorted to variety acts, and the concerts were in decline by the mid-

'50s. The decade saw a draining of musicians away from Melbourne on a short- and long-term basis. Frank Johnson and Len Barnard investigated country tours, Terry Wilkinson, Billy Weston, Geoff Kitchen, Dick McNally were among a great number who moved to Sydney while others, incl. Laurie Parr, Ross Fusedale, and Lin Challen, went further afield, if only temporarily. During this lean period, anything which gave support to jazz therefore assumed unusual importance. The Southern Jazz Society was founded in 1949 by Shirley Wood (later Shirley House), with Will McIntyre as its first president. Len Barnard led the house band, a monthly magazine, the *Southern Rag*, was edited by Tony Standish, who later founded the Heritage record label. The society organized jam sessions, annual balls and ran the 1952 AJC in Melbourne. It provided jazz experience for a number of enthusiasts who were later to make significant contributions to the music, incl. Bill Haesler, Dick Hughes, and Frank Traynor. Above all, it provided a forum for traditional jazz during a period when the music was enjoying comparatively little public support. The society ceased operating in 1958, passing the flame on to the Melbourne Jazz Club.

Towards the end of the '50s there was a resurgence of jazz activity in Melbourne, as always with the emphasis on the traditional styles. This growth manifested itself primarily through the casual dance or club scene, in which local halls of varying sizes were hired, generally by bands, for dances attended primarily by teenagers, who were beginning to constitute an affluent market. An early venture had been opened by dance band leader Mick Walker at Malvern Grove ballroom in 1951, but it was not until the late '50s that these dances became the major entertainment centre for young people. Under the leadership of Len Barnard, Max Collie, Frank Traynor, and increasingly using younger groups like the Melbourne NOJB, the dances and clubs ranged from ephemeral events to durable institutions like Club 431 which ran from 1957 to about 1964, the Esquire in Glen Iris, the Gasworks in Kew, and the Peninsula Jazz Club at Frankston. The clubs were not licensed, and as far as alcohol being on the actual premises in the possession of the 'members' was concerned, this was fairly strictly adhered to. The casual dances were dominated overwhelmingly by traditional jazz. That was the sound that characterized these functions for socializing, entertainment, courtship, for this generation was not much interested in the cabaret and night-club atmosphere, which was associated anyway with an older generation. They wanted more energetic outlets than movies, radio, and TV, and something more regularly time-tabled than the increasingly infrequent and commercially contaminated jazz concerts. The extent to which the casual jazz dances made inroads on the established industry is hinted at in the April–June 1960 issue of *Melbourne Jazz Club News*. It took issue with Bon Gibbins of the Council of Ballroom Dancing who had gone into print with one of the oldest objections to jazz: that the jazz dances were breeding grounds for immorality. The article in reply insisted that the clubs were run strictly, that they were no more inherently immoral than ballroom dancing, and that the real grievance was that the casual dances were stealing custom. Although the emphasis was on traditional jazz, there were some venues which experimented with a broader spectrum of the music, incl. Jazz Centre 44 and Bob Clemens's Downbeat Jazz Club which opened in 1958 with a band that incl. Brian Rangott (gtr) and Stewie Speer. Although in a separate category from the clubs or dances, mention should also be made of The Embers night-club which opened in 1959 in Toorak Road on the site of the old Claridges. Addressing itself to a more sophisticated, wealthy, and older clientele than that which supported the casual dances, it nonetheless signified in the renewed interest in jazz by virtue of the outstanding musicians it employed, incl. Frank Smith, and top-line imported artists like Oscar Peterson and Ella Fitzgerald.

The establishment of the Melbourne Jazz Club perpetuated the tradition embodied in the Southern Jazz Society. While most of the casual dance/club activity was simply oriented to providing a context for teenage socializing, the Melbourne Jazz Club was specifically set up to promote jazz. It grew out of record sessions at the home of Bill and Jess Haesler, with Frank and Pat Traynor also prime movers. The club opened at the RSL Memorial Hall in Church St, Richmond on 6 June 1958 with a house band led by Traynor. The club's rapid growth led it finally to St Silas Church Hall, Albert Park in June 1959, then into the church itself when the hall was demolished in 1960. The club became a centre for traditional jazz in Melbourne. Young musicians gained early experience at its functions—Ian Orr (tpt), Bob Brown (bs), John Hawes (tpt), Gavin Gow (clt), Kevin Goodey (clt), Eddie Robbins (clt)—older musicians knew they could resume contact through the club, interstate visitors and bands could use it as a jumping-off point. In July 1961 it produced its first newsletter and in August 1962, began a magazine which appeared alongside it until the former ceased publication in mid-1963. These publications promulgated information covering the national traditional movement and contributed to its solidarity. In June 1965 the club moved operations to the Musicians' Union premises in Queens Road, then expanded its activities in May 1966 to Mario's in Exhibition St with the advent of 10 o'clock hotel trading, as well as holding Sunday afternoon functions at Frank Traynor's club. Following another shift to the George Hotel in Fitzroy St, St Kilda, the club gradually wound down as the trad boom receded, ceasing operations at Traynor's in 1967. The Melbourne Jazz Club was a fixed point in a dispersed and fluid club and dance scene, and brought together at various times virtually every major jazz musician, particularly in the traditional area, of the post-war period.

The casual dance/club phenomenon set the scene for Melbourne's trad boom. In addition to the continued activity of veteran Frank Traynor, the early '60s saw the rise to importance of a number of new bands and musicians. Some, like the Driftwood JB and the Melbourne Dixieland JB, scarcely survived the boom, but two, the Yarra Yarra JB and the Red Onions JB dominated the traditional field well into the '70s, and the former continued to carry the flag of NO jazz in 1986. As elsewhere, this boom reverberated into the more modern styles of jazz, though at a greatly reduced level. Brian Brown and Alan Lee continued to be significant exponents of progressive approaches. Doug Dehn's Soultet enjoyed a brief burst of activity, as did groups led by Les Patching and Geoff Bartrum (with Roger Sellers, dms). Jazz Centre 44 had reverted to a traditional jazz policy by 1963, but in the same year American entrepreneur Ali Sugarman opened the Fat Black Pussycat at 90 Toorak Road, South Yarra. The Pussycat became Melbourne's centre for jazz experimentation, opening with a band led by Barry McKimm (tpt), and ex-German Heinz Mendelson (ten.), with Brian Fagan (bs) and Barry Woods (dms). The Pussycat also presented visiting musicians incl. Bob Bertles who joined McKimm/Mendelson for some months in 1963, and the Heads (Bernie McGann, John Pochée, Andy Brown (bs) and Dave MacRae) in 1964–65. By May 1965 the Pussycat had become a disco, but later that year Adrian Rawlins took it over and installed Brian Brown on Saturday and Sunday nights and Barry McKimm's trio on Sunday afternoons. With the ebb of the early '60s jazz tide, however, the Pussycat's days were numbered. In early 1966 a rock policy was introduced, and by about April it closed its doors.

The late '60s saw jazz eclipsed by electric pop as the music of the new generation. Jazz venues closed, most of the bands spawned by the trad boom folded, and again, musicians began leaving for Sydney with its more extensive entertainment industry. As in the previous decade, anything that kept jazz alive was that much more important. The liberalizing of licensing hours in 1966 created new opportunities of which jazz was at least partial beneficiary. More directly influential was the establishment of the Victorian Jazz Club in 1968. At the outset the club was run wholly by a collection of committees, each with its own portfolio. Many of the members were musicians incl. veterans of the '50s like Nick Polites, others were part of the new generation that emerged in the trad boom, incl. Geoff Thomas (dms), Allan Leake, Peter Grey, Bob Paul, Don Heap, John Murray. Some of the non-musicians, like Ken Carter and Don Anderson, have remained active in jazz up to the present. Using the Prospect Hill as its base, the club presented

regular functions, using bands on a rotation system. It published its own magazine, *Jazzline*, first under the direction of a committee, then under the editorship of Jim Loughnan from February 1969. In March 1970 it started a second night each week at the Manor House Hotel, and severed connections with the Prospect Hill from January 1971. The '70s also saw the beginning of its own radio programme on 3CR presented by Roger Beilby, from 1974 a newsletter in addition to *Jazzline*, and from 1977 a series of jazz workshops. The club moved from the Manor House in March 1981 and finally settled in at the Museum Hotel, where it was still based in 1986. Over the years, the Victorian Jazz Club has organized balls, commemorative concerts, picnics, barbecues, even sporting events. It is the contemporary manifestation of a distinctive element in the history of Melbourne jazz, the main antecedents of which are the Southern Jazz Society and the Melbourne Jazz Club. For this succession of clubs, jazz has been more than just a form of entertainment; rather, it is the centre-piece of a way of life, with club members being part of an extended family, often (by blood and marriage) literally so. The social functions, the newsletters and magazines, have created an *esprit de corps* which centres on, but goes beyond jazz, creating a national fraternity.

In the late '60s the Victorian Jazz Club created opportunities for new bands. Some, like the Limehouse JB, the Chicagoans, Kansas City Six, enjoyed active but brief lives. Others, notably the New Harlem JB and the Storyville group, both established in 1968, have continued to operate through to 1986. In the '70s the situation for traditional jazz stabilized somewhat after the violent fluctuations of the previous decade. John Kellock's La Vida JB, John Tucker's Yacht Club JB, and Peter 'Poppa' Cass's Dixielanders all enjoyed relatively regular work in addition to the older surviving groups. The '70s also saw some stylistic opening up in the traditional area in Melbourne. The Storyville Jazzmen ventured into mainstream territory that had been rather fenced off in previous decades (though Kenn Jones's long-running Powerhouse residency had also situated itself in

that area). A number of musicians hitherto associated with the traditional style also absorbed rock
influences, notably the groups led by Dave Rankin
(his Rankin File incl. Ian Orr, Ian Coots dms, Ron
Sedgman keyboards, Tom Cowburn vcl/el.bs, and
Graham Davies sax: a blend of musicians from
various stylistic areas).

Brian Brown continued to be the dominant
figure in more modern jazz styles during the '60s,
with his sometime bass player David Tolley pushing
further into electronic experimentation and free
improvisation. Brown's work at the Commune in
the late '70s brought in a new generation of young
musicians interested in contemporary styles and in
mid-decade, there was also a minor bop revival
which threw into prominence musicians like Ken
Schroder, Ray Martin, and Mike Murphy, only
one of whom, however, has remained in Melbourne. It remains true that Sydney offers more to
musicians interested in playing outside the traditional idiom. The brief impetus to contemporary
styles which was hinted at in the foundation of the
Victorian JAS in 1974 failed to materialize when
the society went into abeyance.

In the '80s there has been an increase in the level
of contemporary jazz activity, partly because of
the introduction of jazz studies courses at institutions like the Victorian College of the Arts,
through other institutional support from government funding, and through exposure on community FM radio. Three groups have provided the
main focus for such music: Onaje, Pyramid, and
Odwala. Odwala (named for a composition by
Roscoe Mitchell) grew out of a meeting between
Martin Jackson (reeds) and Jamie Fielding (p.)
when they were taking lessons from Bob Sedergreen. Both had earlier been drawn to contemporary jazz after hearing Brian Brown in the late '70s.
Odwala stabilized with Jackson, Fielding (until he
moved to Sydney in 1985), Barry Buckley, and
Keith Pereira (dms), though other alumni incl.
Stephen Hadley (bs), Jex Saarelahrt, Craig Beard
(vibes), Steve Miller (tbn.). Pyramid was formed
by David Hirschfelder (keyboards) and David
Jones (dms), with Bob Venier (tpt) and Roger
McLachlan (bs), and came to general notice with

*Onaje, 1985, with, l. to r., Gary Costello, Bob Seder-
green, Richard Miller, Allan Browne*

its first album in 1981. Its performance at the Montreux Jazz Festival in July 1983 was so acclaimed
that the group was immediately booked to do a
second concert. The band has since broken up,
however, and David Jones has moved to Sydney.
Onaje is the most durable of the three groups and,
interestingly, consists largely of veterans of earlier
decades of Melbourne jazz: Allan Browne, Dick
Miller, Bob Sedergreen, with Gary Costello and
Derek Capewell both having served on bass. Less
free form than Odwala and Pyramid, Onaje nonetheless works in contemporary areas, but closer to
the mainstream.

In 1986 Melbourne continues to be essentially a
stronghold of traditional-based jazz, albeit more
loosely and generously conceived than during the
'50s. Peter Gaudion's Blues Express and Roger
Hudson's Jazz Lips moved into bop areas, the current New Harlem JB has an eclectic repertoire that
incl. material by Randy Newman. But the basic
lines of force in Melbourne jazz continue to
arrange themselves around the traditional core.
The longest residencies are held by bands like the
New Harlem, Storyville, and such products of the
late '70s as Steve Waddell's Creole Bells and the
Maple Leaf JB. One of the most successful new
groups in the '80s has been Ross Anderson's New
Melbourne JB, formed in 1982. It has performed at
the Sacramento Dixieland Jubilee, recorded for
Roger Beilby's Anteater label, and has worked on

radio and TV. Unlike most other centres throughout Australia, Melbourne has also enjoyed a steady, if small, influx of new traditional musicians in the post-'trad' boom period. Pip Avent (tba) and the late Maurie Dann have made significant impact. Graeme Pender (clt) is a more recent recruit, and even younger musicians are showing interest in and enthusiasm for a style of jazz which is almost completely ignored by the new generation of players in other cities. The strength and vitality of traditional jazz in Melbourne is unique in Australia. No other state has boasted a jazz community which is simultaneously so durable and so elaborately and tightly reticulated. Why such a phenomenon should arise specifically in Melbourne is a matter for speculation, but indisputably it has been largely sustained by the succession of clubs which have maintained a very palpable sense of fraternity embracing every traditional musician active in the post-war period.

Melbourne New Orleans Jazz Band (Melbourne, Vic.)

In 1957 Llew Hird founded his NOJB, one of the first in its style in Australia, with Peter Sheils (tpt), Lou Silbereisen, John Kavanagh (bjo/gtr), Graham Bennett (dms), Nick Polites (clt). The band played at Southern Jazz Society functions and also a Downbeat concert before opening at the Blue Heaven Restaurant, Fitzroy St, St Kilda where

The Melbourne New Orleans Jazz Band, with George Lewis (third from left), London 1961; from l. to r., Kevin Shannon, Willie Watt, Lewis, Graham Bennett, Mookie Herman, Frank Turville, Nick Polites

Paul Marks (vcl) was also working in a skiffle group. Frank Turville and Willie Watt replaced Sheils and Kavanagh, 1957, then Mookie Herman replaced Silbereisen. In March 1958 Hird left, Polites took over as leader, Dave Rankin joined on trombone, and the name was changed to the Melbourne New Orleans JB. They left Blue Heaven and began at Jazz Centre 44, with Paul Marks and his 'folk group' made up of members of the band. In 1959 Rankin was replaced by Charlie Powell, who was in turn replaced in 1960 by Kevin Shannon, who took over as leader in 1961, at about the time the band began to plan an overseas tour. Frank Turville preceded the others to England and was replaced by Geoff Bull, who then left the band in Sydney following its departure from Melbourne in August. Arriving in England on 15 Sept. 1961, they rejoined Turville and played their first gig at the BBC Jazz Club on 21 September. For seven months the band toured the English jazz club circuit and played concerts, recording sessions and TV. In April 1962, Shannon and Bennett returned to Australia. Bennett continued playing and in the '70s moved to Bellingen, NSW, where in the mid '80s he was associated with Brett Iggulden; Kevin Shannon was killed in a motor accident in Melbourne on 3 Mar. 1963. Max Collie (tbn) and Mal McGillivray were flown to England to replace them, joining in time for a five-month European tour. In August 1962 Paul Marks returned to Australia and was replaced by Long John Baldry. A further seven-month tour of the UK finished in April 1963, after which the band broke up. Polites, Turville, Watt, and McGillivray returned ultimately to Australia. Collie remained in England, joining the London City Stompers until forming his own Rhythm Aces, which has since become one of the most successful traditional groups in Europe. Mookie Herman joined Doug Richford's band and went on to other groups before returning to Germany where he died in 1985.

The Melbourne NOJB stimulated a 'school' of jazz based on the music of NO veterans like Bunk Johnson and George Lewis. It was a style which enjoyed considerable vogue in Melbourne, with some ripples in Sydney, during the '60s and which is still preserved by the Yarra Yarra JB. The history of the Melbourne NOJB has been documented by Eric Brown in articles in *Jazzline*, September and December 1973, from which much of the above is derived.

Miller, Richard (Dick)   b. 4/12/44, Melbourne, Vic.   reeds/comp.
First important band was the Red Onions JB which he joined in 1965, remaining until it broke up in the mid '70s. Its Polish tour with Roland Kirk gave impetus to Miller's musical evolution, which has continued under the influence of a range of musicians incl. John Coltrane, Miles Davis, and, locally, Bob Sedergreen. Stylistically Miller is at ease over an unusual range and is equally at home in both the main bands he worked with during the '80s: the mainstream to bop setting of Peter Gaudion's Blues Express and the more contemporary and experimental Onaje. Miller teaches music and is building a significant reputation as a composer, and in 1986 was with Chris Ludowyk's Society Syncopators.

Miller, William Henry (Bill)   b. 22/2/14, Melbourne, Vic.   wbd
Began playing in 1947, and as a musician his most significant activity was organizing the Portsea Trio, with George Tack and Will McIntyre. This group played the Nepean Hotel, Portsea, on holiday weekends 1952–54, and since then at occasional concerts and AJCs. Miller's primary importance, and it is very considerable, lies in having provided a focus for jazz activity and discussion during the emergence of the traditional movement in Australia from the '40s. He developed an interest in jazz as a schoolboy and on arriving in England in 1933 to read law at Oxford, he began to buy from the large record catalogues available. The collection he brought back to Australia in 1938 was one of the most significant repositories of jazz for young Melbourne musicians. It was made accessible through *Swing Night* on 3UZ, for which Miller began writing scripts and providing records, and through sessions which included

Bill Miller, the Playhouse
Theatre, Hobart, for the 32nd
Australian Jazz Convention, 1977

Ade Monsbourgh, 1985

those run monthly by the 3UZ Jazz Lovers' Society, founded also by Miller in January 1941. At the same time he started the society's official magazine, *Jazz Notes*, which he continued to produce for some time after going into the army, until handing it over to Ced Pearce, resuming editorship for a year from February 1944. He was responsible for numerous other publications including a discography of small record labels. He founded Ampersand records which preserved the sounds of many Australian groups of the '40s, of which we should otherwise have had no record. He also started the XX label, specializing in reissues of 'classic' jazz.

Miller was prominent in the administration of the AJCs, and in the year of the first one, 1946, he started the *Australian Jazz Quarterly*. To circumvent paper restrictions, the first issues, going back to 1945, had had different titles; no. 1 under the name *AJQ* appeared in May 1946. Providing, among other things, a forum for discussion and an international record trading mart, the periodical continued until April 1957, Bill Haesler having taken over as editor in December 1954. From the late '50s Miller's professional commitments have diminished his participation but not his interest in the jazz scene.

Monsbourgh, Adrian Herbert (Ade, Lazy Ade) b. 17/2/17, Melbourne, Vic.   (tbn./tpt/reeds/p./ tba/bjo/vcl/recorder/comp/ldr
Brought up in Koyuga near Echuca, Vic., Monsbourgh moved with his family to Melbourne in 1925. He had begun playing a mouth organ and had taken lessons on piano from age eight. In 1932 he heard Clarence Williams on the radio and, with classmate Roger Bell, he developed an interest in jazz. In 1935 he formed the Shop Swingers (incl. Spadge Davies, Ivan Arthur, and Jack Coughlan (bs), cousin of Frank), for local dances. Over the next few years he became associated with the Bell brothers, mainly on banjo. Began trombone and trumpet and in June 1937 had, with Sam Benwell, founded the Melbourne University Rhythm Club while proceeding to his BSc. Continued sitting in with Bell groups, incl. at the Nepean Hotel, Portsea, and joined the band for its Heidelberg TH residency beginning in July 1943. In September, he participated in the Roger Bell/Max Kaminsky recording session and in April 1944 the first sessions under his leadership were released. By the time he joined the RAAF in 1944 he was already being referred to as 'Father Ade', a term which has subsequently caused considerable confusion in discussions of Australian jazz. In 1945 he recorded with

the highly talented but only briefly active Kelvin (Kelly) Smith, and was also discharged from the RAAF, upon which he rejoined the Bells. During the two tours of Europe and England, his work with the Bell band had particular impact, and during the second tour he was presented with a plastic alto by the manufacturer, an instrument which became something of a Monsbourgh trademark back in Australia. He was offered a permanent position in the band of leading English jazz musician, Humphrey Lyttelton. Monsbourgh declined the invitation, returned to Australia, and when the Bell band broke up in 1952, he joined Pixie Roberts in establishing the Pan recorder manufacturing company, an interest which also manifested itself in the *Recorder in Ragtime* album he made in 1956, a unique exercise in jazz on the recorder. In 1952 he won the 'other instrument' category for his work on alto, and the vocal category, in the *Music Maker* poll. He played alto with Len Barnard's band, 1953–55, and in 1954 made a solo multi-tracked record, playing tpt, tbn., clt, alt., p., wbd. Through the late '50s he freelanced, with a relatively long period at the Powerhouse rowing club, and in 1961 became the resident guest musician for the Melbourne Jazz Club, which brought him into frequent association with Frank Traynor's band. Monsbourgh has continued to perform, compose, and record prolifically with each succeeding generation of Melbourne traditional jazz musicians, incl. with blues/jazz singer Margret Roadknight. He has been a regular performer at AJCs since their inception, and frequently plays concerts in other parts of Australia. Also an influential mentor, perhaps most notably of the fledgeling Red Onions JB in the early '60s.

Monsbourgh is one of the most original and influential jazz musicians Australia has produced. His distinctive approach, both in terms of timing, harmonic line and, especially on alto, his timbre, is central to what is widely, if controversially regarded as the 'Australian' or 'Melbourne' jazz style. He has been copied more productively than any other Australian jazz musician, incl. by overseas musicians. He has established a crucial authority in every group he has been associated with,

incl. the Southern Jazz Group with which he recorded in 1950. Among Australian jazz compositions, Monsbourgh's are played more often than the work of any other composer. As a teacher he has fostered and directed several generations of Melbourne traditionalists. In the early days of the second jazz phase during the late '30s and '40s, Monsbourgh's writing and his discussions with other musicians were central in the attempt to distinguish and articulate the roots of jazz. In 1986 Kevin Whittingham, an Australian researcher now resident in Canada, was working on a Monsbourgh biography.

Mooney, Martin (Marty)   b. 18/9/43, Brisbane, Qld   reeds
Began playing engagements in 1963, and joined Nick Boston's band in 1964. Moved to Sydney, 1966, and persuaded Boston to do likewise; in Sydney continued with Boston's new band. Began jamming c. 1970 with other Sydney musicians during holidays in the Snowy Mountains, and this led to the formation of Galapagos Duck, of which Mooney was a founder member. Left the Duck in 1977 to join Dick Hughes's Famous Five at Soup Plus. Left Hughes, 1980, and joined Roger Janes's band; also during the period played with Freddie Wilson at the Cat and Fiddle Hotel in a band which incl. Lachie Jamieson (dms). Mooney toured Australia for three months with Spike Milligan, 1984, with John Callaghan (bs) and American

Marty Mooney, The Basement, 1976, with Tom Hare in the background

Dave Paquette (p.). Following a layoff during 1985, he was freelancing in 1986, incl. as frequent deputizer for Paul Furniss in the Eclipse Alley Five. Mooney is an impressive example of that category of Sydney jazz musician who is so fully formed in terms of the jazz tradition that he is at home in virtually any traditional to bop setting.

Morgan, Graham George   b. 30/4/37, Melbourne, Vic.   dms/ldr
Heard traditional jazz as a schoolboy; introduced to more modern styles by Peter Martin (reeds). In 1957, with Ted Preston Quartet, and opened Jazz Centre 44 leading his own quartet. Joined Channel 9 Orchestra, 1958. To the US, 1962, where he studied with Murray Spivak. Returned to Melbourne then moved to Sydney, 1963, and worked in the night-club scene, incl. leading the Latin Quarter group, 1965. Returned to Melbourne, 1965, and mainly occupied in studio work, incl. for Channel 10, but intermittent jazz activity incl. work with Graeme Lyall quintet. In the US, 1973, and performed with Cleo Laine at Carnegie Hall. Back in Melbourne, has been primarily occupied with the Channel 9 Orchestra under Graeme Lyall. In 1985 he was still with 9, teaching, and with the Peter Sullivan big band. Although mostly active in session and TV work, Morgan has been very important and inspirational as a teacher.

Morrison, James   b. 11/11/62, Boorowa, NSW   tpt/tbn./sax./bs/p./comp./arr/ldr
Born into a musical family, began cornet at seven and formed his first dixieland band while still a schoolboy. In secondary school started a big band with brother John (dms). Began piano, 1974, and formed a quartet. With the Young Northside Big Band, made considerable impact at Monterey, 1979. Passed through the jazz studies programme at the NSW Conservatorium, 1979–80, and joined Don Burrows as well as forming his own small group in the early '80s. In 1983, began teaching at conservatorium, became MD for the Pan-Pacific Music Camps, and, with John, formed the Morrison Brothers Big Bad Band, many members of which were fellow graduates of the conservator-

ium. Represented Australia at the Olympic Jazz Festival, Los Angeles, 1984; played concerts with Tommy Tycho and also performed the Haydn Trumpet Concerto with the Conservatorium Symphony Orchestra. Played Expo '85, Japan, and spent three months in the US in 1985. A regular performer at the Don Burrows Supper club in the Regent Hotel, and at Soup Plus. Morrison's mastery of the established bop conventions and his flamboyant presentation have made him a significant popularizer of jazz in the '80s.

Munro, Charles Robert (Charlie)   b. 22/5/17, Christchurch, NZ;   d. 9/12/85, Sydney, NSW
reeds/fl./cello/comp./arr/ldr
Born into a musical family, he began piano at seven, sax at nine and cello at 11. As well as playing in the family dance band, he was conductor of the Christchurch Boys' High School Orchestra at 13, and became a professional musician from 17. Worked in New Zealand, incl. under Maurie Gilman, then played on ships travelling between New Zealand, Australia, and Vancouver, with Linn Smith. Settled in Sydney in 1938, joining Myer Nyman at the Glaciarium, and replacing Bert Mars at the Backstage Club with Wally Parks in 1939. Was prevented from joining Frank Coughlan at Romano's because of being articled on the *Niagra*. In 1940 he was back ashore, working with Dud Cantrell, with Sam Babicci at Rose's, then for six months with Tiny Douglas at the State Theatre.

Charlie Munro, 1978

After joining the army he was posted with the American-Australian concert party (the '50-50' entertainment unit) on tenor, spending six months in New Guinea. Following the group's disbandment in June 1944, Munro was placed in charge of the 7th Div. concert party, which after six months in Perth was absorbed into the orchestra of the 1st Australian entertainment unit under Eddie Corderoy. Munro later became its conductor. The orchestra was broken up in 1945, and he then formed a Liberty Loan unit (which incl. Billy Weston) to tour NSW and Vic. Discharged in 1946, he joined Wally Norman on alto at the Roosevelt, having deputized there in late 1945. Although a night-club group, there was strong jazz interest among the members, which incl. Sid Beckwith, Al Vincer and Django Kahn (a.k.a. John Hodgson), and when Munro heard the bop records imported by Norman in 1946, he immediately assimilated the style. In 1950 he left the Roosevelt to join Bob Gibson, and began four more years formal study of the cello. With Gibson he became prominent in the jazz concert scene, and made recordings with various participants incl. Les Welch. In 1954 he began an association with the ABC dance band, at that time under Jim Gussey, which was to last until its break-up in 1976. Beginning as lead alto, he later composed and arranged for the band, often applying the modal concepts which he absorbed from Bryce Rohde. The association with Rohde lasted through the early '60s until the latter's departure for the US in 1965, and involved workshops, concerts, and recordings. Through this period Munro was also working with groups of his own, incl. Mark Bowden and George Golla, and fellow New Zealand trombonist Bob McIvor (b. 9/1/42, a session player who later featured on John Sangster's *Lord of the Rings* recordings). Munro also played with Don Andrews's Castillian Players, and was approached by Chico Hamilton to join his band on cello. The experimental urge which had found an outlet in the Rohde workshops found expression through the late '60s in a variety of projects. In 1966 he assembled a group incl. Ron Gowans, Dave Rutledge, Bob Iverson, Ken Brentnall, Marie Francis, Ed Gaston and Bowden, to

perform the Seiber/Dankworth *Improvisations for Jazz Band and Orchestra* with the Sydney Symphony under Dean Dixon. Studying Eastern music well before it became fashionable, he recorded *Eastern Horizons* in 1967, an album which John Clare coupled with *Here's Bryce* (on which Munro was also present) as the two 'most successful jazz recordings' in Australia. In 1970 he recorded his ballet *Count Down*, a work which included atonal and free-form excursions. Through the '70s he remained in comparative obscurity, emerging in 1979 to perform with Bruce Cale. In the '80s he ran a workshop (Jack Thorpe p., Mark Bowden, and Wayne Ford on cello) which made broadcasts, and led his own quintet at the Earlwood-Bardwell Park RSL. At the time of his sudden death from a cerebral haemorrhage he was working with Georgina de Leon's Lucy Brown Quartet. Through his whole career Munro maintained an extraordinary breadth of outlook and was at the forefront of a succession of jazz developments in Australia— bop, modal, free form and eastern experiments. As an arranger, he showed a profound grasp of the total context, including the medium, of the music. His intelligence was one of the most comprehensive in Australian jazz, yet wholly without pretentiousness. He was one of the closest things this country has had to a jazz-based genius.

Munro, William Herbert Colyer (Bill)   b. 1/10/27, Adelaide, SA   tpt/ldr
Joined J.E. Becker's Adelaide Drum and Fife Band, 1937, graduating to the Banjo Band on banjo-mandolin, 1938, and began trumpet, 1940, playing in the Junior Military Band. First jazz work was with Colin Taylor, Bob Wright, Bruce Gray, playing 'Mood Indigo' and 'Playmates' for an Adelaide High School function, 1943. Munro was in Malcolm Bills's band from its foundation in 1944 to its dissolution, by which time he had joined the Southern Jazz Group, with which he performed at the 1st AJC, 1946. Left the SJG to join Bruce Gray, mid-1950, and has remained associated with Gray to the present. Played in Billy Ross's quartet at the Carousel in Glenside and the Burnside TH, 1956–57. Formed his Jazz

Bill Munro, 1986, with Bruce Gray in background

Six (Rod Porter, John Malpas, Wright, Glyn Walton, Jim Smith wbd) for the 1957 AJC, and kept the band together for occasional gigs, incl. the Carousel. Later re-formed the band (Ernie Alderslade tbn, Gray, Norm Koch, Don Knispel dms, Ron Williams bs) for ABC work. In 1962, joined the University JB, which became the Campus Six, Munro remaining with the band until its dissolution in 1976. As he had with Ross, Munro demonstrated a command of more progressive styles in broadcasts he made in 1963 as a member of Maurie Le Doeuff's sextet and octet. In the late '60s, played in the Neville Dunn Big Band. Replaced the late Roger Swanson in Gordon Coulson's Climax JB, 1977–83. Since retiring from his job with Australian National Railways in 1984, Munro has increased his freelancing, incl. with the Barossa Valley Vintage JB, Wog Fechner's (reeds) Southern Vales JB, the Jazz Jesters, and the Eric Bryce (p.) Quintet. In 1985 he joined the Jubilee Stompers. In 1986 he continues to work regularly in Bruce Gray's groups, and with Dave Dallwitz, with whom he renewed association following the latter's resumption of jazz activity in the early '70s. Munro has been at the centre of the jazz movement, and one of its most complete and poised musicians, throughout the whole of its most significant history in Adelaide, working with virtually every band that defined the music: Malcolm Bills, the Southern Jazz Group and other Dallwitz groups, Bruce Gray's bands, the Campus Six. Notwithstanding his venerable status, his playing remains undiminished in its vitality.

Murray, John Philip   b. 5/3/41, London, England   tbn./vcl
Family arrived in Yarrawonga, Vic, 1953; moved to Melbourne, 1959. Murray joined a trombone trio at Toorak Teachers' College, with Brian Hanley and Don Stanton. Hanley switched to trumpet and formed the Chicago Seven, which Murray joined, 1959–64. With Storyville All Stars, 1968–81. In 1986, with Ross Anderson's (bs) New Melbourne JB at the Bridge Hotel, Richmond, and the Tower Hotel, Camberwell.

Napier, Lloyd Warren   b. 10/8/45, Perth, WA   cnt

Exposed to jazz through his father, Don, who played reeds, and through the rehearsals of the Riverside Jazz Group, held at his home. Napier's Wild Bill Davidson-inspired playing has been heard in most traditional bands in Perth, incl. the New Era JB, the Blue Note Jazz Group, the Great Western JB. In 1986 he was still with the Lazy River JB of which he was a foundation member, and playing freelance work with small pick-up groups.

Nash, Alan T.   b. 14/10/22, Melbourne, Vic.   tpt

Joined Coburg School Band at age nine, later entered the Coburg City Band under Tom Davidson. Joined the Melbourne Centenary Brass Band, 1936. Abandoned trumpet for lack of dance band opportunities in 1937, resuming in 1939 to play with brother Wally's (p.) band at the Galleone. Nine months later joined Clarrie Gange at the Glaciarium, the band incl. Al Vincer and Neville Maddison. At Plaza coffee lounge, 1940, with a band incl. Tom Davidson, Charlie Lees, Benny Featherstone, and in Davidson's radio orchestra, 1941. Drafted into the army, Nash served in the '50-50' (American/Australian) entertainment unit, then in September 1944 joined the Army Concert Orchestra under Eddie Corderoy (alongside, *inter alia*, Maurie Le Doeuff and Charlie Munro); later served in the Seventh Division Concert Party under Munro. Following discharge, 1946, Nash very briefly played with Tom Davidson, then, following three months at the Brisbane Tivoli, moved to Sydney. Joined Jim Gussey's ABC Dance Band, 1947, and also made a reputation in the jazz con-

certs into the '50s. During the same period he worked in Sydney dance halls/night-clubs, with John Best at Segar's, Leo White at Prince's, at other venues with Ralph Mallen and Denis Collinson, under several leaders at the Stork Club, and in 1953, replacing John Bamford at the Colony Club. From the late '50s Nash has not performed in jazz settings, confining his playing to various kinds of session work, though he has recorded occasionally with jazz groups, such as Len Barnard's. In 1986 he had retired from playing, and was secretary of the Sydney branch of the Musicians' Union, an office he was initially elected to in 1976.

Nelson, Michael Raymond (Mike)   b. 31/7/49, Perth, WA   keyboards/comp./arr

Until returning to Perth after a period in England, 1974–79, Nelson's work had been primarily in rock/pop settings, though in England he formally studied at Trinity College and the International Film School. Back in Perth he has been active in concert and jazz club work, backing visitors incl. Keith Stirling, Milt Jackson, Mark Murphy, John Dankworth, Bernie McGann. In 1983 worked on film music under ABC commission and with WA Symphony Orchestra. In 1985, lecturing in the jazz courses at Mt. Lawley and leading his group Four on the Floor.

Nettelbeck, Theodore John (Ted)   b. 4/1/36, Streaky Bay, SA   p./comp./ldr

From 1955–60 with Bruce Gray's All Stars, incl. for the ABC radio series *The Evolution of Jazz* in which he demonstrated a sympathy for traditional styles which is unusual in musicians with progressive preferences. In 1960 at The Embers night-club in Melbourne with Frank Smith, an important in-

Ted Nettelbeck, 1984

fluence and for whom he wrote one of his compositions, 'Not Only In Stone'. Nettelbeck took over as leader of The Embers group, with Graeme Lyall, Alan Turnbull, Darcy Wright, through 1961. From 1962–65 in Europe and England with various groups, incl. for Louis Bonnet's (a.k.a. Matt Flinders). From his return to Adelaide until 1969 led his own quartet with Bob Jeffery, Trevor Frost (dms), Dave Kemp, and since 1970 has freelanced incl. with Billy Ross's trio, Bruce Gray's sextet, Schmoe & Co., Barry Duggan. In the mid '80s led his own trio with Laurie Kennedy and Michael Pank. Has accompanied numerous visiting Americans incl. Mel Tormé, Buddy Rich, Phil Woods, Mark Murphy, Milt Jackson. Nettelbeck has also been influential as a teacher; with WEA, in the Diploma of Jazz Studies course, SA CAE 1981–83 and at the Elder Conservatorium in 1982. 1959–69 he was a professional musician, but he is now also a lecturer in psychology at the Univ. of Adelaide.

Newcastle (NSW)

As one of Australia's largest provincial cities, Newcastle has registered many of the nodes and antinodes in the history of Australian jazz. These have been amplified to some extent by virtue of the proximity of Sydney, just over 170 km away, providing a pool of potential visiting performers able to augment the corps of local musicians. Jazz in Newcastle has been sustained by a combination of a small but fiercely enthusiastic jazz fraternity, and inspirational visitors from Sydney.

The first phase of jazz during the 1920s left little distinctive impress in Newcastle, and by the time its main pre-war dance hall, the Empire Palais, opened in 1929, the fashion was dying nationally. Nonetheless, during the next major phase Novocastrians were well to the fore, with the Newcastle Hot Jazz Club one of the first of its kind in the country when it functioned briefly from 1934. The 2HD Hot Jazz Club followed in 1936, and indeed the succession of clubs dedicated to swing, hot jazz, or modern music during the late '30s was so rapid and so confusingly reported that it is difficult to determine precisely the identity and duration of each. This much is clear, however: the mainstays of these early clubs, and frequent office-bearers on their committees, included Athol 'Happy' Sutherland, Horrie Greenwood, and bass player Jack Sinclair. Sutherland in particular was a dedicated and well-informed student of jazz at a time when many, including members of the various clubs, simply equated a 'swing' club with a medium for socializing to the accompaniment of any modern music. Indeed, one of the features of the Newcastle jazz clubs into the '50s was the periodic splintering away of smaller associations impatient with the commercial influences beginning to dominate the parent organizations. Sutherland, Sinclair and Greenwood led the way here when they dissolved the Hot Jazz Club functioning in 1938 and formed a new 'select' club with membership limited to 30. These early clubs were the primary focus of hot music in Newcastle well into the '40s. It was a perennial lament that the commercial dance venues showed virtually no comprehension or encouragement of jazz. The clubs drew regularly on Sydney for their live performances, with Frank Coughlan, Stan Bourne, Dick Freeman, Maurie Gilman, Reg Lewis, Barbara James, Dud Cantrell, making considerable impact on the local enthusiasts. Although the local dance bands provided little inspiration for aspiring young jazz or swing players, the club functions helped to give a start to several musicians who later established significant reputations in Sydney, incl. Jack Crotty (tpt), who later joined Frank Coughlan, Al (a.k.a. Mick) Vincer (dms/vibes) who became very active in the night-clubs when they were 'unofficial' havens of progressive jazz thinking, and Jack Maittlen, later a Sydney bandleader.

The post-war jazz boom gave impetus to the clubs which continued to function intermittently, and Newcastle also enjoyed its own modest jazz concert era. Ron Brown was an important promoter, and mainstays of the Newcastle concerts were Sydney leaders Bob Gibson, Wally Norman, Frank Marcy, Bob Limb, with bands made up of varying combinations of local and Sydney musicians. Among the former were Dick Jackson, the blind reed player at that time resident in Newcastle, and members of bands which sprang into being in connection with this new jazz phase. Some of these, like Harry Tabernacle and his Australian Jazz Band, were little more than opportunistically re-titled dance bands. The Steely City Seven, however, was an authentic jazz group inspired by concerts given by Graeme Bell, and probably the first of its kind in Newcastle. It began as a house band for the Steely City Jazz Club in 1948. Although the club dissolved next year, victims of the 'jivers and boppers' according to a local mouldy lament, the band continued to function under Jack Sinclair's leadership, and include Holly Butler (reeds), who later freelanced in Sydney. The early '50s saw another resurrection of a local jazz club, with Ray Scribner (tpt) a leading spirit. Again, purist frustration with creeping commercialism led to the formation within the body of a Backroom Jazz Club, having a reported membership of two. In 1954 the Hunter Valley Jazz Club was formed with a house band that included Tony Howarth, Scribner, and Bill Boldiston, later replaced by Rod Lawliss (clt). Although both club and band appear to have become dormant by the late '50s, they signified a reserve of energy ready to be released when times were more propitious.

Such times arrived in the early '60s. Newcastle presented a microcosm of the proliferating scene in the capital cities. In March 1960 the recently reactivated Newcastle and Hunter Valley Jazz Clubs amalgamated. Other clubs were established under the aegis of the WEA and the YMCA. The Harbourside Jazz Club was established in January 1963, and its house band, which included Bob Henderson, became one of the busiest bands in the area around Newcastle. Other groups were formed, including several working in post-traditional styles, and the Univ. of Newcastle presented recitals and boasted its own group. Tony Howarth ran jazz workshops for a period, and in addition to other musicians who had been active earlier, the boom brought other names into prominence, including Jack McLaughlin (reeds) who has remained a steadfast local proponent of the New Orleans style ever since, and Paul Leman (dms) and Rod Travis (reeds), who have gone on to work in Sydney groups. The AJC in Newcastle in 1964 brought its usual flow-on of jazz interest, and a new Newcastle Jazz Club was formed in its wake, but the national jazz slump of the late '60s saw the virtual disappearance of the music locally.

The establishment of the Maryville JB in 1971 marked the beginning of Newcastle's most recent continuous jazz phase, and established that phase as traditional in musical character. Founding members of Maryville were Tom Tyler (bjo), Peter Buckland (clt), Eric Gibbons (tbn./ vcl), John Vernon (bs), Leman and McLaughlin. Harry Cantle replaced Vernon from 1972–74 until leaving for Brisbane, upon which Vernon rejoined. Bob Henderson was added briefly in 1973, but in the mid-'70s, economics reduced the band mostly to five pieces, with Buckland and Gibbons as the front line. Through 1976–78, Roger Graham (tpt) who had moved from Sydney, and Charles Pope (p.) were added. Leman moved to Sydney and was ultimately replaced by Ron Hogan who remained until 1982, when he was replaced by Peter Young. From the late '70s the band worked regularly as a five piece as Graham, Pope, and Tyler left, the latter replaced on banjo by Cantle who had returned from Brisbane. Since 1984, Cantle has moved to bs, Greg Griffiths has joined on banjo, and Ron Hogan has returned on drums. Work has been intermittent however. In 1979 the Silver Bell Quartet was formed as a more economical offshoot, with Gibbons, Tyler, Warren McCluskey (bs), John Wilson (cnt). Wilson was a Harbourside Six alumnus, and McCluskey had worked with the La Vida band in Sydney where he had taken lessons from Don Heap and Jack Lesberg. Buckland and Griffiths replaced Wilson and Tyler respectively in 1984.

Since the '30s jazz activity in Newcastle has waxed and waned in response to forces emanating from elsewhere, primarily Sydney. Since the early '70s the work of the Maryville JB has been the central focus of the Newcastle scene. During that time a new jazz movement has appeared in Sydney, but with the emphasis upon the more contemporary styles. For once this seems to have transmitted few ripples to Newcastle. The city's traditional scene has contracted towards the mid-'80s, but at the same time the progressive initiatives in Sydney have made little impression. The Maryville JB has enjoyed a uniquely long life for a Newcastle group, but its lapse into comparative inactivity and the negligible infusion of young blood into the local scene, make the future of jazz in Newcastle a considerable enigma.

The New Harlem Jazz Band, 1972, with, from l. to r., Bill Morris (tba), Chris Farley (bjo), Doug Rawson (p.), Richard Opat (dms), Chris Ludowyk (tbn.), Ian Fleming (clt), Ian Smith (tpt)

New Harlem Jazz Band (Melbourne, Vic.)

In 1967 Ian Smith formed the Royal Harlem JB for the Geelong Jazz Festival; in 1968 the name was changed to New Harlem JB. The personnel stabilized as follows, with each musician's tenure indicated: Smith tpt (1967–79), Chris Ludowyk tbn. (1968–82), Ian Fleming clt (1967–73), Rube Croft p. (1967–69), Richard Opat dms (1967–78), Chris Farley bjo (1967–c. 74), Jeff Parkes tba (1967–72). Subsequent turnover in personnel was: tba, Bill Morris (1972–); bjo, Cam Crofts (c. 1974–77), Maurie Dann (1977–84), Chris Farley (1984–85), Rod Evans (1985–); dms, Bob Wood (1978–81), Richard Opat (1981–); p., Doug Rawson (1969–75), Neil Orchard (1975–), with an interruption while Chris Somerville replaced Orchard in the early '80s; tpt, Sandro Donati (1979–); reeds, Bob Gilbert (1973–). When Smith left, Ludowyk became leader, and when he left in 1982 he was replaced by Pat Miller, making the New Harlem a two-reed group. In addition to band members singing, Pippa Wilson became vocalist in 1979. The band has been one of the most visible groups in Melbourne, with numerous residencies incl. at

Athol's Abbey from 1975, Smacka's Place from 1976, Alexanders from 1981. In 1986 the band was based at the Railway Hotel, Windsor. It has played numerous concerts, incl. as support for Turk Murphy, and with the Storyville band in tributes to Fats Waller, Duke Ellington, Louis Armstrong. It appeared in the movie *The Rise and Fall of Squizzy Taylor* and recorded the title music for *The Last of the Knucklemen*. The group made its first recording in 1970 and has since made a number of others, incl. an album with Alex Frame as guest. The New Harlem JB was established primarily as a vehicle for novelty material of the '20s and '30s, and less famous jazz classics with emphasis on the early work of Ellington. It has also presented original compositions by band members, notably Orchard and Morris. Its strength was in the cohesion of its ensemble work, involving often elaborate but crisp arrangements. Changes in personnel and instrumentation, especially in the '80s, have inevitably altered the band's character; its current stylistic base is broader and its repertoire more eclectic, embracing material not normally associated with jazz.

Tony Newstead's Southside Gang c. 1950, with David Ward (tbn.), Tony Newstead (tpt), Don Reid (dms),
Ray Simpson (gtr), Will McIntyre (p.), George Tack (clt), Keith Cox (bs)

Newstead, Anthony (Tony)   b. 1/10/23,
Melbourne, Vic.   tpt/ldr
Played piano as a child and took up trumpet in
1940, with a few lessons from Frank Coughlan.
Played with various musicians incl. Don Banks,
Keith Atkins (clt) and Charlie Blott, until he
joined the RAAF in 1942. Posted to Port Moresby in
New Guinea, he became involved with Sid Bromley, Will McIntyre, and Don Reid, then later back
in Australia played in the RAAF jazz band with
Bob Cruickshanks at Tocumwal. In 1946 he joined
the Varsity Vipers of which the alumni also incl.
Ken Ingram (tbn.), Nick Polites, Frank Milne
(p.), Colin Taylor (wbd), Will McIntyre, George
Tack, Keith Cox. Newstead's Bix Beiderbecke-
influenced style was recorded with McIntyre's
band in July '46, with Atkins, Reid (suitcase),
Frank Mardell (bs) and Eric Washington (tbn.),
and he began leading bands for Eureka Hot Jazz
Society functions. President of the Melbourne
University Rhythm Club, 1947, and leading his
Jazzmen to deputize for the Bell band during the

latter's absences from the Uptown Club, using
various musicians incl. Geoff Kitchen, Keith Cox,
Ian Pearce, Rex Green, and Laurie Howells. When
Bell left for Europe, Newstead took over with his
Uptown Band, using a nucleus of Tack, McIntyre,
Ingram, Cox (bs), Reid (dms), and Ray Simpson
(gtr). About this time, began using the word
'Southside' in the band name. He ran the Southside
Club at the Maison de Luxe from Feb. to April
1948, leading his Southside Gang, which also con-
tinued at the Uptown Club until it closed in late
1947 after a falling out with the Eureka Youth
League. In 1949 his band was deputizing for Bell
at Leggett's during the latter's tours with Rex
Stewart. The Southside Gang (a.k.a. Downtown
Gang) was one of the most acclaimed bands in
Melbourne during 1950–51. It played numerous
concerts, incl. the Cavalcade of Jazz for the fund-
raising concert for the Melba Memorial, May
1950, and brief residencies at the Powerhouse
Rowing Club (using David Ward tbn., and Toni
Lamonde vcl) and the Katherina, St Kilda. In

1952–53 it was virtually dormant, and Newstead was freelancing, but the Southside group reunited occasionally through 1954–58. In 1958–60 Newstead was engaged in post-graduate study in London, with some months in New York in 1959, when he played with Cecil Scott, Zutty Singleton, and Don Frye. Back in Melbourne by January 1960, and subsequently worked with the Powerhouse and Storyville bands and occasionally resurrected the Southside name. In 1970 he was seconded to the World Bank in Washington and was active in local jazz. Back in Melbourne from 1974, he freelanced, then in 1977 moved to Hong Kong as assistant general manager of the Hong Kong Telephone Co., and led his own group. Returned to Melbourne, 1984, and established his own telecommunications consultancy, but has played no regular work since 1985.

Nicholson, Ross Edwin   b. 8/2/40, Perth, WA   reeds/fl./arr
Took piano lessons as a child and became interested in jazz through the North Cottesloe beach broadcasts by Kevin Conroy. At 14 began clarinet lessons with George Caro and in 1955 formed a group with Jim Beeson, Leon Cole, and Ross Donaldson (dms) playing dances. He attended the Sunday afternoon jazz concerts at the Youth Australia League, and in 1956 he deputized for a period in the Riverside band. Began attending AJCs in 1958, incl. Melbourne 1960, as a member of the Westport JB. Posted as a teacher to Geraldton, 1962–65, and upon returning to Perth began about a decade with J.T. and the Jazzmen. In the mid '70s he spent two years with Abe Walters's night-club group, where his repertoire was invaluably extended. Following two more years with Will Upson's big band at Pinocchio's, Nicholson joined the Swan City Jazzmen, remaining to the mid-'80 excepting for 1981 when he lived in England. At the same time, his broad stylistic sympathies bring him extensive freelance engagements, and there are few if any established bands in Perth with whom he has not worked. Nicholson's decision to remain in Perth has resulted in inadequate recognition on a national scale. Apart from being one of the most ubiquitous musicians on the Perth jazz scene since the '60s, he is one of the most original traditional/mainstream stylists currently working in Australia.

Nolan, Colin James (Col)   b. 16/4/38, Sydney, NSW   p./bs/comp./arr
First jazz work was with Doc Willis in the Bermu-

Ross Nicholson, 1987

Col Nolan in the Rock's Push, Sydney, 1976

da JB at the Macquarie Hotel, 1959–60. With the rock group the Dee-Jays, 1961–62, and during the same period became a regular leader of trios and quartets at El Rocco, with various personnel incl. John Sangster, Gerry Gardiner, Ron Carson, Stewie Speer, Warren Daly. This activity continued into the late '60s; in the meantime Nolan had been with Ray Price during this period, and was leading his Soul Syndicate at the Celebrity Room (1968–69), continuing to use this group name at other later venues incl. Jasons in the early '70s. Foundation member of the Daly-Wilson Big Band. From the '70s has continued to be in great demand on a freelance basis, and has led his own groups, often with Errol Buddle, at established jazz venues like the Rocks Push and Soup Plus. With Galapagos Duck, 1979–80, and frequent performer at local festivals. A pioneer in Australia of the Hammond organ in jazz. A player of greater stylistic versatility, possessing to an unsurpassed degree that compulsive but indefinable capacity called 'swing'. In the '80s, continues to lead small groups, frequently with drummer Harry Rivers.

Norman, Wally   b. 14/3/19, Melbourne, Vic.   tpt/tbn./ldr/arr

Began in brass band at nine and entered dance band profession in 1939 with Jay Whidden at St Kilda Palais, moving to Sydney to join Frank Coughlan at the Bondi Esplanade. Extremely active in dance band/night-club scene: 1940 with Jim Davidson's ABC Dance Band, Ernest Ritte at Rose's, Dick Freeman and later Abe Romain at the Trocadero; 1943 with Monte Richardson's radio group, Craig Crawford at Prince's restaurant, and Bert Howell at Prince Edward theatre. Began at the Roosevelt restaurant in 1944, taking over as leader in 1945 for a long period, during which most of Sydney's top progressive musicians passed

Wally Norman (tpt), 1945, with Don Burrows (clt) and Al Vincer (vibes)

through the band. In 1945 Norman was on the historic Regal-Zonophone records under George Trevare. By 1947 he was known as a leading pioneer in Sydney in fostering interest in bop, through journalism, record sessions, and concert performances which ran into the '50s in Sydney, Newcastle and Brisbane. During this period he also continued night-club work, leading at Christy's, 1950, with Les Welch, 1951, then leading, 1952, at Sammy Lee's. In 1954 he began working with visiting concert packages: MD for tours Gene Krupa, Ella Fitzgerald, Artie Shaw, Buddy Rich, Louis Armstrong, Mel Tormé. In 1955 his band toured Australia and New Zealand backing Buddy de Franco, then Norman spent a 12-month musical study tour in the US. In 1956, MD for the Rex Hotel chain; in 1957 led the band at Chequers and arranged for Les Welch's recording band. Retired from performance in 1960 and formed a theatrical agency which he was still running in 1986, as well as transcribing arrangements on an occasional freelance basis.

O

Orr, Ian Magee (born John McGill Magee)
b. 5/12/39, Paisley, Scotland tpt
Migrated to Australia, 1948. Began playing in a dance band c. 1954. Introduced to jazz by sister Helen and future brother-in-law Bill Haesler, and listening to the Frank Johnson band. Began playing jazz at the 1956 AJC in Melbourne, and sat in at jazz functions during 1957. Through the late '50s active in the casual dance scene as sideman and leader, working with, *inter alia*, Eddie Robbins (clt), Kuzz Currie (dms), Kevin Goodey (clt). Regularly performed at AJCs, including with Frank Turville in re-creations of the early King Oliver repertoire. With Max Collie from 1960 until departing for Europe in March 1961, sailing on the *Patris* as a member of a jazz band.

In London, formed the Yarra Valley Stompers (including Goodey and John Cahill bs) and played with Ken Barton's Jazz Band. Returned to Australia in 1965 and joined Eddie Robbins. Through the late '60s freelanced with various bands incl. the Storyville group and Mabel's Dream. In c. 1970 began a long association with the jazz/rock/blues groups led by Dave Rankin (Rankin File, the Divers, etc.), incl. for the very successful Lemon Tree residency. This broadened his stylistic interests, and Orr took lessons and participated in workshops with Barry Veith's big band, though he continued also working in more traditional groups such as the Maple Leaf Jazz Band. In c. 1981, with Steve Miller (tbn.), established a hard bop jazz venue at the Star and Garter. In 1987 had been with the Storyville group since 1983. Orr is one of that group of musicians (including John Hawes tpt) who emerged during the casual dance era, and continued to develop musically, but who for various reasons have received little recognition outside of Melbourne although they have constituted an essential pool of jazz talent since the '50s.

Oxley, Edwin (Eddie) b. 1929, Romsey, Vic.
reeds/fl.
Began on sax as a schoolboy in Bendigo. To Melbourne in 1945 to study clarinet and piano at the conservatorium. At 18 joined the ABC Dance Band and from the late '40s was involved with the emerging bop movement in Melbourne, playing with numerous small groups incl. the Don Banks Boptet and the Teddy Preston Quartet, as well as with the Freddie Thomas's big band. Formed small groups for ABC broadcasts and country tours, using other pioneer bop players of the day incl. Preston, Ron Rosenberg, Ron Terry or Ken Lester (bs), Bruce Clarke or Brian Rangott (gtr) and Billy Hyde. Studied in England (1955) and the US (1960). Staff musician at Channel 9 Melbourne, 1961–70, and with the ABC Show Band under Brian May from 1977 to its demise in 1982. Since then, session work has been supplemented by teaching and theatre work.

Parkes, Frederick William Richard (Fred)
b. 12/5/31, Melbourne, Vic.   clt
Started clarinet in 1949 and played his first gig in
September. Formed a quartet playing local church
dances until 1955. Joined Len Barnard's band at
Mentone LSC, which involved a stylistic change
away from Goodman towards Johnny Dodds, an
adaptation which probably contributed to Parkes's
strikingly distinctive style. Joined Kenn Jones's
Powerhouse band until 1966. 1966–75 withdrew
from music apart from occasional gigs and recordings, then returned via Robin McCulloch's Datsun
Dixielanders. Joined the re-formed Powerhouse
aggregation for its Victoria Hotel residency and
Sacramento appearance. In 1982 with Peter Gaudion's Blues Express, and in '83 with Ross Anderson's New Melbourne JB. Has taught clarinet in
State schools since 1983.

Pearce, Ian Philp (Nog)   b. 22/11/21, Hobart,
Tas.   p./tpt/tbn./comp./arr
Started piano as a child and began listening to jazz
on radio and records with his elder brother Cedric
and near neighbour Tom Pickering, in the mid-
'30s. In late 1936 or early 1937 he took up trumpet
to participate in private sessions with the other two
and with Rex Green on piano. Remained with this
quartet, which became the Barrelhouse Four, until
posted to the mainland during the war. Until then,
also worked with local dance bands and participated in jam sessions at Fouché's Stage Door. Following his army enlistment, 1942, he played very
little music, but heard Graeme Bell on a broadcast
in Darwin, giving him his first inkling of other jazz
activity in Australia. Stationed back in Hobart
during his last year of service, he resumed as trumpet player with Green and Pickering. Following

his discharge in 1947 he moved to Melbourne to
study at the conservatorium under army auspices,
and began attending the Uptown Club to hear
Graeme Bell, with whom he recorded in 1947. Dental problems led him to give up the tpt, and when
the Bell band returned from its first European tour
Pearce joined the band on tbn., while working
with Tony Newstead's band. Because of his final
examinations he declined an invitation to go with
Bell on the second tour in 1949. In 1950 he went to
England with Don Banks and Ivan Sutherland, and
joined the Mick Mulligan band, tbn. then p. His
experiences in this band are anecdotally recorded
in George Melly's *Owning Up* (Penguin), and
Pearce remained with Mulligan until the band
broke up following a serious road accident involving its touring bus. Pearce went on to play casual
jazz work with, inter alia, Bob Mickleborough and

Ian Pearce, 1984

Sandy Brown, and supported himself in various day jobs. In 1955 he returned to Australia with his family (he had married in England) to work in a bookshop. Resumed association with Tom Pickering, first as intermission pianist at the latter's Town Hall dances, then replacing Keith Stackhouse in the Pickering band. The two have co-led a group ever since. In the late '50s Pearce also formed his own sextet in a more mainstream style, to play ABC broadcasts produced by Ellis Blain, an enterprise which led to a series of recordings produced by Jack Smith of the ABC for Swaggie. Pearce's attitudes to music have been in part documented in Mike Williams' *The Australian Jazz Explosion*. Since rejoining Pickering, his jazz career has followed the same path; in 1986 the two were still working together, mainly in hotel residencies. With Tom Pickering, Ian Pearce is one of the founders and preservers of jazz in Tasmania.

Perth

Perth's geographical isolation, with relatively few visiting musicians until the '80s, and the low level of interstate movement of the locals, is a major determinant of its jazz history. The city's isolation also resulted in a comparatively late start for jazz. Veterans of the '20s, Roy Delamare, Ken Murdoch, Merv Rowston, agree that the music only gained momentum in the '50s.

Bert Ralton's visit in the '20s gave Perth its first contact with live performance billed as jazz, and there is evidence that some aspects of Ralton's music rubbed off on local bands such as that of Charles Sheridan at the Piccadilly Ballroom. Tommy Stratton also developed a reputation as a 'hot' reed player with Fred Nice at the Youth Australia League dances. Theo Walters arrived from the east and formed the Knickerbockers in 1930, but the evidence is that this band entertained a vaudevillean view of jazz. Nonetheless, some foundations were being established for later jazz performance through the massive Youth Australia League band in which many later jazz and swing musicians received their tutelage.

The limited opportunities there had been in the

'20s for jazz performance were reduced to a negligible level as Perth suffered the effects of the Depression. Whatever jazz was being played was in private sessions, such as at the Roe St brothels in the '30s, where an easygoing atmosphere and the presence of parlour pianos were conducive to after hours blowing. The Westralian Modern Music Club, founded in 1934 by Ken Murdoch and Merv Rowston, also provided some focus for swing interests. Some musicians began to develop reputations within the dance band profession as jazz stylists, including Roy Delamare, Jim Riley, 'Skippy' Alexander (reeds) and Bill Hendrie (gtr), in addition to others dealt with elsewhere in this volume. But jazz as such had virtually no public forum. An attempt to establish a promising venue, a night-club called the Miami, foundered for lack of public support.

The Second World War opened the city to new influences. Military postings brought in outsiders and sent locals elsewhere to discover, as John Regan did, that the eastern states were far ahead of Perth in the development of popular music. Visiting American servicemen created a demand for the more frantic music required for jitterbugging, and also constituted a market for more extravagant night life. At the American Red Cross centre Merv Rowston led, then directed, a group which catered to a taste for hotter forms of swing, and with which American musicians jammed. The situation also 'pressure-cooked' younger players like Keith Whittle, who were not only being given unusually extensive opportunities to play, filling gaps created by enlistment and conscription, but in addition, to play a more jazz-based music than had been possible in those circumstances before the war. There was an American Service dance band led by Johnny Turk, which gave the Perth musicians unprecedented experience of dynamics and aggressive articulation.

The end of the war found the same charged situation as in other capitals. Local taste remained relatively conservative—a rhythm club inaugurated on 27 May 1945 survived for only the one meeting. But the general level of jazz consciousness and competence had been raised, and continued to be fostered by an increase in its representation on radio, which had been airing live and recorded swing since the late '30s. The programmes of dance band musician and record collector Ken Murdoch (1907–86), provided many young musicians with their earliest taste of the music on 6KY where he was programme manager during the war, and later on ABC.

In the immediate post-war period, the prevalent flavour of the traditional style in jazz was dixieland of the Condon variety, and this was to be sustained through to the '60s in a succession of bands whose lifetimes overlapped and who shared personnel. Unlike the modern styles which centred on particular musicians, the history of traditional jazz in Perth is the history of bands rather than of individuals. These bands ranged from essentially

cabaret-style groups with a dixieland tinge to a more mouldy posture, with a preponderance of the former. It remains true to the present that conditions in Perth have favoured bands that can leaven the jazz with commercial material, to a greater extent than in other parts of the country, so that the line dividing a jazz band from a cabaret or show band, is not as firmly fixed in the public eye. The Jazz Stars operated from about 1947–52, incl. a very successful run at Maylands TH, on a jazz-tinged commercial programme, and subsequent groups which have effectively operated the same formula are the long running J.T. and the Jazzmen, and the Swan City Jazzmen, the latter, however, with a firmer jazz base. J.T. and the Jazzmen was founded by John Thornton in the early '60s, but leadership was taken over by Dave Way (tbn.) for

The West Side Jazz Group, 1946, with, from l. to r., on the ground, Bruce Wroth (reeds), driver, Harry Child (manager), Arthur Downie (p.), Col Goldsmith (bs), Dick Hatton (tbn), on the vehicle, Keith Hounslow (tpt), Cliff Keeley (dms)

most of its life, which continued into the '80s. At various times, most jazz musicians in Perth have worked in the band, and although strongly oriented to cabaret material, the group had a strong jazz flavour when it enjoyed the presence of musicians like Don Bancroft and Ross Nicholson.

Since the '40s there has also been a line of bands more exclusively jazz oriented, beginning with the West Side Jazz Group which grew out of private practice sessions in 1946. Its personnel incl. Bruce Wroth (clt), Dick Hatton (tbn.), Colin Goldsmith (bs), Bing Throssell and Arthur Downey (p.) and Tom Bone (dms), with Keith Hounslow on trumpet before he left for the east in the late '40s. Many of these musicians became members of the subsequent groups which maintained a traditional jazz presence in Perth. Tom Bone, for example, played piano in the Alvan St Stompers, who were active from 1949–50 with Phil Batty (tpt), Rod Wyatt (clt), Vince Davies (dms), Vince Holmes (gtr/bjo) and founder Brian Foster (tbn.) Several of these in turn became the nucleus of the West Coast Dixielanders through the '50s, which then yielded personnel to the Westport JB from 1958 through the '60s, exchanging some of its members with the Riverside Jazz Group in 1961. Veterans of this succession of bands are still active in two of Perth's established traditionally oriented groups, the Corner House JB and the Lazy River JB.

Later styles of jazz have not enjoyed the same continuity. Unlike Sydney and Melbourne, Perth did not develop an early uncompromising bop movement. In the post-war period, progressive jazz moved from a base in the swing style to a West Coast approach which was heavily watered down for public cabaret consumption. In the mid-'50s John Regan assembled a progressive group which, under his own name and also as the Moderniques, continued to operate with fluctuations in personnel and instrumentation until 1970. Until the early '60s, however, Perth could not sustain a continuous post-traditional activity, so that musicians interested in the style had to be content with non-jazz work, the occasional concert, or leave for the east. There was, however, a local version of the jazz concert phenomenon, although never on the same scale as in the other mainland capitals. In 1947 Sam Sharp and Harry Bluck initiated the Jazz Jamborees, annual concerts which presented a similar variety of material as was being offered through the coming decade throughout the country, with the proportion of material marginal to jazz increased to an extent consistent with the relatively low level of local jazz activity. The drawing power of the jazz billing, however, led to similar ventures, like the Playhouse concerts and the Sunday sessions at the Youth Australia League in the '50s, as well as dance band contests which attracted both jazz groups and straight dance outfits. Sam Sharp organized concerts in addition to the Jazz Jamborees, often with ambitiously augmented orchestras playing elaborate semi-symphonic jazz material, as well as using more conventional big bands with leanings towards the progressive style of Stan Kenton.

The passing of the concert era left a hiatus in the public visibility of jazz although in the traditional area the line was preserved through the succession of bands noted above. There was also some new activity in more modern styles, beginning towards the end of the concert period and involving such musicians as Bill Clowes (p.), Bill Tattersall (dms), Theo Henderson (reeds), Guido Bartolomei (a.k.a Guy Bart) (dms), George 'Rocky' Thomas (tpt) and Frank Smith (bs). Thomas graduated from the

Frank Smith (bs) and Rocky Thomas (tpt), 1957, with Bob Cochran (dms), and Ron Morey (valve tbn.)

Condonesque Riverside Jazz Group and formed several groups, incl. the Modern Jazz Quintet, pushing the music towards a bop style as far as local tastes and capabilities would allow. He subsequently moved to Sydney and went into club work where he is still active. Smith was an energetic organizer who later opened a record shop in Claremont which became a meeting place for visiting and local musicians. He later had a prominent role with the Perth Jazz Club, and the Festivals of Perth, for which he promoted jazz functions up until his sudden death in 1981. Notwithstanding the efforts of these musicians, however, the next big boost to modern styles of jazz in Perth came with the arrival of Bob Gillette followed by Keith Stirling. Gillette was the first fully committed bop/post-bop player heard in Perth, and he and Stirling became the nucleus of a flurry of progressive jazz in the early '60s which was virtually the equivalent of similar movements in Sydney, Melbourne and Adelaide. In a succession of night-clubs/coffee lounges—the Shiralee, the Melpomene, and the Hole in the Wall—Stirling gathered groups of younger musicians such as Bill Gumbleton to consolidate the work of other locals like Bill Clowes, creating some public awareness of jazz other than traditional and swing. In addition to Clowes, the younger musicians who served their apprenticeship at this time continue to be the basis of the modern jazz scene.

During the same time, Perth enjoyed its share of the trad boom. In 1962 the University Jazz Club and a new Perth Jazz Club were established, the latter based in the Tabu coffee lounge, then moving to the Shiralee. In addition to the well-established Westport and Riverside jazz bands, several new groups briefly flourished, providing a start for some of the younger musicians still active in the '80s. The Traditionalists, the Varsity Five, the South Side Five, were among the groups that appeared at club functions, and the popularity of the music was boosted by the film *It's Trad Dad* which opened in July 1962, and the visit of Kenny Ball in October. In a city so isolated and so rarely visited by touring bands, Ball's concerts had greater impact than in the east. The membership

of the Perth Jazz Club increased dramatically, and subsequent traditional jazz bands in Perth continue to bear an unusually strong imprint of the English 'trad' sound.

Following the national slump in jazz activity in the late '60s, Perth was suddenly invigorated in a way unique among Australian capital cities. Before, and independently of the beginning of the Jazz Action Society movement in the east, a new Perth Jazz Club (later changed to Society) was founded in March 1973, very much as a consequence of the work of Don Mead, with Ivan Oliver as foundation president. The club established a house band, the Tony Ashford Quintet and operated initially from various premises until beginning a long association with Hernando's Hideaway. Although its musical brief extended to all styles, the club was particularly valuable in providing support for more contemporary styles of jazz which could scarcely survive in the normal commercial context. The club has continued to operate to the present. Although others have sprung up at different times (the East Perth Jazz Club in 1976, the Jazz Club of WA in 1984), the Perth Jazz Club/Society established in 1973 has had the most profound effect on the local scene. Subsequently, Perth has enjoyed the benefits of the national enfranchisement of jazz that began in the late '70s, with jazz studies courses established under Pat Crichton at Mt Lawley College (later incorporating the conservatorium), the development of student orchestras, incl. the West Australian Youth Jazz Orchestra, participation in the national jazz co-ordinator programme, representation of jazz at the Festival of Perth, and hosting the Australian Jazz Convention in 1979. Perth now enjoys the stimulus of frequent visiting musicians and bands, and in 1983 the first WA Jazz Festival was held in York. In 1985 the festival presented 18 bands, all but one of them local. Although some of these were assembled just for the occasion, the volume of activity is impressive. It would require an analysis of much broader issues than just the local entertainment industry to account for the current scale of jazz performance in Perth, from the evolution of civic attitudes to the economic boom of the late '70s, though even this would leave some mystery as to why jazz in particular should have been the beneficiary of these circumstances. The fact remains, however, that outside of Melbourne and Sydney, Perth currently enjoys more jazz activity across a broad range of idioms than anywhere else in Australia.

Philpott, Ron   b. 21/11/45, Sydney, NSW bs/gtr/arr/comp
Began guitar in 1956; formed a traditional band with Ed Wilson, 1959. This lasted c. six years, gradually evolving into a big band and incl. Warren Daly. From 1965 to the early '70s, he worked commercially incl. touring with a duo floor show with his wife Caron. Began writing for the Daly-Wilson band (1971) and joined Judy Bailey (1974). Also with Doug Foskett's quartet (1974–76). In the late '70s played with The Last Straw at Pinball Wizz, and with Kindred Spirit. Began teaching at NSW Conservatorium in 1979 and in the '80s worked with Jack Grimsley's Channel 10 Orchestra, prepared jazz curricula for secondary schools with Paul McNamara, and has taught and played in schools. In 1986, still associated with Judy Bailey's groups.

Pickering, John Pearce   b. 16/7/30, Adelaide, SA   tbn./bar. sax
Began trombone in 1947 to form a band with Bob Harper (clt). In 1949 Pickering formed the South City Seven which, with almost identical personnel, played at the 1950 AJC as the Cross Roads JB. In the early '50s with Bruce Gray playing tuba. Played with the Black Eagles throughout its lifetime and subsequently a mainstay of the bands active in the Taverns, incl. the Adelaide All Stars. From 1968–69 worked with the Vencatachellum Jazz Peppers, which incl. a number of Black Eagles veterans, playing material inspired by King Oliver's groups. Through the '70s active in traditionally based bands, notably the Pioneer JB, a brief period with the Captain Sturt band led at the time by John van der Koogh (tpt), and most recently with the Adelaide Stompers. Pickering's Ory-based style became nationally known through his regular performances at AJCs, for which he has acted as official parade marshall since 1972.

Pickering, Thomas Mansergh (Tom)   b. 8/8/21, Burra, SA   reeds/vcl/ldr/comp.
The family arrived in Hobart in the early '30s, and with near neighbours Ian and Cedric Pearce, Pickering developed an interest in jazz. In 1936 he was offered a clarinet as an incentive to pass his examinations, his choice of instrument inspired by Benny Goodman. Took lessons from Alex Caddy and began playing privately with Ian and Cedric. Over the next year, Rex Green became involved and the group ultimately assumed the name the Barrelhouse Four, from Jess Stacy's composition 'Barrelhouse'. In 1937 Pickering became the founding president of the 7HO Swing Club and also began tenor at about this time. At Fouché's Stage Door with the Barrelhouse Four and with other musicians, and also recruited as 'hot' soloist as well as session player into the band of Max Humphries (who was later active in Qld). The Barrelhouse Four was identified as the one jazz band *per se* in Hobart, and as such played in the pit for the revue *Red, Hot and Blue* in 1940, and broadcast for the ABC in 1941. In 1941 Pickering was transferred by his employer to Westbury, to be replaced in The Barrelhouse by Melbourne saxophonist Sel Chidgey. Shortly after Pickering returned to Hobart, the band was broken up by wartime postings. (The subsequent careers of Ian Pearce and Rex Green are sketched elsewhere. Cedric Pearce

Tom Pickering, early 1980s

continued to be Pickering's chosen drummer until a stroke forced his retirement from music. Up to that point however he recorded with Pickering and with groups led by Ade Monsbourgh and Roger Bell. Cedric was also a frequent contributor to jazz magazines, and edited *Jazz Notes* during 1941–44. He died in 1982.) During the war, Tom Pickering was stationed locally except for a short period on the mainland where contact with the Bell circle directed his attention to dixieland as opposed to swing. In 1945 there were reconstitutions of a Barrelhouse band with different personnel incl. Milton Driscoll (sax.) and Kay (a.k.a. Yvonne) Staveley, Tony Bell (p.) and Ron Roberts (bs). The original personnel, excluding Cedric Pearce, attended the 1st AJC. In 1947 Ian Pearce and Rex Green moved to the mainland, and Pickering formed a trio with Keith Stackhouse (p.) and Cedric Pearce, later Col Wells, then Ron Roberts and Benny Cuebas. This band, with the usual fluctuations in personnel incl. the addition of Geoff Sweeney in 1949, twice went into the finals of the Maples Parade competitions on Melbourne radio (1949, 1950), played in alternation with Ray Watson's band at the town hall in 1949, and began its own weekly functions at the 7HT Theatrette, so that it was extended to Wednesday nights, using Nick Moore instead of Stackhouse. In the mid-'50s they moved to the town hall, where Ian Pearce came in as intermission piano with Sweeney on vibes, and sometimes Alan Brinkman. The Town Hall dances ran to the end of the '50s, after which Pickering went into cabaret, leading a band at the Stork Club in 1960 broadcasting with Ian Pearce's sextet. Pickering became the inaugural president of the Hobart Jazz Club in 1961. Throughout the '60s Pickering played residencies with Ian Pearce and Max Sweeney through to the mid-'70s. In the meantime the ABC broadcasts which had begun under Ellis Blain passed into the control of Jack Smith, and in 1970 he produced the first of a succession of albums of bands co-led by Ian Pearce and Pickering for release on Swaggie. Mike Colrain now came in on drums; Col Wells was temporarily incapacitated by illness and ex-Brisbane trumpet player Bruce Dodgshun replaced him. In the mid-'70s Wells

resumed, and Oscar Smith (bjo/gtr) and Graham Ranft (bs) began working with the band which stabilized with that personnel for the next decade, except for 1979 when Pickering was replaced by John Broadley and Alan Brinkman in alternation while he was travelling overseas on a Churchill Fellowship for jazz studies. He suffered a heart attack following his return in July, then rejoined the band, which continues to be the most active jazz band working in Hobart. Pickering has always maintained and extended a broad repertoire which incl. his own compositions. In 1982 he was made Member of the Order of Australia for his services both as a jazz musician and as parliamentary librarian. He has also written extensively, both for scholarly periodicals incl. *Quadrant*, and fiction, for which he has received literary awards. In 1984 the Pearce/Pickering band played a concert in Salamanca Square which was shown on national television in *The Burrows Collection* series.

Piercy, Peter (The Judge)   b. 10/8/27, Pt Lincoln, SA   p.
Studied piano from seven to 12, then at the SA Conservatorium in Adelaide. Became interested in jazz in late teens. From the late '40s played in RAAF dance band and on Bill Holyoak's swing/jazz concerts at Adelaide TH and the Tivoli, in his own groups and those of Maurie Le Doeuff, Paul Thomas, and Ian Boothey (like Piercy, Boothey has subsequently settled in Sydney where, in addition to club work and jazz freelancing, he organized the ambitious Australian Jazz Orchestra in the '70s in which Piercy was also involved). Piercy also worked on the Adelaide dance circuit, incl. with the Unley Palais Orchestra, 1953. To Sydney c. 1954, playing night-clubs incl. André's and Chequers. Toured Australia and New Zealand with the all-black American revue, Harlem Blackbirds, 1955, and was much influenced by the show's arrangements by MD Gene Kee. The revue incl. comic Pigmeat Markham whose famous 'Hear come de judge' routine led to Piercy's nickname. Piercy joined John Bamford's big band at the Sunday Ironworkers' Hall sessions (1957). In Nov.

1958 he played the Billy Eckstine show, the first of many Sydney Stadium concerts, incl. jazz, rhythm and blues, and rock, in which he performed. Through the '60s and '70s he was mostly active in club and commercial ventures, and was inactive for a period following a street bashing in Kings Cross. Freelanced in jazz contexts, and in 1976 joined Paul Furniss's band at the White Horse Hotel, Newtown. Led a trio for the Jumpin' at the Jubilee Concert (1983), and joined Terry Rae's (dms) Orient Jazz Express (1984), with Dart McRae (reeds), Dave Colton (gtr), Tony Buckley (bs). In March 1985 the Stormy Monday unit from 2MBS-FM honoured Piercy with a concert, *The Court's In Session*, with Piercy leading Merv Acheson, Colton, Buckley, Alan Geddes, and Sally King (vcl). Piercy is an example of the authentic and adaptable jazz musician who, over many years, has helped to sustain the music but who, in choosing to freelance, receives almost no recognition outside the profession.

Pochée, John Kenneth   b. 21/9/40, Sydney, NSW   dms/ldr

A left-hander playing on a right-handed kit, he began his career with Dave Levy at the Mocambo in 1956. The two worked together at the El Rocco in 1957, and in 1958 Pochée abandoned journalism to become a professional musician. He spent 1959–62

John Pochée at the Sydney Musicians' Club for the Jazz Action Society, 1977, with Ron Philpott in background

in Melbourne playing with Dave MacRae, Keith Stirling, Graeme Lyall, Tony Gould, Chuck Yates, and Joe Lane who was an enthusiastic influence. Following an interval in Sydney, 1963–64, with Levy, Bruce Cale, Bryce Rohde and Lyn Christie, he returned to Melbourne to join the Heads at the Fat Black Pussycat in 1964, a band name which, with Bob Bertles, maintained an intermittent existence through the late '60s back in Sydney. Following work with Don Burrows, Pochée spent a brief time in Adelaide leading a trio. Back in Sydney by 1966 he led a trio (Yates, Mike Ross) at the Mandarin Club for two years, then became involved again at the El Rocco with, *inter alia*, Bobby Gebert, Keith Barr, Serge Ermoll, Bernie McGann. During the late '60s, Pochée worked in less jazz-oriented areas, for two years as house drummer at Chequers night-club, and in 1969 and 1970, touring with Shirley Bassey. In 1971 he became MD for the vocal group, the Four Kinsmen, with tours in Australia and the USA. Has worked with most of the important progressive groups of the '70s and '80s, incl. those of Gebert, Roger Frampton, recording and touring SE Asia with Judy Bailey, and playing with The Last Straw since its foundation in 1974. Although primarily influenced by bop and post-bop styles, Pochée maintains a very broad stylistic sympathy and has been one of the most durable and active drummers of that young generation of modernists who rose to prominence from the late '50s.

Polites, Nicholas (Nick)   b. 2/7/27, Melbourne, Vic.   clt/alt.

With Manny Papas's band (with Doc Willis), 1945–46, and played with these two in the Jazz Appreciation Society Band at the 1st AJC. With Allan Bradley's Rhythm Kings, 1947–50, John Sangster's JB, 1950, and Frank Johnson's Fabulous Dixielanders, 1951–56. Joined Llew Hird in what became the Melbourne NOJB under Polites's leadership. Following the band's dissolution in England, Polites returned to Australia via New Orleans, the first of several visits. With Yarra Yarra JB, 1964–66 and 1971–73, with various other groups between those two periods. Formed the NO Stompers, 1975, and in the early '80s was playing at the Auburn Hotel where his band had started in 1977. Melbourne AJC trustee, and regular AJC performer. A businessman until retirement in 1970, Polites is director of an ethnic welfare organization.

Pommer, Rolph   b. 24/11/14, d. 1980, Ipswich, Qld   alto

After his education at the local Christian Brothers' College, Pommer moved to Brisbane in 1932 and became active in dance bands. The advent of swing in 1936 gave him the opportunity to develop a hot style in performance, and he attended the first meeting of the Brisbane Swing Club on 15 November. In 1937 he took over from the Premier Dance Band, in which he had been playing, with his own Swing Men at the Bishop Island Sunday dance. Following casual work he started at Prince's ballroom under Coughlan in 1940, taking over as leader mid-year, then in July left for Sydney joining Tut Coltman. During the war he played in the RAAF, with Giles O'Sullivan at the Booker T. Washington Club. With Reg Lewis's band at Grace Bros. Ballroom in 1944, and, following his discharge, went into the Roosevelt under Wally Norman in 1945. In July 1945 he played on the controversial George Trevare recording sessions, his only commercially released material. Regarded

as among the country's top alto players, his quintet was reviewed as one of the best bands of 1945. During the late '40s he worked with Frank Coughlan, with Frank Smith under Marsh Goodwin (tpt) at the Surreyville dance hall, with Bill Cody's (dms) swing shows at Segars, and was active in private jam sessions organized by Bill Mannix (sax). In 1949–51 at Surfers Paradise Hotel replacing the recently deceased Jim McLaren in Jack Goldner's group. After a brief period living in Wollongong, NSW, he returned to Sydney and worked with Ralph Mallen's band before going into retirement in the late '50s. Resumed playing in 1959, but only intermittently active from this time on. Some work for Sammy Lee's night-clubs in Sydney was followed by a return to Brisbane where he sat in occasionally with Sid Bromley at the Oxley Hotel. He returned to Sydney in 1963, but health problems, incl. increasing deafness, led to diminished musical activity. He retired in about 1970. That Pommer's work has been so sparsely preserved is one of the great misfortunes in Australian jazz, and no small indictment of the national sensibility. He was idolized by Frank Smith, himself an under-recorded legend. Willie Smith, one of the most accomplished altoists in American jazz, was in Australia during the war, when he heard Pommer, whom he described as 'The best white alto player in the world' (quoted by Merv Acheson, *QR*, April 1980).

Rolph Pommer, 1945

The Port Jackson Jazz Band (Sydney, NSW)

This band was started by trombonist Jack Parkes in 1944—from the beginning, trumpeter Ken Flannery was the star member. By 1948, Flannery was the only remaining original, Ray Price and Jimmy Somerville being the most significant of the newer arrivals. In March 1948, a series of concerts at the Conservatorium of Music made the PJJB a household name in Sydney, and triggered the jazz concert era. Following this great success, they undertook a country tour to Brisbane and return, but it was a financial disaster. The band broke up in Brisbane. Flannery reformed the band in Sydney, Somerville formed a rival band, the Jazz Rebels, and Price temporarily forsook music, then studied

The Port Jackson Jazz Band, probably 1948, with, l. to r., unknown vcl, Jim Somerville (p.), Duke Farrell (bs), Ken Flannery (tpt), Bob Cruickshanks (clt), Ray Price (gtr), Clive Whitcombe (dms)

the bass violin and joined the Sydney Symphony Orchestra. Somerville rejoined Flannery in 1949 and the band resumed its successful concert routine. Personal tensions still existed, and Flannery handed leadership to Somerville before leaving the band completely. He rejoined later in 1949, but by the end of 1950 the band had disappeared from the scene. 1952 and 1954 saw brief revivals of the principals under the name Ray Price & His Dixielanders, but it wasn't until October 1955 that the Port Jackson Jazz Band really got back together again, and once more became 'Sydney's band'. The publicity generated led to Price being sacked from the SSO, and he devoted his energies to the PJJB full time. Flannery left to join the ATN-7 orchestra, and Bob Barnard came up from Melbourne to play with the PJJB, and the trio at the Macquarie Hotel. Bob returned to Melbourne

in June 1958; his place in the trio was filled by John McCarthy, and Flannery was able to rejoin the full band again. From 1955 to 1962, the PJJB was Sydney's leading group, with a long Sunday night residency at the Ling Nam, two ball seasons at the Empress ballroom, and the trio or quartet at the Macquarie and Adams Hotels five sessions per week. They shared the bill with Dizzy Gillespie and Coleman Hawkins at Sydney Stadium in 1960, and with the Brubeck Quartet at the Adelaide Festival of Arts in 1962. In September that year, disagreements between Price and other members of his quartet—John McCarthy, John Costelloe and Dick Hughes—led to the breakup of that four, and of the seven-piece PJJB. Since then the band has reformed, without Price, for particular gigs such as the Katoomba Festival in 1965 and the Jazz Convention at Balmain in 1975. Under Dick Hughes's leadership it played a season at the Stage Door Tavern. In 1978, Flannery and Somerville took the Port Jackson Quartet, later expanded to a Quintet (with John McCarthy) into the Mosman Rowing Club, Saturday nights. In 1985, Ray Price's health permitted his return to music, and the PJJB was featured at the Regent Hotel and the Sydney Town Hall. It also shared the bill on different occasions with the Woody Herman All Stars and the Glenn Miller Reunion Orchestra.

Entry compiled by Jack Mitchell

Porter, Rodney Crawford (Rod)   b. 9/3/35, Adelaide, SA   reeds
Studied piano from age 10, and inspired to take up clarinet after hearing Bruce Gray. With Richie Gunn's Collegians (1950), Doug Whitrod's Jubilee JB (1951), and became president of the University of Adelaide Jazz Club in 1956 when he also took up alto. As an undergraduate, played with the university jazz band, which incl. John Melville (tpt), Colin Nettelbeck (p.), Mick Drew (dms) and Ron Williams (bs). Played with Roger Hudson's Eumenthol Jazz Jubes, 1957, and played at 1958 AJC with the University JB. In England, 1959–62, with negligible playing. Back in Adelaide, the trad boom provided many opportunities: with the Adelaide All Stars, then the Campus Six until 1976, as well as other, short-lived, bands. With Dave Dallwitz groups since 1973, and led the Friends of Jazz, which evolved from the Campus Six, 1978–79. During the '80s has continued on an occasional basis with Dallwitz, but apart from casual work, has been comparatively inactive musically.

Price, Henry Thomas (Harry)   b. 17/6/30, Melbourne, Vic.   tbn./vcl
Taught by Cy Watts in 1949, led the Melbourne JB, 1949–52. Joined Neville Maddison at Leggett's, 1952–53, and in 1955 joined bands led by Russ Jones and ex-Adelaide sax player Ian Boothey. With the Kenn Jones Powerhouse group, 1958–66, and from c. 1959–62, with Frank Gow's band. With Bob Harrison's Firing Squad at the Keyboard Club 1967–69, then with Alan Eaton

until 1971. During the '70s toured with visiting performers incl. Johnny Mathis, Kathryn Grayson, Des O'Connor, Manhattan Transfer. With John McGaw's Jazz Experience, 1977–81, then with Kenn Jones's re-formed Powerhouse group at the Victoria Hotel. In 1984, with the New Melbourne JB and the Camelia Quartet.

Price, Eric John (Rick)   b. 12/9/30, Portsmouth, England   tpt/vcl
Studied piano as a child, and began trumpet in his teens. Worked in jazz clubs and pubs in southern England, then 1949–50 in RAF as a musician. To Brisbane in 1974. Formed Bebop Revival Band, incl. Clare Hansson, Frank Tyne (reeds), Jim Howard (dms), Ian Cocking, which represented a departure from the usual Brisbane scene. Continued to lead various bop-oriented groups and freelances in more traditional/mainstream settings, performing for Qld JAS, the Cellar Club.

Price, Ray   b. 20/11/21, Sydney, NSW   bjo/gtr/ldr
Began on drums, then guitar with lessons from Charlie Lees. Played in this family's band, incl. a tour of New Zealand, 1938. Assistant editor of *Tempo*, 1938–39. Served in the army, 1940–43, and played for a time in the Booker T. Washington Club. With Les White on 2UW, 1945 and with Merv Acheson at the Syncopation Swing Club, 1946. By July 1947 was in the Port Jackson JB, playing various engagements incl. dances at Air Force House until, discovering the management barred blacks, Price pulled the band out without notice. Price organized various concerts for the Port Jackson band, incl. at the conservatorium and a tour of NSW and Qld, 1948, with personnel Ken Flannery, Dick Jackson (reeds), Jim Somerville, Clive Whitcombe (dms), with Duke Farrell joining mid-tour. Bad weather and maliciously circulated accusations of communist sympathies led to the collapse of the tour in Brisbane. In 1949 Price began studying bass at NSW Conservatorium (he previously had discontinued bassoon study because of asthma). In June 1950 he joined Sydney Symphony Orchestra under Eugene Goossens on

Ray Price Quintet, early 1970s, with, from l. to r., Col
Nolan (p.), Paul Simpson, Laurie Bennett (dms), Mike
Hallam, Ray Price

bass, later working with the National Ballet and
ABC orchestra. He formed the short-lived Dixielanders (Flannery, Somerville, Jim Shaw dms,
Jackson, Billy Weston, Nellie Small vcl) in 1952,
revived them in 1954, then reformed the PJJB,
1955–62. An ultimatum regarding his jazz activities led him to be dismissed from the SSO, 1956 (a
circumstance flavoured retrospectively with some
irony when Constantin Silvestri, subsequently
guest conductor of SSO, praised enthusiastically
Ray Price's group at the Macquarie Hotel). In 1956
Price formed a trio (Bob Barnard, Dick Hughes)
for the Macquarie Hotel, moving to the Adams
with John McCarthy having replaced Barnard.
John Sangster was added on trumpet to be replaced
by John Costelloe in 1959. The quartet toured
Queensland for the Arts Council in the early '60s,
and was booked for the Australia Hotel. Because
of management antagonism arising after the contract had been signed, however, the band found
itself in the unusual position of being paid but
not permitted to play. The Port Jackson Band
and the quartet both broke up with the depar-

ture of Hughes, McCarthy and Costelloe, 1962,
and Price formed a new quartet with Col Nolan,
Cliff Reese, Pat Rose (reeds). In the mid-'50s the
group began its school concert/lecture recitals,
which continued until 1980. When the Adams
Hotel residency finished in 1966 these recitals
became Price's main activity, with a succession of
Quartets and Qintets [*sic*]. He also acted in other
entrepreneurial roles, incl. as MD for the Waratah
Jazz Festival, 1973. He retired with ill health in
1982, moving to the Port Macquarie area where he
organized a jazz festival, 1985. He also resumed
playing on an occasional basis for PJJB reunions
from 1985. Price has been one of the most vocal
and ubiquitous jazz evangelists in Australia, both
as writer and performer. His school tours gave a
whole generation of children their first exposure to
Australian jazz, incl. a great many musicians
currently keeping traditional jazz alive. The band
alumni incl. most traditional musicians working in
Sydney since the '60s, and many who have also
made an important contribution in more modern
areas.

Primmer, Sterling   b. 5/5/31, Albury, NSW   p.
Took lessons from Roy Maling at age 17 while
living in Sydney where, with other beginners Colin
Bates (p.) and Jimmy Shaw, he followed the Ralph
Mallen band. Played in Bill Townsend's jazz band
before moving to Canberra (1951) where he joined

Sterling Primmer, 1980

Bruce Lansley's group. Travelled in England, Europe, and Canada, 1953–56, where exposure to top bands was inspirational. Returned to Canberra and rejoined Lansley, with whom the Sydney Frank Johnson (tpt) also played for a period. Worked with Greg Gibson 1957–58, and later with a quartet (John Hamon bs, Terry Wynn, Lansley) which won the Canberra Jazz Club's Band of the Year Competition, leading to a number of TV appearances. Through the '60s and '70s primarily led small groups for restaurant work. In Sydney, 1981–82, with Noel Crow, then leading a trio. Since returning to Canberra, briefly with the Fortified Few, and casual freelancing.

Properjohn, Gregory E. (Alf)   b. 10/3/49, Hobart, Tas.   dms/ldr

Began drums in 1966 and first performed publicly with Ian Pearce at Hobart AJC, 1967. To Melbourne, 1969, working in commercial settings and in a rehearsal group led by Ken Schroder. Returned to Hobart, joined Sullivan's Cove JB, 1972–77, then led groups in Tattersall's Hotel, 1977–81. Was invited guest at Allan Leake's Jazz Parties in Melbourne in 1978 and 1979. Has

Alf Properjohn, 1986

worked with visiting musicians, incl. Kenny Davern, Herb Ellis and Barney Kessel, Penny Eames, Johnny Nicol and Col Nolan, Bob Barnard. Played school tours for the Arts Council in 1980 and 1981 for JAS, of which he became musical co-ordinator in 1980. In addition to casual playing, Properjohn was with Peter Webster's Integrations in 1984. Percussionist in performances of Claude Bolling's Suite for Flute and Piano and percussion tutor at Hobart College of Music 1978–80. Founder member of the Contemporary Jazz Society in 1981 and is the first Tasmanian jazz co-ordinator.

Qua, Christopher George (Chris)   b. 14/11/51, Orange, NSW   bs/tpt

Son of pianist Pat Qua. Learned trumpet from Harry Berry, and began bass at school. First paid gig at 15, with Alan Lee. Began jamming c. 1970 with other Sydney musicians during holiday seasons in the Snowy Mountains, and from this became a founder member of Galapagos Duck.

Chris Qua, The Basement, 1976

Left the Duck, 1980, and over the next three years worked freelance and regularly with Johnny Nicol, Tom Baker's Quartet, and with the trio of ex-English drummer Stuart Livingston. Toured with Daly-Wilson big band (1983) and resident with Su Cruickshank's group at the Brasserie until 1985. In 1986 still freelancing and doing some teaching.

Qua, William Everest McKay (Willy, Quill) b. 31/5/53, Orange, NSW   dms/reed/fl.

Son of pianist Pat Qua. Began at 11, and played first gigs on drums while at school. Began working with different groups, incl. Roger Frampton's, at the Rocks Push in early '70s. With Galapagos Duck, 1973–76. Formed Quill's Folly, 1976, disbanding the group in 1981. Has freelanced extensively, worked with Vince Genova and the group Nebula (with Indra Lesmana), and toured with visiting musicians incl. Jimmy Witherspoon, Herb Ellis/Barney Kessel, Mark Murphy, and five tours with Georgie Fame. Ran the Quill Club, a late night jazz venue in Petersham, 1984–85. In 1986 was active on a freelance basis.

Record Labels

The major recording companies do not record a lot of jazz, and when they do, they look first to the well established 'safe' names. Less well-known bands and younger musicians can gain exposure only by producing their own records on the custom pressing labels of the big companies, or by appearing on one of the minor labels which specialize in jazz records alone.

This problem was even more apparent in the '40s. Not only was there but one major company (albeit producing five different labels), but that company, owing to the wartime shellac shortage which coincided with an increased demand for all types of entertainment, could sell every record it could press. Consequently it was not inclined to experiment with untried jazz talent. The increasing number of jazz fans had to look elsewhere.

In Melbourne, collectors Bill Miller and C. Ian Turner were recording, mainly for their own interest, the local jazz musicians, such as Graeme Bell and Ade Monsbourgh. It wasn't until wartime controls were eased in 1946 that materials became available, allowing the commercial release of some of these recordings.

AMPERSAND was the first specialist jazz label to appear. Owned by Bill Miller, a Melbourne solicitor, it eventually ran to 36 × 10″ and 5 × 12″ issues recorded by Australian jazzmen, as well as a reissue series of 5 × 10″ discs of American material. Another Miller label was called XX—a sly reference to the doubtful legality of some of the issues which were dubbings of early American jazz recordings on long defunct labels such as Gennett and Paramount.

In Adelaide, Bill Holyoak brought out the MEMPHIS label (which remained the most attractively artistic label of them all) to feature the Southern Jazz Group. Only eight issues appeared on Memphis.

Soon other collectors and organizations were following in the footsteps of Miller and Holyoak, bringing out discs which featured bands and/or styles not to be found on Ampersand or Memphis. The following is believed to be a complete listing of these specialist labels in the 78 rpm format. The owner of the label is named in brackets, and then the number of issues on each label.

BLUE STAR (Ron Wills) (3)
CIRCLE AUSTRALIA (Ron Wills) (6)
DUEL DISC (Lyall Richardson?) (1)
ELMAR (Norman Linehan) (6)
GEORGIA (Loel's Music Store) (2 or 3)
JAZZART (Clemens Musical Serivce) (c. 34)
LIBERTY (Liberty Music Shop) (1)
ROSEVILLE (Lyall Richardson) (5 or 6)
SOUTHERN CROSS (Victorian Musicians' Union) (1)
SWAGGIE (Graeme Bell) (5)
VARSITY (Melbourne University Jazz Club) (1)
WILCO (Ron Wills) (25)
ZENITH (Ross Fusedale) (2)

Other 78rpm labels which were advertised, but cancelled, were:
CLUB (Frank Johnson?)
JAZZ BAND (Frank Owen Baker)
JELLY ROLL (C. Ian Turner)
VOX (unknown)

The ending of the 78 era coincided with a falling off in the public interest in jazz, and the above listed labels had all been discontinued by the end of

1952. Apart from Swaggie and the Ron Wills labels, they had limited distribution, often from agents otherwise unconnected with the music or record retailing business. These labels not only spread the jazz message around the country, they preserved for us a treasury of music that demonstrates how the pioneers of both revivalist and bop in this country sounded better than can any printed page. LPs have reissued material from at least six of the above labels, plus some titles recorded, but never issued, by them. More can be expected.

Jazzart continued into the microgroove era, issuing 4 × 10″ LPs, but the label was only a memory by 1954. It is understood that the entire catalogue will soon appear on cassettes.

Swaggie lapsed when the Bell Band returned to Europe in 1950. Nevill Sherburn bought it in 1954—starting cautiously he built up a catalogue of 11 × 10″ LPs and 36 × 7″ EPs. Two of these 45s were in fact folk music of British origin, but it is not unreasonable to regard Swaggie not only as a specialist jazz label but as Australia's best such. Since the early MG days, Nevill Sherburn has put out more than 100 × 7″ and 200 × 12″ LPs, featuring local and overseas bands in both reissues and original recordings. Swaggie records are well known in overseas collecting circles, and are highly regarded for the quality of programming and production.

With some 30 LPs to its credit, Jazznote is our next most prolific specialist label. This is owned by a Melbourne consortium headed by Allan Leake, but most issues are produced co-operatively with the band involved.

44 is unique, not because its name is a number (there are others such), but because it is the only Australian jazz specialist label produced by a major company. Managed and/or owned by Horst Liepolt, production, artwork and distribution are undertaken by Polygram. Thirty LPs and three 45s were issued before Liepolt moved to New York and the project ended.

A host of other labels have seen the light of day, most with only one or two issues to their credit. Amongst those with at least five LPs in their catalogues are: Batjazz; Cumquat; Jazz and Jazz;

Rain Forest and Seahorse. Many bands now produce their own records and cassettes, often using no named label, or that of the custom pressing service of a major company, but it is to be hoped that there will always be individuals interested enough to produce jazz records on their own specialist labels. Without them we cannot understand and appreciate the infinite variety of Australian jazz.

Compiled by Jack Mitchell

Red Onions Jazz Band (Melbourne, Vic.)

In 1960 schoolboys Brett Iggulden, Allan Browne, and Bill Howard discovered jazz through Iggulden's father's records. Iggulden and Howard took lessons on trumpet and trombone respectively from Ade Monsbourgh, and with fellow pupils Kim Lynch (tba) and John Funsten (clt), and Browne, who was learning drums from Norm Hodges, they formed a band with the addition of Felix Blatt (bjo) and John Pike (p.). The band started a residency at a dance in Beaumaris in 1960 and at the Oxford Club and Edithvale LSC in 1961. In June 1961, appearing on Channel 2, the members were forbidden the use of the current band name, the Gin Bottle JB, because of the teenage audience, and they renamed themselves the Red Onions JB. The youth and exuberance of the band were amplified by the members' adoption of long hair and Edwardian style dress. When much of the Melbourne scene was dominated by NO mannerists, the Onions stood out with their high energy and sometimes theatrical image, enhanced by featuring Iggulden's sister, Sally, on washboard. On the crest of a jazz wave, the band quickly achieved cult status. They began a residency at their own jazz club, the Onion Patch, in 1962, made their first record and Gerry Humphries (clt) joined. In 1963 a Red Onions Fan Club was formed and the band worked major clubs incl. Memphis, Esquire, and the Melbourne Jazz Club, with concerts in other states. In February 1964 Ian Clyne took the piano chair which had been vacant for about a year, and with the assistance of Bill Haesler the band negotiated a long-term recording

The Red Onions Jazz Band, 1968, with, l. to r., Conrad Joyce (bs), Richard Miller (reeds), Allan Browne (dms), John Scurry (bjo), Brett Iggulden (tpt), Rowan Smith (p.), Bill Howard (tbn.).

contract with W & G. In November their opening night at the Opus Club attracted more than 2000 people.

In 1965 John Scurry replaced Rainer Brett, who had been on banjo for some time, then in September a split created two bands. Humphries, Lynch, and Clyne broke away to form a pop group, the Loved Ones, and Iggulden, Howard, and Browne recruited Dick Miller, Rowan Smith (p.), and Bill Morris (tba) into the Onions. This personnel remained stable for five years, and represented the band at its peak. In 1966 they began a two-year residency at the Royal Terminus Hotel, Brighton, and continued recording and TV work. At this time their repertoire was beginning to embrace big band material from the '20s and '30s, in particular from Luis Russell and Duke Ellington.

In June 1967 the band left for England, preceded by Dick Miller to organize advance publicity. The Onions played in England and toured Europe in packages which incl. Georgie Fame and Roland Kirk. They appeared at the Polish Jazz Festival and on Eastern bloc TV. In March 1968 they arrived back in Australia and appeared on the first direct TV transmission to Japan by satellite. They played residencies at the Commodore in Sandringham, the Village Green, the International at Essendon, the Prospect Hill, and ran their own club at Danny's. By the time they left for their second overseas tour, Conrad Joyce, who replaced Bill Morris, had switched from tuba almost exclusively to bass, thus increasing the suppleness and drive of the band. During their second tour of Europe in 1970, they played the International Jazz Festival in Hungary, and upon their return to Australia in December, Rowan Smith remained in England to study art.

Early in 1971 the Onions began residencies at the Prospect Hill and the Waltzing Matilda Hotels, and appeared regularly at the Victorian Jazz Club functions. In 1972, with some members feeling an urge to develop, they cancelled all work and began rehearsing more advanced material. In Feb. 1973 they resumed performing, but found themselves prisoners of the old public image which had earlier been so important to their success. Although they continued to diversify with, for example, electric instrumentation, they found it necessary to revert to more traditional approaches for most of their work. It was clear that its members could not progress freely within the constraints of the Red Onions' image. Except for occasional reunion performances such as in 1977 at Smacka Fitzgibbon's restaurant, and in 1983 as a benefit for Bill Howard who had been burnt out in the disastrous bushfires of that year, the Red Onions ceased operating around 1974–75.

Brett Iggulden later moved to Bellingen, NSW, where he runs his own business and plays casually with other musicians in the area incl. Graham Bennett. Bill Howard spent a long period in musical retirement, but resumed playing in the mid '80s, and has been Bob Barnard's 'tour' trombonist since the death of John Costelloe. John Scurry and Conrad Joyce have remained active on a casual basis, the latter also with Allan Leake's Swing Shift in 1985. The subsequent careers of Allan Browne and Dick Miller have been sketched elsewhere.

Much of the foregoing has been derived from a history of the band prepared by Roger Beilby, and published in part in *Jazz Down Under*, March 1975.

Reese, Alexander Clifford (Cliff)   b. 27/10/24, Brisbane, Qld   tpt/comp./arr
Began on drums in East Brisbane Juvenile Brass Band, then switched to Eb brass bass then to cornet. With George Craitem at the Bellevue Hotel (1940) then joined the Cocoanut Grove band (1942) where patronage by US servicemen encouraged a hot swing approach as well as featured dixieland segments. Also worked through the war years with Billo Smith's older style dance band, with US service musicians incl. Frank Thornton, and in broadcasting in groups led by Max Humphries, Eddie Coburn, and Joe Allen. With Joe Allen's Four Aces at Redcliffe, 1946–47, and played Brisbane Swing Club functions. Founder member of the Canecutters, remaining until departing from

Brisbane. In 1948 led the Hot Jazz Society's house band and played in radio 4BK's swing group (Jack Thomson, Arthur Howard tbn, Maurie Dowden p., Eric Wynne reeds, Allen, Humphries). To Melbourne, 1951, to join Col Bergerson at Ciro's, then with Stan Bourne at State Theatre, followed by work with the Tivoli orchestra. In 1952 he began six years in the RAAF with its band at Laverton, then returned to Melbourne, working with Lee Gallagher's *In Melbourne Tonight* TV orchestra, at the Tivoli, with Denis Collinson, and backing visiting tours. To Sydney, 1960, to join Graeme Bell at Sylvania Hotel, then joined Ray Price at Adams Hotel, 1964. Rejoined Bell for his new All Stars, 1968 to the mid '70s, after which he joined the Harbour City JB of which he was still a member in 1986. Reese has also freelanced in session work, has played in symphony orchestras and is nationally eminent in brass band work.

Reeves, Adolphus Francis (Splinter)   b. 9/1/24, Jarrahdale, WA; d. 23/1/87, Tweed Heads, NSW   reeds/ldr
Began on concertina at six, switched to alto in 1935. In the pit orchestra at His Majesty's, Perth, for the Will Mahoney Show in 1939. To Melbourne, 1941, where his slim frame earned him his nickname, and joined Frank Coughlan at the Trocadero in 1942, followed by work at the Dugout with George Watson. In about late 1945, Reeves began two years with Frank Arnold at the Palm Grove and in 1946 joined Craig Crawford's band at the Storklub, taking over as leader in 1947. In Sept. 1945 he had led the winning band on the P & A Parade competitions on 3KZ, with Charlie Blott, Lin Challen, Don Banks, and Alf Baker (gtr). These musicians, and others incl. Storklub colleagues Bob Young, Errol Buddle, Ken Brentnall, became part of the nucleus of the bop movement of the late '40s in Melbourne, with Reeves one of the leading lights. Played with Doug Beck and Brentnall at the first of the New Theatre concerts promoted by the Modern Music Society in Aug. 1947, and formed his Splintette in 1948, recording for Jazzart in February 1949 with Jack

Splinter Reeves's Splintette, Delphic Nightclub, Melbourne, 1949, with, from l. to r., Ted Preston, Splinter Reeves, June Carey, Jack Williams, Charlie Blott, Bruce Clarke, Ken Lester

Williams (tbn.), Ted Preston, Lin Challen, Bruce Clarke and Charlie Blott. In Aug. 1949 the Splintette played on Rex Stewart's first Australian concert and subsequently on Bob Clemens's Downbeat concerts. Reeves was also playing with Geoff Brookes's group (incl. Preston, Blott, Clarke) at the Copacabana, and in October his Splintette recorded again, with Ken Lester replacing Challen, and June Carey (vcl). In 1950 he participated in the Boposophical Society sessions at the Galleone and in July made his Sydney debut at a Celebrity concert with Blott. Through the next two years the local bop activity diminished, although Reeves continued to lead his group for concerts, broadcasts and in cabaret, and he recorded two more sessions, in June 1952 with Keith Hounslow, Geoff Kitchen, Ron Rosenberg, Bud Baker and Blott, then in 1953 with Hounslow, Lou Silbereisen, Blott and Stan Walker. In 1954 he left Melbourne,

first touring New Zealand with the Folies Bergères, then moving to Surfers Paradise, Qld where he worked for many years with Stan Bourne. In 1955 he won the tenor category in the *Music Maker* poll. During the early '60s he was based in Townsville, Qld, and also wrote for *Music Maker*. From 1964–70 Reeves was back in Melbourne with the Channel 9 orchestra, then to Sydney, 1970–81, from whence he worked on cruise ships for two years, then mostly club work with very occasional jazz gigs. Following some ill health retired to northern NSW, where he died of injuries sustained in a road accident.

Regan, John Norman (Jack)   b. 19/7/26, Fremantle, WA   dms/ldr
Inspired to take up drums in 1940 by Merv Rowston, who gave him lessons. Following cabaret work with Roy Jenkins, Regan entered the RAAF,

John Regan, c. 1970

was posted to Melbourne, Adelaide, and Darwin where he received his discharge after playing in a service band. Back in Perth, he worked in dance bands before forming his own group in 1954 with Jim Beeson, Ralph Filmer, George Caro (reeds) and George Franklin. Beeson, Filmer, and Caro were at different times replaced by Bruce Halliday, Alan Cole, and Keith Loder, and this band worked as the Jack Regan Quintet or the Moderniques. The band was a pioneer of West Coast jazz in Perth, playing concerts and cabaret engagements. Through the '60s the size and character of the band altered, with Loder's commercial vocals assuming greater importance. The band ceased functioning c. 1970. Regan retired from music until joining Barry Bruce's Chicago JB, for the Ocean Beach Hotel. Following this residency, Regan has remained active on a casual basis and since 1982 has been with Fine and Dandy.

Riley, Ernest James (Jim)   b. 13/11/14, Fremantle, WA   p./vibes/bjo/arr
Began piano in 1922; first engagements in 1929 for silent movies. With Ron Moyle, 1937–38, and the Lyric Trio (Merv Rowston, Ken Murdoch), subsequently augmenting the latter to the Swing Five for the ABC. At Cocoanut Grove during the war, then in Sydney as pianist/arr for Frank Coughlan at the Trocadero. Except for nine months in England, where he played with Nat Gonella, Riley remained at the Trocadero, taking over as leader in 1951, until 1952. Moved to Qld, playing on the Gold Coast with, *inter alia*, Charlie Lees, and on cruise ships. Returned Perth, 1958, working in radio and

TV and serving as secretary of the Perth branch of the Musicians' Union. Recently based again on the northern coast of NSW, in clubs and repertory, then returned to Perth, 1985. Apart from his work in the Trocadero, Riley was regarded as one of the most able exponents of swing-based jazz in Perth during the war.

Riverside Jazz Group (Perth, WA)
Founded in 1954 in Perth by Don Thomson (tbn.), with Frank Thomas (tpt), Brian (a.k.a. Brett) Lockyer (clt), Bob Dixon (dms), Frank Herbert (gtr) and Roy Coates (bs), with John Van Oyen (p.) added later, subsequently replaced by Phil Batty and Coates by John Bartlett. In 1960 the band attended the AJC in Melbourne. American trumpet player King Fisher, son of novelty band leader Freddie 'Schnicklefritz' Fisher, found the style of the Riverside so congenial that he returned with them to Perth. Personnel changes left the Riverside as: Fisher, Lockyer, Barry Bruce, Thomson, Bartlett, and Will Dower (dms). During 1961–62 the band played television and jazz club engagements at the Shiralee. Attended the AJC in Sydney, 1962, to great acclaim. The band stayed on in Sydney, playing under Fisher's name, but gradually devolved in 1964–65 as members developed careers elsewhere. Bruce returned to Perth, the others remained in Sydney except for the nomadic Fisher who retired from music and continued travelling. Lockyer later began a long association with Mike Hallam, Thomson worked with various groups incl. Graeme Bell's and the Harbour City JB, Bartlett has been active on a freelance basis and Dower went into TV session work, returning to the jazz scene in 1986 with Noel Crow's band.

Roadknight, Margret   b. 16/7/43, Melbourne, Vic.   vcl/gtr/kalimba/tambourine/ldr/arr
Began singing and playing guitar in early '60s in the folk scene, then introduced to the classic blues repertoire by Frank Traynor, with whom she began singing in the mid '60s. Also encouraged by members of the *Black Nativity* troupe touring Australia. From the late '60s, ran folk music courses for the Vic. Council of Education. In 1971

Margret Roadknight, Soup Plus Restaurant, 1978, with Dick Hughes (p.), Mal Jennings (tpt)

recorded *The Odds Are Against Me*, highly acclaimed both locally and in the US. Toured extensively for festivals, revue and theatre companies, and in education programmes. In Sydney worked with Judy Bailey in the early '70s, and on ABC TV. To the US, 1974, on a study grant from the Australia Council, then again in 1977. Resident at Soup Plus through the late '70s. Roadknight has freelanced extensively with jazz groups and led her own jazz/blues/folk groups. Dave Dallwitz has written material for her, and she has extensively recorded, incl. with Dutch Tilders. In 1986 she was doing mainly solo work, and engaged in promotional enterprises.

Roberts, Donald Bruce (Pixie)   b. 4/12/17, Melbourne, Vic.   reeds
By 1934 Roberts was learning reeds, mainly alto, and began playing with Ron Huntington's trio. Joined Bob Clemens's band in 1936, and worked with other dance band musicians Lou Silbereisen and Bud Baker. These three began playing at Saul's coffee lounge in 1940 with Baker's brother Harry and Norm Bickerton (dms). The Bell brothers heard them, and the two groups converged in what became, with changes, the nucleus of the Bell band. Roberts remained with the Bells until the band broke up in 1952, with occasional interruptions incl. a period in the navy in 1945. In 1952 he started a recorder factory, Pan Recorder Co., with Ade Monsbourgh and in the mid-'50s worked with

Max Collie. He joined the 431 Club band in 1959, but from the '60s reduced his musical activity. Resumed playing in 1984 and has since performed in Bell band reunion concerts.

Roberts, John Charles   b. 28/9/38, Wollongong, NSW   multi-instrumental, mainly tpt/comp.
Moved to Sydney in infancy. Began trumpet at 16, played trombone in Ryde Brass Band. Began sitting in at Sydney Jazz Club functions and played first gig with Mick Fowler, c. 1956. In 1958, with Pat Qua's JB, on Sydney Jazz Club and AJC committees. Over the next few years led his Jazz Bandits and worked with the Jazz Pirates, the Royal Georgians, and during vacation seasons with the Melbourne University JB at Lorne, Vic. In Canberra, 1965–68, where he played with Greg Gibson, then in Melbourne, with Mabel's Dream Orchestra before moving to England c. 1970. Upon return to Sydney became the regular trumpet player with Bill Haesler's band. From 1975 has worked with Adrian Ford, first with his Ragtime Orchestra, then with the Big Band whenever it is assembled. In 1977–78, with the Abbey JB, followed by freelancing. Briefly with Merv Acheson, 1983, and with Tony Gardner's big band, 1984–85. Primarily a freelance musician with daytime commitments as a medical practitioner, Roberts has been a stalwart of the AJCs since attending his first in 1957.

Rohde, Bryce   b. 12/9/23, Hobart, Tas
p./comp./arr/ldr
Began piano at 11. Heard jazz on V-Disc while serving in New Guinea and following discharge worked with Alf Holyoak in Adelaide, 1949. From 1949–51, led trios (incl. 'Jazza' Hall gtr, Milton Hunter bs) at Bill Holyoak concerts and played solo for the ABC. Hearing Jack Brokensha led Rohde into progressive areas. To Canada with Brokensha in 1953 to join Buddle, and with the AJQ, 1954–59. With Ed Gaston, recruited George Golla and Colin Bailey for a quartet which worked Sydney venues incl. Bel Air café/club and El Rocco. Organized Jazz Goes to College concerts, 1960, at Univ. of NSW and Univ. of New Eng-

land. The quartet backed Dizzy Gillespie at his Oct. 1960 appearances. Rohde took the quartet to US with Frank Thornton (ten.) replacing Golla; toured with Kingston Trio, April to May 1961, then Thornton, a swing-based player, left as Rohde began experimenting with the Lydian Concept of Tonal Organization. Management problems led to the quartet's dispersal and Rohde based himself in California for further study, meantime playing with Frank Phipps (brass) and Dick Knutson (bs) in a Mill Valley restaurant. Returned Sydney, 1962, and established a workshop with Charlie Munro, Bruce Cale and Mark Bowden (Bowden is a highly rated session drummer (b. 29/5/27) who began in theatre and night-club, but apart from work with Judy Bailey, John Sangster et al. at the El Rocco, has rarely been regularly visible in specifically jazz contexts). The Rohde workshop quartet recorded the album *Corners*, a major innovative work. Rohde played concerts, broadcasts, and continued workshop experimentation until 1965, when he recorded *Here's Bryce* with his quartet augmented by young sax player Sid Powell, whose sudden death about a year later was a significant loss to Australian jazz. Rohde acted as MD for Katoomba Jazz Festival in 1965, then departed for US where he continues to be musically active. One of the major experimenters in Australian jazz, he continued a line of innovation stretching back to Alf Holyoak, taken up from Rohde by Charlie Munro, and continuing in the '80s in the work of Bruce Cale.

Rose, Jon b. 19/2/51, London, England vln/p./cello/various 'invented' instruments/comp./arr/ldr
Began violin in childhood. Later worked in rock, classical, and jazz, before concentrating on experimental improvization. Worked as an audiovisual technician at the Royal Academy of Music, London, and with various multi-media performance groups. To Sydney, 1976. Started Fringe Benefit records for experimental music, 1977, undertook the jazz studies conservatorium course, 1977–78, and gave solo performances on improvised instruments at different venues incl. Pinball Wizz, Watters Gallery, Sculpture Centre, Austra-

Jon Rose

lia Music Centre. Formed his Relative Band, 1980, presented its debut at The Basement. Regularly performed internationally: at ICA London (1980), Moers International New Jazz Festival, Germany (1981), solo tours in Europe and the US (1982), the Festival d'Automne, Paris (1983). In Australia has played the Sydney Festival (1982), and has toured in other states with the constantly changing Relative Band (1984). In 1982 he played a 12-hour solo marathon, and produced the first International Relative Band Festival, augmenting his group with performers from overseas. With the multi-media group Slauterhaus, 1983. Rose has also worked in institutional contexts: musician-in-residence at La Trobe Univ. (1981), and at Canberra School of Music (1984); artist-in-residence at Praxis, Fremantle WA (1983). In 1985 he was a prize-winner in the first 2MBS Radiophonic Composition Competition. In 1986 he gave farewell concerts at The Basement and the Performance Space, with Sandy Evans, Jim Denley (fl.), Chris Abrahams, Louis Burdett (dms), Jamie Fielding (keyboards), before leaving to play in Houston for the New Music American Festival, proceeding then to a residency at Kunstlerhaus Bethanien, a contemporary performance centre in Berlin. Many commentators would question the inclusion of Rose in a book on jazz, yet if improvization is integral with the music, this seems the least unsatisfactory way of situating him. Most of the received performance categories do not apply to Rose's work. His music, his instruments, the format and context of his performances, and even his curriculum vitae, do not fit into the conceptual structures which have evolved to accommodate and discuss jazz, though jazz is unquestionably one of the traditions feeding his work. Many have found his work difficult of access, though it is not inherently so. It is rather that those conventional categorical structures obscure what is happening, incl. the presentation of a very considerable sense of humour: the advertisement for his 'Relative Band plays cricket', a musical performance based on the rules of cricket, invited the audience to 'voice loud opinions at anything that happens . . . which is not "fairgo", or is getting too close to "Art"'. Rose said, 'I like music to continually attack my preconceptions.' He applies this axiom radically, and is a musical improviser in a deeper sense than are most conventional jazz musicians.

Ross, William John (Billy)   b. 3/3/34, Adelaide, SA   dms
Took lessons from Bob Foreman in 1950 and began sitting in with local traditional bands. They were not receptive to his style, and he moved towards more mainstream musicians, joining Ian Drinkwater (with John Bermingham bs, Graham Schrader, Roy Wooding gtr) in about 1953. Led a band briefly, then in late '50s with Bruce Gray, with a spell at the Paprika in 1958 with Lew Fisher. In 1960 at The Embers in Melbourne for six months under leadership and influence of Frank Smith. Back in Adelaide played the Cellar during its most virile jazz policy. Also playing cabaret and night-club work, arriving at the Cellar after the gig. From the mid '60s, more night-club work, incl. a year at the Trocadero. In 1974 at Wrest

Billy Ross at The Embers, 1960, with Ted Nettelbeck (p.) and Buddy Rich (vcl)

Point Casino, Hobart, then in Sydney working casually with Col Nolan and Errol Buddle. Returned Adelaide, 1974, played for two years at the Pizza Palace, then the Paprika. The opening of the Creole Room in 1977 gave short-lived opportunities, after which he retired from music for about two years. Mal Badenoch, as MD for an entertainment centre in Surfers Paradise, invited Ross to work there, with, *inter alia*, veteran sax player Les McGrath. Back in Adelaide, 1980, with Alan Hewitt, and by the mid-'80s was resident at the Gateway Hotel on North Terrace. Ross's career demonstrates the difficulties facing a committed progressive musician in Adelaide (and to varying extents throughout Australia). Nonetheless, his work at the Cellar in particular established his importance, and his reputation spread nationally through the peripatetic musicians who worked with him there. Ross is one of the pillars of post-traditional jazz in Adelaide.

Rowston, Mervyn (Merv)   b. 1907, Perth, WA   dms/vibes
Began drums 1926, in 1927 joined Oscar

Merv Rowston, c. 1934

Mayerhofer's band, then with Bill Naughton at the Kit Kat Tea Rooms. In 1928 with Harry Cross in the Capitol pit band, later with Charles Sheridan at the Piccadilly, Merv Lyons at Temple court, Miss Darcy, Colin Smith, and from April 1932 with Fred Nice, doubling on xylopone. In Theo Walters's Knickerbockers, 1930. In England, 1934–35, where he took lessons from Max Abrams. In 1937, led the Dixielanders and a trio with Jim Riley and Ken Murdoch for the Miami night-club. Throughout the war, he led broadcasting groups for the ABC, worked at the Cabarita and at Cocoanut Grove, 1942, and MD at the American Red Cross Centre from 1943. Post-war, Rowston was active in the cabaret/dance band area, incl. a long association with Roy Delamare. Inaugurated the annual Drummers' Convention in 1949 or 1950 and was foundation president of the Perth Drummers' Club in September 1958. Australian organizer of the International Association of Modern Drummers established in America. Active as journalist and influential teacher. In 1982 he recorded a tribute to the Embassy with the Will Upson Big Band, and in 1983 was still freelancing several nights each week.

Rutledge, David Down (Dave, Dagwood)
b. 18/10/23, Sydney, NSW   reeds/fl.
Began his career during the war, establishing a reputation as a 'take-off' man in swing/jazz-based groups playing the night-club/restaurant scene: with Wally Norman at the Yankee Doodle Club for allied NCOs (1944) and with Reg Lewis at the Modern Music Club and Bondi Esplanade (1945). After the war he continued to work in the same settings, incl. with Norman at the Roosevelt (late 1945), Leo White at Prince's (1949), Chequers in the early '50s, with Ron Gowans at Segar's Ballroom (1955) and with Julian Lee at the Ling Nam, 1956–57, taking over as leader. He was also active in broadcasting bands, with Leo White on 2UW and with Monte Richardson's Persil unit in the mid '40s, and on *Jazz For Pleasure* (1958) with Don Burrows, Terry Wilkinson, Freddie Logan, Ron Webber. This band, the Australian All Stars, played the Sky Lounge dances and recorded the *Jazz for Beachniks* albums. Rutledge also played with Terry Wilkinson at the Silver Spade Room in the Chevron Hotel from late 1960 or early 1961. He was also active during the '50s in the Sydney Stadium concerts, backing Frank Sinatra in 1955. He toured with other imported musicians incl. Stan Kenton, Buddy Rich, Sammy Davis Jr., Gene Krupa, Mel Torme, Shirley Bassey, Henry Mancini. From the late '60s until 1986 Rutledge has not been visible in jazz settings, being primarily active in TV studio work.

Ryan, John Andrew   b. 8/6/42, Sydney, NSW   bs
Learned violin in childhood, later took bass lessons from Chic Denny. In teens played in rock groups (incl. Lonnie Lee's), then introduced to jazz by Dave Levy. Replaced Bruce Cale at the Mocambo, c. 1963, and played the Mandarin Club with Tony Esterman. In England, c. 1964–71, where he freelanced, worked with Humphrey Lyttelton, played commercial work incl. residency at Talk of the Town and recordings with Cat Stevens. Following return to Sydney, toured SE Asia with rock groups, then played club work with the Four Kinsmen. Began playing with Nancy Stuart shortly after the latter's resumption of performance in the early '70s, and was her regular bass player until her death. Subsequently with the Harbour City band for a period; in 1986 freelancing.

Sangster, John Grant   b. 17/11/28, Melbourne, Vic.   tpt/vibes/dms/perc./comp./arr/ldr

Sangster first came to national attention playing trumpet at the 1948 AJC and in 1949 began leading his Jazz Six (incl. Ross Fusedale, reeds), a band which performed at the Melba Memorial fundraiser in May 1949. Worked in groups associated with the Bells, incl. Ade Monsbourgh's Late Hour Boys and Kelly Smith's trio (with Monsbourgh on p.), and joined the Bell band on drums for their second overseas tour beginning in 1950. Following the break-up of that band in 1952, Sangster remained with Graeme Bell, on the Korea/Japan tour and in Brisbane in 1955 for a night-club residency where he taught himself vibes, subsequently his main instrument. Returned to Sydney, 1956, and played with Bell at Bennelong Hotel. Left Bell in 1958 and worked on a freelance basis until joining Ray Price on trumpet in the early '60s. Sangster's increasingly progressive interests emerged in an association with Don Burrows, and in his experimental work on the *Jazz Australia*

John Sangster, c. 1952

album released in 1967. In November 1966 he recorded for the ABC an experiment in total improvisation with taped material, piano (Bob Gebert) and bass (Ron Carson), with himself on drums. With Burrows for Expo '67 in Montreal. Through the late '60s Sangster worked with electric groups like Tully, and the Nutwood Rug Band. He continued to extend his jazz range in performances over a six-year period at the El Rocco with musicians incl. Dave MacRae, Graeme Lyall, George Thompson (bs), Alan Turnbull, Col Nolan. Played at Expo '70 in Japan with Don Burrows, and played in the pit for *Hair* for two years with, *inter alia*, Keith Stirling and Freddie Payne (tpt). From the '70s most of Sangster's activity has been in film, TV, and radio. In 1973 he recorded the first album in one of the most ambitious enterprises ever undertaken in jazz recording. *The Hobbit Suite* was the first issue of a series of jazz suites inspired by Tolkien's writings, finally resulting in six LPs under the general title of *Lord of the Rings*, together with material recorded under the title *Landscapes of Middle Earth*, the whole consisting of Sangster compositions in styles ranging from ragtime to electronic and aleatory. In the '80s, Sangster has continued to record his own compositions, performs at occasional concerts, and is writing his autobiography.

Schrader, Keith Graham (Graham, Gramps)
b. 15/5/33, Adelaide, SA   p./arr

Began at eight with later lessons from Bryce Rohde. Jazz activity began with the Holyoak swing and jazz concerts through the '50s. Later played at the Cellar, but most of his work has been in the night-club/restaurant setting, and with the Penny Rockets rock group. With the broadening of the jazz scene in the '80s, he has found work with bands led by Don Armstrong (reeds), Ken Way (bjo), Phil Langford and Peter Hooper, and leading his own trio.

Schroder, Ken b. 5/10/47, Melbourne, Vic. reeds/fl./comp./arr/ldr
Began playing jazz, 1962 with fellow university students. Throughout the '60s he worked with various groups, including the rock/pop band Nova Express with Derek Capewell, Graham Morgan and Ian Hellings (tpt). Also received valuable advice from Frank Smith. In the '70s he became involved with a young group of bop revivalists in the band Bebop Inc., before leaving for London. Following his return Schroder worked in various settings, incl. writing the music for the film *Exits*. In 1982 he participated in the artists-in-schools programme and in 1983 travelled under an Arts Council Grant to Germany where he studied with Herb Geller. Since returning he has formed a group with Allan Browne, Gary Costello, Jex Saarelahrt (p.), Ian Hellings, Margaret Morrison (vcl), sometimes augmented to a nine piece with Peter Martin (reeds), Bob Venier, and Doug de Vries (gtr), and also teaches.

Scott, Colin Richard Winston b. 20/3/41, Kojonup, WA gtr
Started guitar in 1958. Participated with various groups, notably that of Bill Clowes, during the efflorescence of progressive jazz in the mid '60s at the Shiralee and the Hole in the Wall, and established a reputation, with Ray Walker, as one of the foremost modern guitarists in Perth. Later with Walker in a trio, and has enjoyed some renewed jazz exposure through the Perth Jazz Club/Society, incl. with the Tony Ashford Quintet in the '70s.

Scott, Craig Blakefield b. 20/10/56, Sydney, NSW bs
Began bass in his teens, with lessons from Cliff Barnett. Turned professional, 1975, and undertook the jazz studies programme at the NSW Conservatorium in the late '70s. Worked with groups led by Dave Levy and Steve Brien, and spent three years with Keith Stirling's quartet/quintet, from 1979. Toured with Mike Nock, 1980, with Joe Henderson, 1981, and studied under Todd Coolman and Ed Soph at the Conservatorium. Joined Don Burrows, Dec. 1981. With John Bostock Quintet, Lorraine Silk's group, and Sydney Jazz Quintet, 1982. With James Morrison's quartet and the Big Bad Band, 1984, and later that year went to the US for further study with Coolman and Rufus Reid. In 1985 was still with Burrows, with the Sydney Jazz Quintet, and active with Morrison, Silk, Julian Lee, Steve Giordano, and as a resident bs player at the Regent Hotel. Has backed numerous visiting musicians incl. Bobby Shew, Phil Wilson, Joachim Kuhn, and Lee Konitz. Has taught for the JAS since 1981 and at the conservatorium since 1984.

Sedergreen, Robert Alexander (Bob) b. 24/8/43, Acre/Haifa, British Palestine p./synth./tpt/comp/ldr
Moved to UK in 1947 and trained classically 1949–50. To Australia 1951, and with the Southern City JB 1959–60, then the George Jury Quartet in 1961.

Bob Sedergreen, 1987

His style was developed by his work with the Fred Bradshaw Quartet (1962–70), Ted Vining Trio (1971–72), Alan Lee's Plant (1973), Brian Brown's Quintet (1974) and Quartet (1977). In the '80s he has been working with the Australian Jazz Ensemble, Onaje, and Peter Gaudion's Blues Express. Backed visiting performers incl. Richie Cole, Phil Woods, Dizzy Gillespie, Milt Jackson, Bobby Shew, David Baker, Jack Wilkins and Mel Lewis, and has played numerous festivals incl. Moomba (1974–79), Perth (1975), Adelaide (1976), Sydney (1976–78), and the International Jazz Festival (1978). Extensively recorded. One of the most creative and emotionally intense pianists in Australia, a player of great power, Sedergreen is involved in teaching both privately and institutionally in addition to his performing activities with Brian Brown, Onaje, McCabe's Bones, and Blues Express.

Sharp, Samuel (Sammy)   b. 23/10/19, Jerusalem, d. 16/1/79, Perth, WA   tpt/tbn./ldr/arr
Self-taught from 12 until beginning formal lessons at 17, then worked dance bands incl. those of Ron Moyle and Colin Smith. To England, 1934, via engagements in India and Ceylon, with Wally Hadley (bjo/gtr). Worked at the highest professional level, with Geraldo, Ambrose, Lew Stone, Harry Roy, and with Henry Hall on the *Queen Mary*. Studied arranging and composition, worked in films and for the BBC. His wife, Leah, a violinist, worked with Ivy Benson. Returned to Perth, 1946 and worked with an ABC band incl. Sylvia Caporn (p.), Ray Le Cornu (dms) Roy Coates, Maurice Lawrence, Jack Pateman, 'Skippy' Alexander, Mick Bartlett (saxes). He introduced vigorous and innovative approaches to the local popular music scene, replacing Bill Kirkham's orchestra with his own at the Embassy shortly after his arrival, inaugurating with Harry Bluck the Jazz Jamborees in 1948, and touring rural areas with his bands. He assembled ambitiously voiced orchestras on a scale unprecedented in the history of popular music in Perth, such as his Jazz Fantasia aggregation in Aug. 1951, consisting of 35 pieces with 20 string players and 70 voices, for a concert at the Capitol, and incl. such authentic jazz musicians as Horry King, Jim Beeson, Keith Whittle, and Brian Bursey. Sharp's music often attracted censure from more purist jazz enthusiasts, and it seems clear that he inclined to a musical lushness if not bombast. Nonetheless, he energized the local scene, increased the currency of the notion of jazz,

Sam Sharp's orchestra, probably late forties, Embassy Ballroom, Perth; from l. to r., brass: Ted McMahon (tbn), Ray Taylor, Norm McMahon, Laurie O'Keefe (tpts), reeds: Jack Pateman, Mick Bartlett, George Caro, Ken Murdoch, John McCaughlan, rhythm section: Keith Whittle (dms), Harry Pender (gtr), Jim Beeson (p.), Roy Coates (bs), Sam Sharp at microphone

provided platforms for jazz musicians and, on the testimony of his colleagues, was himself a convincing swing-oriented jazz musician.

Sharpe, John Bruce (Sharpie)   b. 12/12/41, Parkes, NSW   tbn./tba
Played piano as a child and first played for a church dance in 1959 in Sydney where he was a university student. While an undergraduate, formed the Swingtet (incl. Paul Furniss) and attended Sydney Jazz Club workshops on drums. Began trombone in 1963, then moved to Canberra, 1964, and intensified practice by joining Canberra City Band. Joined the Cavaliers Dance Band, 1965, and became a founder member of the Fortified Few, remaining with the band until its dissolution in 1983. He then joined the Double T Jazz Ensemble formed and led by Tony Thomas, moved to the Jerrabomberra JB until it disbanded, and in 1986 was playing tbn./tba with Pierre Kammacher's Hot Four.

Shaw, James (Jimmy)   b. 5/6/26, Sydney, NSW   dms
Started drums 1945 and joined Chullora Electric Car Shops Orchestra, 1947. Studied under Alard Maling at NSW Conservatorium on a scholarship (1948–49). Through the '50s, active in restaurants/night-clubs: with Gaby Rogers at Romano's (1951), Paul Lombard at the Celebrity and Les Welch at Sammy Lee's (1952), Percy Winnicke at Prince's restaurant (1953). Also played jazz concerts featuring both local and imported musicians incl. Eddie Shu (1954), Mel Tormé (1955 and 1961), Ella Fitzgerald (1961), and with Don Harper's quintet as support for Dave Brubeck (1963). Worked in the big bands of Wally Norman for town hall concerts, Billy Weston at the Gaiety, and John Bamford at the Ironworkers' Hall; also with Ray Price's Dixielanders and the Port Jackson JB in the mid '50s. Since the late '50s was active in session work with Bob Gibson, Jack Grimsley, Julian Lee, and in the TV orchestras of Bob Limb (1961–64), Eric Jupp (1961–66) and Bob Young (1961–64). With Alan Pennay trio at Top of the Cross restaurant, Sydney, and at Mt Thredbo

(1967–69). With Wally Johnson Orchestra at the Central Coast Leagues Club (1970–74). To Adelaide, 1974, playing Paprika night-club, and backing touring packages incl. Kamahl at the Festival Theatre. With Schmoe and Co. as support to Herbie Mann, and with Glenn Henrich Quintet as support for Dave Brubeck in early '80s. Played on the cruise ship *Fairstar* (1982–84), then freelanced in Adelaide with Schmoe, Penny Eames, Bruce Gray, Dick Frankel, and played Murray River jazz cruises. Teaching privately, and has been learning trumpet since 1982. Resident in Macclesfield, 50 km south of Adelaide, until returning to Sydney in 1986.

Sherburn, Nevill Louis   b. 24/12/30, Melbourne, Vic.   p./bjo/wbd/ldr
While at Royal Melbourne Institute of Technology (1946–48), was leader of the Swing Wing of the Music Society, presenting jazz record sessions, and involved in organizing the college JB and jazz concerts. Took piano lessons from Graeme Bell, 1948–49. Attended his first AJC in 1948. Formed his Rhythm Kings in 1950. Secretary of 1956 AJC in Melbourne. Editor and publisher of *Jazz Notes*, July 1960 to December 1962. Most important contribution to Australian jazz is as producer of Swaggie records since 1954.

Silbereisen, Louis Colin (Lou, Baron)
b. 25/2/16, Sebastopol, Vic.   bs/tba
In Queenscliff, Vic., 1918–35, heard his first jazz on records for silent movies. To Melbourne, 1935, and became involved in jazz through Melbourne Swing Club and the sessions at Fawkner Park Kiosk. Began bass to join a band playing written arrangements. Met Pixie Roberts, Bud Baker and, later, the Bells. 1940–45 in England and Middle East with RAAF. Returned to Australia and joined Graeme Bell at the Uptown Club, where he also took up tuba. Remained with the Bell band throughout its most significant period, winning the *Music Maker* Poll in bs/tba category in 1952. When the group broke up in 1953, withdrew from music until 1976, when he resumed casual playing. Currently learning piano.

Smith, Frank   b. 30/7/27, Sydney, NSW;
d. 18/2/74, Melbourne, Vic.   reeds/ldr
First became prominent in the night-club and concert scene in the immediate post-war period. With
Reg Pedersen and his Colossal Casuals (1946), and
at Surreyville ballroom under Doug Cross then

*Frank Smith with his Embers group, 1960; front to rear:
Ted Nettelbeck, Billy Ross, Ivan Videky, Barbara Virgil,
Frank Smith*

*Frank Smith with Billy Antmann's group from Sammy
Lee's, c. 1950; from l. to r., Frank Smith, Ron Gowans,
Dick McNally, Billy Antmann.*

Marsh Goodwin, and working with Raymond
Cray (1947). With Col Anderson at the Bondi
Esplanade and Johnny Best at Segar's Ballroom
(1948), Gaby Rogers at Romano's (1949), then
joined Billy Antmann (p.) at Sammy Lee's,
Christy's, and the Trocadero in succession
throughout 1950. Left the Trocadero for the
Roosevelt (1951), joined Warren Gibson's Metronome dance circuit (1952), then returned to Romano's under Bela Kanitz (1953–54). During these
years he had also been active as a sideman in the
jazz concerts, incl. with Ralph Mallen, Enso Toppano, the Joe Singer Trio, and Jack Allan, with
whose group he recorded with Rex Stewart in
1949. Smith worked again with Mallen in 1954, but
went on primarily to freelance. Was active at El
Rocco during its early years, with, *inter alia*, Dave
May (p.), Cyril Bevan (dms), Cliff Barnett. In Aug.
1959 he played the opening of The Embers in Melbourne, leading Frank Thornton, Mike Nock,
Chris Karan, Peter Robinson (bs), with Thornton
shortly replaced by Billy Weston. The Embers was
destroyed by fire on Nov. 1959, but Smith again
led a quintet when it re-opened in Feb. 1960, remaining until Ted Nettelbeck took over leadership
in 1961. Throughout the '60s Smith was primarily
involved in various forms of studio work, including in TV orchestras, and freelancing with, *inter
alia*, Bruce Clarke with whom he was associated in
a jingle workshop. In 1971 Smith established his
own music production company which wound
down in late 1972/early 1973. He joined the band
at Wrest Point Casino in Hobart, Tas. early in
1973, but had returned to Melbourne at the time of
his death. A benefit for his family on 21 April 1974
gave some testimony of the enormous respect and
affection in which Smith was held. As an inspirational figure, he was one of the most important
musicians in the history of Australian jazz and one
of the handful to achieve legendary status. Generous with encouragement and advice, he is remembered by scores of players of all instruments as the
most influential musician in their lives. He exercised this influence primarily through informal
discussion and after-hours jam sessions; in addition to being one of the most gifted exponents of

'modern' jazz saxophone of his time, he imparted a general attitude to music, a sense of commitment, which is recalled as being at least as important as specific matters of theory and execution. While a great many musicians have been better known to the lay public, none has been more admired and more affectionately remembered by that more discriminating body composed of his or her peers. The sparseness of Smith's recorded output is one of the greater misfortunes in Australian jazz.

Smith, Ian Leslie (Smithy)   b. 23/4/48, Bletchley, England   tpt/tba/dms/wbd/vcl/some p. and bjo
Immigrated to Tas. 1951, to Vic. in 1955. Began drums at school, 1962; lessons from Hal Boyle of the Melbourne Dixieland Band with which he began sitting in in 1963. Played in a rock band, 1967. Began trumpet in 1966 and formed New Harlem Band 1968. Regular gigs, recording, and concerts incl. support to Turk Murphy 1978. President 28th AJC, 1974 Melbourne, and produced documentary films on this and the 30th, 1976 Brisbane. Played 2nd trumpet in Tom Baker's San Francisco JB at Sacramento, Calif. in 1978. Since leaving New Harlem in 1979, has worked with various groups incl. two years on drums with Maple Leaf.

Smith, June Elizabeth   b. 9/6/30, Leeds, Scotland   tpt/vcl
Her mother was with Ivy Benson's band. At 16 June began trumpet professionally with Blanche Coleman and in the late '40s with Benson as lead trumpet. Based in London, early '50s, working with dance bands and as solo act. Retired from music to raise children, returning via casual work after the family moved to Melbourne in 1961. Joined a commercial group, Division Five, from which her husband Lew formed Maximum Load in 1970. To Perth, 1974, vocalist with Will Upson, in Barry Bruce's Chicago group and his Middle of the Road with Lew, and with bands ranging through a broad jazz spectrum, often through the Perth Jazz Club/Society. Regarded, with Helen Matthews, as the city's most important jazz vocalist.

Smith, Lewis Arthur (Lew)   b. 29/4/30, Pickering, Yorkshire, England   reeds/fl.
Son of dance band leader Arthur Smith, Lew started on violin at age five, drums at 11, and sax in 1946. Following National Service played at a Butlin's resort then six months in 1950 playing on the *Mauretania* between Southampton and New York, where he was inspired by American musicians incl. Dizzy Gillespie and Charlie Parker. Early '50s toured the provinces in England, incl. with John Dankworth's band as first alto. From 1955 he was resident in London, with Denny Boyce (with wife June in the trumpet section), then at the Palladium. Lew and June migrated to Melbourne in 1961, Lew serving briefly in a band with the RAAF, then into pit work while teaching music for the Education Department. Played informally with Alan Lee and Brian Brown and worked at The Embers with 'Boof' Thomson, Gary Hyde, Ted Preston. Through most of the '60s, with Channel 9 and playing casually. In 1970 he formed a commercial group Maximum Load. To Perth, 1974, as lecturer at a teachers' college, and began to work for Perth Jazz Club/Society. Has worked in a broad range of styles, incl. Barry Bruce, Mike Nelson's Four on the Floor, Will Upson's Big Band, the Nookemburra with Don Bancroft, and several years with the Lazy River JB. In 1984 he began lecturing in jazz history at Mt Lawley College.

Somerville, James Anquetil (Jim)   b. 14/11/22, Sydney, NSW   p./ldr/comp./arr
Began his professional career at Mosman Rowing Club, 1941. Worked in a variety of settings, incl. writing music for a season of Molière, 1944, and played a black pianist in Saroyan's *The Time of Your Life*, 1943. Led groups at the 2KY Swing Club Functions and in 1944 led a group incl. Ade Monsbourgh, Kelly Smith, Ray Price, Tom Pickering, for the Sydney Arts Ball. Began a long but discrete association with the PJJB, 1947, leaving in 1948 to form his Jazz Rebels: Marsh Goodwin (tpt), Doc Willis, Wally Johnson (ten.), Merv Acheson, Clive Whitcombe (dms), Lennie Evans (bs), Harry Shoebridge (gtr) with Georgia Lee and

Jim Somerville, Bellevue Hotel, Paddington, 1977

Nellie Small (vcls). Rejoined the PJJB, 1949, becoming leader in 1950. During his years with this band he was active in other groups in the nightclub/restaurant scene, incl. with Jack Maittlen at the Roosevelt, 1952–53, as well as being ubiquitous on jazz concerts. With Ray Price's Dixielanders, 1956, with whom he recorded under the name Jed Sullivan for contractual reasons. Through the '60s and '70s Somerville's main activity was freelancing in club and studio work, incl. a long residency at the Paddington-Woollahra RSL with John McCarthy. Since the late '70s he has been more visible in jazz context, incl. with the Harbour City JB. He played in New Orleans during a 1980 visit, deputized for an ill Graeme Bell for a tour, 1984, and joined Ken Harrison's Compass, 1985. Somerville has demonstrated a longstanding interest in jazz, his style retaining freshness by virtue of continuous development. His articles on jazz in *Music Maker*, 1945–46, demonstrated unusual insight for the time, and he has freelanced and deputized in most of Sydney's important traditional to mainstream groups incl. those of Mike Hallam, Dick Hughes, Alan Geddes, Ray Price.

Southern Jazz Group (Adelaide, SA)

In 1945 a band began rehearsing at the home of Errol Buddle's parents in Adelaide. As the Southern Jazz Group it presented its loose '40s dixieland style on a Bill Holyoak broadcast late in 1945 with Dave Jenkins (tpt/ldr), Dave Dallwitz (tbn.), Buddle and Bruce Gray (reeds), Lew Fisher (p.), George Browne (bs), Claude Whitehouse (dms) and John Malpas (gtr). Dallwitz took over as leader in 1946, implementing changes which left the band as Dallwitz, Gray, Fisher, Malpas (now on bjo), Bill Munro (tpt), Bob Wright (tba), Joe Tippet (dms). With a clear idea of the sound he wanted, Dallwitz coached the band to a style based on the classic jazz of the '20s, the first consciously revivalist group in Adelaide. Fisher remained behind when the band went to Melbourne for the first AJC in 1946, and the absence of the piano, Tippet's use of washboard instead of drums on that occasion, and Wright's deft, springy attack, gave the band a lightness that impressed all present. The 1st AJC inspired the SJG to further refinement, and it remained the dominant traditional band in Adelaide during its lifetime, much of which was recorded on Bill Holyoak's Memphis label. The personnel remained stable (though they recorded with such guests as Ade Monsbourgh and Tom Pickering) until Tippet's professional commitments led to Kym Bonython replacing him through 1949, followed briefly by Maurice Kelton then Bob Foreman. By 1950 the band was disintegrating under the stress of conflicting musical interests, Gray in particular wishing to introduce changes in approach and personnel. He left in 1950, taking with him Malpas, Wright and then Munro, all of whom became part of his All Star Jazzmen. There was a dispute over the ownership of the name SJG and the last time it graced a recording was in June 1951, when only Dallwitz remained of the 1946 band. The SJG to a large extent determined the subsequent history of traditional jazz in Adelaide, not only for the sound it presented as a band, but also for the subsequent work of its alumni. The band was held at the 1946 AJC to have a distinctive character, to which a two-beat feel and tuba foundation were essential, and this

The Southern Jazz Group, Tivoli Theatre, Adelaide, probably 1948; l. to r., Bob Wright, Bill Munro, Lew Fisher, Dave Dallwitz (tbn.), Alan Spry, sitting in (gtr, seated), John Malpas (bjo), Bruce Gray

The Southern Jazz Group welcoming home Graeme Bell following the Bell band's first European tour, 1949; l. to r., Bob Wright, Dave Dallwitz, Joe Tippet, Bruce Gray, John Malpas, Bill Munro, Lew Fisher

has remained a perennial feature of Adelaide jazz. The conceptual clarity of Dallwitz's direction and the authoritative playing of Bruce Gray were also crucial to the band sound. Although they broke up for all practical purposes in 1951, the surviving members of the SJG have played occasional reunion concerts, most recently at the 40th AJC in Ballarat in 1985.

Spedding, Graham Menzies (Speddo)
b. 25/11/33, Sydney, NSW   reeds/vcl
Began clarinet in 1951 and joined the Zenith JB 1953–55. With the Southern Cross JB for its tour to Moscow, 1957, and joined the Black Opal JB, 1958–60. With the Jazz Pirates, 1960–62, then with Graeme Bell's All Stars, 1963–67. Left Australia, 1968, first working in South Africa, then to England where he played with the bands of Joe Daniels, John Parker, Eric Silk, and the Black Bottom Stompers. Spent four years playing US bases in Europe and the Middle East, and returned to Sydney in 1975, joining Ray Price until 1977. Freelanced, 1977–80, while he ran a restaurant, then joined the Abbey JB, of which he was still a member in 1986, as well as playing deputizing and freelance engagements. Spedding's original and forceful style generates an intensity in the rather angular tradition of Ade Monsbourgh and Fred Parkes, two important Australian influences on his work. In addition to his work in bands of which he has

Graham Spedding, Foresters Hall, Croydon, 1959

been a regular member, his appearances at AJCs have led to many memorable jam sessions which have helped make his name recognized and respected throughout Australia.

Speer, Stewart (Stewie)   b. 26/6/28, Melbourne, Vic; d. 16/9/86, Sydney, NSW   dms
Began on trap drums at 13 and worked in brass bands. With Doc Willis, then in Sydney, 1948–52, when he did no playing. From 1955–58 with Max Collie, and at the Downbeat Jazz Club, 1955–56, with Brian Rangott, Keith Hounslow, Kevin Hocking (p.) and Ron Terry (bs). Founder member of the Brian Brown quartet from 1956 to its disbanding in 1960, and joined Kenn Jones's Powerhouse band in 1958. To Sydney, 1961 to play with John Holman at Chequers night-club, and became active at El Rocco with John Sangster, Judy Bailey, Rick Laird (bs), Errol Buddle, George Golla, Lyn Christie. In 1965 with Col Nolan trio (with Ron Carson), and in 1966 at the Latin Quarter. In June 1967 returned to Melbourne to join the jazz-influenced rock group, Max Merritt and the Meteors, but two weeks later was involved with other band members in a road accident, suffering serious injuries which kept him out of music for six months and subsequently seriously limited his mobility. Spent 1970–80 with the Meteors based in England, though making five trips to Australia. Returned briefly to Melbourne then settled in Sydney in 1980. For some time did little work: one gig at The Basement with Keith Stirling, Bob Bertles, Tony Esterman, Jack Thorncraft, in 1980, another, with Barry Duggan, Ray Martin, Bob Gebert, in 1981. More recently became active freelancing, notably at Soup Plus with McGann, and with his own group incl. Jimmy Sloggett. Notwithstanding frequent periods of local inactivity and a long absence, Speer was one of the most admired and respected musicians in Australia. In an interview, Ted Vining called him 'the best drummer this country ever produced'.

Staveley, Kay   b. 13/10/23, Hobart, Tas.   vcl
Introduced to jazz singers by husband Ron Roberts. To Melbourne, 1942, at Leggett's ball-

Stewie Speer with Max Collie's Jazz Kings, 1958; from l. to r., Graham Coyle, Lou Silbereisen, Roger Bell, Stewie Speer, Pixie Roberts, Max Collie

room, in St Kilda coffee lounges, in concerts with Bob Gibson and for broadcasts. With Fred Russell at Admiralty House, an officers' night-club, where she heard Frank Coughlan. Returned to Hobart, 1944, and toured Tas. with the RAAF show *Tiger Follies*. In 1949 reached the finals in Maples Parade competitions. Began singing with Tom Pickering in 1950, and with Ian Pearce when he returned from England. Has worked and recorded with Pearce/Pickering groups since, and in 1973 organized the first annual reunion of the band. Has also worked with the Jazzmanians and in 1985 was still playing guest spots with Pearce/Pickering.

Steeper, Neil J.   b. 16/10/38, Melbourne, Vic. tpt/p./vcl/valve tbn./ldr
Began piano 1948, cornet 1952, playing in school orchestras. Formed his first band, Steeper's Stompers, for school dances in 1954, and in 1955 led the

group for town hall and various rhythm/jazz club concerts. Revived the name for a band he formed while at Dookie Agricultural College, Vic. Retired from playing, 1960–61, then formed a new Steeper's Stompers while studying at the Univ. of New England, NSW, in 1962. In 1963 formed the Olympia Jazz Ensemble which played functions in the New England area. Moved to Canberra where he founded the the Fortified Few in 1966, but left the band later in the year and ceased playing until re-forming the group with John Sharpe in 1969. Steeper was also involved in the re-establishment of the Canberra Jazz Club, and in 1969 formed Clean Living Clive's Goodtime Palace Orchestra which played at the Dickson Hotel until being replaced by the Fortified Few in May 1970, after which it played at the Back Bar, Hotel Canberra. In the US, 1971–73, playing various venues. Reformed Clean Living Clive's group in 1974 and became Vice President of the Canberra Jazz Club. In 1976, with Gordon Reed and Margaret Moriarty he opened Clean Living Clive's Amusement and Refreshment Hall, handing the business over to Reed in 1977. To Sydney, 1978, where he organized a band with Stan Allworth, which, as Stanley's Steamers, began Saturday morning sessions at the Frisco Hotel, Woolloomoolloo in November (with Ken Tratt, 'Kipper' Kearsley, Viv Carter). Following Allworth's departure in 1981, with Kearsley moving to trombone and Terry Fowler (tba), the band continued at the Frisco as the Raucous Arousal Brass Band, later with Dave Rankin replacing Kearsley. Steeper was president of the Sydney Jazz Club, 1981–83, during which he was instrumental in rationalizing the club's legal status. Following the ending of the Frisco residency in 1985, and apart from some casual work, Steeper has been retired from music.

Stirling, Keith Alexander   b. 5/1/38, Melbourne, Vic.   tpt/ldr/comp.

Both parents played music, and Stirling began at eight, practising secretly on his brother's trumpet. Later took lessons from Freddy Thomas, with encouragement from Brian Brown, Stewie Speer, and Len Barnard. In the late '50s he began playing Downbeat concerts and sitting in with Brian

Keith Stirling, 1979

Brown at Jazz Centre 44, joining the Brown quintet in 1960 and co-leading a quintet with Peter Hall (tbn.). In the early '60s Stirling began hitch-hiking back and forth to Sydney to play at El Rocco, where he co-led a group with Bob Bertles. In Adelaide by 1963, playing at the Havana Motel with Billy Ross, Ron Carson, and Bob Gebert, then the four would go on to the Cellar, playing until around 3 a.m. Keith Barr and Bob Bertles came in as co-leaders later in the year. In 1964 Stirling was in Perth working with Frank Smith (bs), Bill Gumbleton, Theo Henderson, Jim Cook, Bill Tattersall, Peter Hall, at the Melpomene. Led a quintet (Smith, Gumbleton, Henderson, Tattersall) in 1965 at the Shiralee and then the Hole in the Wall until at least late 1966, when Alan Hale and Don Bibby replaced Tattersall and Smith respectively, and Tony Ashford also worked in his groups. Stirling returned to Sydney in the Moscow Circus band, 1968, and played Chequers where he backed, *inter alia*, Carmen McRae, Stevie Wonder, Tony Bennett, Lou Rawls, and jammed with Art Blakey. Returned to Melbourne c. 1970 where, apart from recording with Bruce Clarke, he was primarily involved in studio work. To Sydney, c. 1974, and spent four years teaching in the jazz studies programme at the conservatorium. With David Martin's Big Band, early 1978, then to the US on a grant from the Music Board of the Australia Council. Studied at the Univ. of Louisville in a Jamie Abersold clinic, in Chicago, and in New York, where Lee Konitz was an important mentor. Returned to Sydney, 1979, and formed a quartet which played support for Dave Liebman's tour and an Arts Council concert in Canberra in November. Enlarged the band to a quintet which was active at Jenny's Wine Bar and The Basement, 1980–83, as well as playing concerts incl. the Qantas International Jazz Festival (1980), the Festival of Sydney, the Manly Jazz Carnival, and the Southern Cross Jazz Club in Canberra (all in 1981), and support for Johnny Griffin (1982). The group was televised direct from The Basement in 1982. The main personnel of the group has been Steve Brien (gtr), Jay Stewart (p.), Craig Scott (bs), and drummers Barry Woods, Matt Dilosa, and

Ron Lemke in succession. Stirling has also played with other groups, incl. Richard Ochalski's Straight Ahead, Daly-Wilson Big Band, in pit and session work, incl. for *Hair*, and in a performance of Don Banks's *Nexus* with Don Burrows and the Sydney Symphony Orchestra at the Sydney Opera House. He has played with visitors, Miroslav Vitous, Mike Nock, Richie Cole, the Toshiko Akiyoshi-Lew Tabackin Big Band, and several tours with Georgie Fame. In 1982 Stirling was one of the teachers in the summer jazz clinic set up by Greg Quigley. Stirling's peripatetic career placed him at the centre of the energetic modern jazz movement in the early '60s in four major centres, and helped to establish his reputation nationally. Always a highly respected exponent of bop and modal trumpet and flugelhorn jazz styles, in the '80s his work developed a depth and poised intensity which Stirling attributes to his being introduced to Nichiren Shoshu Buddhism by Ernestine Anderson during her tour of 1981. He is arguably the most substantial and original post-bop trumpet player in Australia in the mid '80s. In 1986 he was resident in Newcastle, NSW, playing mainly club work.

Stott, Alan Bruce   b. 11/4/47, Melbourne, Vic.   tba/bs/flug./comp./p./ldr
Began in the Heidelberg Brass Band and in 1966 joined the Silver Leaf JB, from which he founded the Limehouse JB 1968–72. He joined Poppa Cass's Dixielanders on piano in 1974, then began freelancing. In 1977 Stott began an association with Neville Stribling, with whom he appeared at the Sacramento Jazz Jubilee in 1982. In 1982 was with the Dave Hetherington trio (with John Withers, bjo), which, with the New Harlem JB, provided music for the film *The Rise and Fall of Squizzy Taylor*. In 1986 he was freelancing and leading his own group, the Hottentots.

Stribling, Neville   b. 12/2/36, Melbourne, Vic. Reeds/comp./ldr
Began clarinet in 1952, and in 1953 played a Downbeat concert at Melbourne TH, on Jimmy Wood's 3AW Rumpus time, and was sitting in with Frank Johnson at the Maison de Luxe. During the '50s he listened to the Bell, Johnson, and Barnard bands, and was strongly influenced by Ade Monsbourgh. Joining the Melbourne Jazz Club house band in the late '50s gave impetus to his development. Has become a member of the established traditional jazz population in Melbourne. Has led his own Jazz Players for concerts and recordings and worked with Monsbourgh and Barnard in the Australian JB, which has recorded, and played the Sacramento Jazz Jubilee. In the '80s is freelancing as well as participating in recording projects with Monsbourgh and Dave Dallwitz. Lives in Euroa, Vic.

Stuart, Nancy   b. 6/6/21, d. 22/3/85, Sydney, NSW   vcl/ldr
Began singing at parties in her teens, then local dances. First important professional work was in Melbourne, at the Trocadero with Bob Bell (1947), and at Leggett's ballroom. Retired from singing during the '50s to raise a family, then resumed, singing at small clubs in Sydney's western suburbs, incl. Smithfield RSL, Fivedock RSL, Drummoyne Rowing Club, Western Suburbs Soccer Club. This was invaluable experience in terms of understanding the need to communicate with an audience, and in smooth and unostentatious presentation. She made her first impression on the jazz scene in 1973 when she sat in with Wally Temple's (dms/reeds) band at Cricketers' Arms Hotel, Surry Hills. The warmth of her work, based heavily on the style of Maxine Sullivan, assured her immediate popularity. Played the 1975 AJC, Balmain, to great acclaim, with Paul Furniss, and joined Ray Price during the late '70s. Led her own band, the Gutbucket Five (incl. Temple) at the Bondi Icebergers' Club through the late '70s, and in 1977 was invited to record with her own choice of musicians. These incl. blind pianist George Herrman, who became a regular accompanist until his death, 19 Dec. 1982. Stuart also played a lengthy residency at Red Ned's, Chatswood, and regular Opera House Boardwalk concerts. She played the Manly Jazz Carnival, the Festival of Sydney, the Don Burrows Supper Club (and appeared on *The Burrows Col-*

lection), with Bob Barnard, and played support for tours by Acker Bilk and the Dutch Swing College. In the '80s she regularly sang in the choir of John Colborne-Veel's jazz mass, incl. in the week before her death. The following week the Roman Catholic mass became her memorial service, a remarkable testimony to the general respect and affection in which she, a non-Catholic, was held.

Swain, Murray Gene   b. 26/7/39, Nth Fremantle, WA   tpt/el. bs/ldr/arr

Bought a trumpet to give music lessons as a student teacher in 1959 and began working with a band incl. Dick Hatton and Theo Henderson in 1960. Posted to Geraldton from 1961–63 he played with Frank Harrison (p.) and John Twycross (reeds), and started a big band. Back in Perth, withdrew from music until resuming cabaret work, forming a trio (with Bill Prideaux p., Vic Denford dms) which, with the addition of Twycross, became the Swan City Jazzmen in August 1967. Kevin Findlay (tbn.) joined, 1968, replaced by Neil Slaughter, 1972. In 1969, Denford and Prideaux were replaced by Don Slaughter and Ron Spowart; 'Kipper' Kearsley (tbn./tba) joined in 1971, replaced by Rusty Brookes in 1974 the band remaining stable (Swain, Brookes, Twycross, Spowart, and the Slaughter brothers), until the late '70s. Subsequent members have incl. Ross Nicholson from 1979, Carl Anderson (bs), Ian Daniel (dms) 1979–80. In 1985 it consisted of Swain, Nicholson, Tony Eardley (p.), John Atkinson (bjo), John Kerr (bs) and Bert Tiambeng (dms) with the usual variations imposed by other commitments. Swain's significant career is as leader and playing member of this group, which has become a fixture as a traditional jazz-based cabaret/dance band active in Perth and throughout rural areas of WA. Swain is also active as a broadcaster, in jazz journalism and has established his own record label, Jam Records.

Swanson, Roger Andrew   b. 16/8/37, d. 22/2/79, Adelaide, SA   tpt

Began trumpet at 15, formed a group in 1955 during his National Service, followed by the Pioneer Hall JB with Ian Bradley. With Dick Frankel's Jazz Disciples, 1962–72. During the '60s and '70s also in semi-commercial groups, Lovable Brass and Brass Buckle. Swanson also worked in the progressive Bottom of the Garden Goblins, with Schmoe, Ted Nettelbeck, and Gary Haines (dms), in Frank Buller's Big Band and as lead tpt in Neville Dunn's Big Band. From the late '70s, with the Pioneer JB, the Unity JB, and various pick-up groups. Like Charlie 'Chook' Foster, Swanson was one of the few trumpeters in Adelaide who was convincing in a range of styles. At the time of his death he was with Ron Flack's Unity JB, and Ken Way's Hillside Dixielanders.

Sydney

Like Melbourne, Sydney has set the standards and determined the directions of various movements in Australian jazz. To a large extent, therefore the opening discussions which have traced the evolution of jazz attitudes have simultaneously disclosed developments in Sydney's jazz history, and less need be said on that general subject in this essay than in those covering other centres. At the same time, however, the scale of activity in Sydney makes it the most difficult city to deal with in terms of taking note of all those musicians who have sustained the music. In each individual entry I have attempted to include reference to other musicians who have not elsewhere been remarked. Even so, a dismaying number continues to slip through the net, particularly so when we come to the burgeoning of young talent from the late '70s. The problems of determining what and who is significant in the '80s are discussed in an introductory essay. The number of musicians currently working in Sydney, and the intimidating comprehensiveness of their instrumental command, make the problem especially acute. The size of Sydney's jazz population means that many who would have been outstanding in another city have been swallowed up. Furthermore, the general level of musicianship is so high that there are many musicians in the city who go relatively unnoticed, but whose capacities are considerably greater than musicians who, being based elsewhere, have gained notable reputations;

this volume itself reproduces the anomaly, but with limited space, that is unavoidable.

Jazz in Sydney is not only distinctive in degree, but in kind. In terms of musical interchange, it is the most fluid scene in the country. An unparalleled level of freelancing gives the picture a bewildering diversity. Each entry pertaining to a Sydney musician is therefore likely to be more skeletal than those for other regions; most subjects are, in addition to regular band membership, busy as freelancers, deputizing on short- and long-term bases. This richness of interchange has had a reflexive relationship with the stylistic character of the local music. The lines of stylistic demarcation (traditional, modern, etc.) are less clearly defined than anywhere else, not only within the jazz spectrum, but across the whole musical range. There are enclaves of purist exclusivism, but these are exceptions rather than norms, unlike the case in other centres. The jazz landscape is also diversified and enriched by the huge pool of studio and session musicians, members of which regularly but often unpredictably surface in a jazz setting. All these characteristics have had a self-perpetuating effect, in that they help to make Sydney a magnet for musicians throughout Australia, who in turn multiply the diversity of talent. As a statistical exercise, it is useful to note the number of musicians listed in this book who are resident in Sydney but were born elsewhere. Sydney is the jazz capital of Australia, with its particular strength lying in post-traditional styles. Again, however, even that reservation has less point than would its equivalent observation in, say, Melbourne; the stylistic blending robs the distinction between traditional and modern of much of the clarity it would retain elsewhere. As a familiar example, most of Graeme Bell's Sydney alumni have worked in bop and later styles, and without experiencing any disorientation.

As in other parts of Australia, the distinctive character of Sydney's jazz began to emerge during and after the Second World War. Hitherto, the main distinction had been one of degree: Sydney has had more night life, with more musical opportunities, than other cities. Through the '30s Sydney developed a reputation as more musically advanced than Melbourne, in relation to the standards being set in the US. This circumstance and some of its corollaries have been discussed in introductory essays. At the same time, however, the war years represented the same watershed as in other cities, as a new generation with new attitudes emerged into sudden prominence. Many faces and institutions receded during this period. The main public 'hot' music venue of the '30s, the Ginger Jar (by now renamed again the Oriental) closed in 1942. Its band leader Jack Spooner moved to a new cabaret, the Rex, but was scarcely visible as a performer in the post-war period. Many pre-war band leaders and musicians retired, moved out of Australia (as in the case of Jim Davidson shortly after the war) or moved into more entrepreneurial roles—Dick Freeman was one of many who became involved in the management of dance circuits. The reasons for change were much as they were in other cities, including notably the gaps left by enlistment (which also encouraged the formation of all-woman dance bands), and the American presence, which gingered up audiences and musicians. One immediate effect of the Americans was a fertilization of the night-club/cabaret scene. The opening of Romano's new restaurant in 1938 was fortuitously timed in this respect; the Roosevelt, originally restricted to allied officers, was opened to the public in 1944, and became one of the most popular night spots in Sydney. The Golden Key re-opened in 1946, but had previously established itself as a night-club for servicemen. American tastes also gave further stimulus to swing and hot jazz performance, and various 'jamming' venues appeared during and immediately after the war. As a record session club, the venerable Sydney Swing Music Club was little affected, but progressive forms of hot music were played at the 2KY Jazz and Swing Club (from 1943), the Baltimore in Pitt St (1943), jitterbug championships (as at Leichhardt TH, c. Jan. 1944), the California coffee shop (from 1946), Actors' Equity Club (from 1947), Ellerston Jones's Rhythm Club (founded Jan. 1948). A number of clubs were opened specifically for, or generally aimed at, the American market,

A jazz/jitterbug session at the 2GB auditorium, Sydney, c. 1946. Music by the Port Jackson Jazz Band, with, from l. to r., Bob Rowan (tbn.), Doug Beck (gtr), Ken Flannery (tpt), Clive Whitcombe (dms), Wally Wickham (bs), and pianist Jimmy Somerville (obscured)

incl. the Yankee Doodle Club and the Booker T. Washington Club, and these gave Sydney musicians the opportunity to mix with and play for audiences with well-informed and up-to-the-minute jazz tastes.

In the years immediately following the war, Sydney experienced the national intensification of jazz activity. Bop enjoyed some exposure with Wally Norman's record sessions in 1947 being an early focus. The Harbour City Six was one of the first Sydney bands to present small group bop in public, Ralph Mallen was an important early big band leader in the Stan Kenton mould, and a succession of clubs sprang up to foster progressive styles: the Stan Kenton Society (1949), the Society of Modern Music (1951), the Australian Jazz Club (1954). Traditional styles were promoted by Jack Parkes (tbn.), who established the very important Port Jackson JB (1944), the Midway Stompers (1947), and the Riverside JB (1948). Tony Howarth (tbn.), who was later involved with early Sydney Jazz Club activities, and in the late '70s has been an important organizer in the Parkes/Forbes area of NSW, formed a traditional group as early as 1947. Duke Farrell's Illawarra Jazz Gang (1949), the West Side Stompers, with Bill Boldiston (clt), the Pacific Coast (a.k.a. East Coast) JB, Frank Johnson's (not the same as the Melbourne bandleader) Harbour City Dixielanders (1950), were other early traditional groups whose activity has subsequently been obscured by the extensive documentation on Melbourne bands in that style during the late '40s to early '50s.

These disparate jazz enterprises were brought together in the jazz concerts, which attracted enthusiastic audiences from their outset in 1947. Much of the music presented on these concerts was played by professional musicians who were employed in the night-club and ballroom scene, and who welcomed this new music forum as a chance to play relatively unfettered jazz. Following a brief slump during the fluid and therefore uncertain period immediately after the war, Sydney's night-club activity picked up and became, in many respects, an extended workshop for professional musicians with progressive jazz interests. Many

Wally Norman in the Roosevelt Restaurant, 1944

clubs were fly-by-night operations, but some premises became institutions, even if under a succession of names. Romano's and the Roosevelt had comparatively long histories under those names. Sammy Lee's, opened in 1949, became the Flamingo (1951), the Sheridan (1954), the Pigalle (by 1960), and briefly, the Diamond Horseshoe, under which fading sign it currently stands derelict in Oxford St opposite Centennial Park. Christy's reopened under the more famous name Chequers in 1951, and the Orchid Room changed its name to Andre's within a year of opening in 1954. The Hayden, Golds', the Stork Club, the Silver Ash, were other night-clubs/restaurants which constituted a sort of woodwork from which Sydney's progressive jazz musicians emerged to play the jazz concerts from the late '40s through the '50s. The concerts in some ways represented in microcosm one of the distinctive features of the Sydney jazz scene in that they presented countless combinations and permutations of musicians, both amateur and professional, with interests ranging from 'righteous' jazz to relatively commercial cabaret material. A man who led a small progressive group one night would be likely to be playing in the section of a Glen Miller-style big band the next, and on another occasion filling in in a more or less *ad hoc* dixieland group. Many of the perennial leaders were also regular sidemen, and familiar faces in the early period included Ralph Mallen, Bela Kanitz, Les Welch, Ron Falson, Wally Norman, Enso Toppano, Johnny Best, Billy

Weston, Ron Gowans, Jack Allan, Gus Merzi, Jack Brokensha. Later, Frank Marcy, Bob Gibson, Bob Limb, Pat Caplice, Lee Gallagher, Keith Silver, Joe Singer, and at various times, the Mannix brothers Ron and Bill, were prominent. In a concert situation, vocalists were particularly popular; Georgina de Leon and Edwin Duff had explicit jazz credentials, and Norman Erskine and Larry Stellar leaned more heavily towards cabaret/variety material. The more traditional styles of jazz were presented by various groups, including many of those referred to above, with the Port Jackson JB as the main standard-bearers as a group, and Jim Somerville frequently leading bands assembled on a short-term basis through the traditional to mainstream range. Other important participants in the Sydney concerts have been acknowledged under individual entries.

As the concerts lost momentum in the mid-'50s, most of the musicians disappeared from jazz settings into television and radio, restaurants, licensed club', teaching, copying and arranging, various forms of production and promotion, and into the night-clubs, although for various reasons this last category was also soon to languish. Concerts did continue, though with greatly diminished frequency and altered character, the main change being in favour of imported performers: Gene Krupa (1954), Buddy de Franco, Frank Sinatra, Winifred Atwell, Ted Heath (all in 1955), Stan Kenton (1957), and from 1960 to 1965, Dave Brubeck, George Shearing, Sarah Vaughan, Dizzy Gillespie, Jonah Jones, Mel Tormé, Ella Fitzgerald, Kenny Ball, the Eddie Condon package, Thelonius Monk. Most of these were presented at the Sydney Stadium.

The mid-'50s witnessed considerable changes in the entertainment industry, two important factors in Sydney being the advent of TV and of late night hotel trading. Night-clubs were to feel the strongest effects of these changes, but TV in particular, and the arrival of the long playing record, had more far-reaching effects, including the gradual winding down of the venerable Sydney Swing Music Club.

If jazz ceased to be a broad based public recrea-tion, it nonetheless has never disappeared completely in Sydney. Its continuity was ensured by a determined minority for whom the music was inherently rather than commercially compelling. Mainstream to modern venues were established at the Sky Lounge, c. 1956 or 1957, where, in the interests of dancers, the emphasis was on relatively well-established mainstream music, the Biltmore, Club 11 (which opened 13 Jan. 1957), and the Mocambo at Newtown. Overshadowing all of these, however, was El Rocco, which became not only the centre of the modern movement in Sydney during its lifetime, but one of the seminal venues in Australia's jazz history. Apart from activities sketched elsewhere, its reputation spread with such compulsion that many musicians regularly commuted from Melbourne simply to play or listen at El Rocco for a weekend. The interest which it generated led to the establishment of other venues such as the short-lived Cellar in Liverpool St, and the Parramatta Modern Jazz Club. Traditional jazz found a similar focus in the foundation of the Sydney Jazz Club in 1953, under the aegis of which several bands, notably the Paramount JB and the Black Opal JB, and hundreds of enthusiasts, were able to satisfy their appetite for the music in the years preceding the boom of the early '60s.

The spread of the 'jazz pub' venue also provided an outlet for the sudden explosion of activity from around 1960. Merv Acheson's residency at the Criterion was an early and durable example. These pubs generally favoured the more extroverted 'good-time' flavour of traditional jazz, and among the scores of pub venues which came and went at short order, the Macquarie in Woolloomoolloo, Adams, and the Criterion, were virtual institutions. Later in the '60s but in due time no less celebrated, were the Brooklyn and the Orient in George St across the road from Ironworkers' Hall where the Sydney Jazz Club held its functions, and the Windsor Castle in Paddington. The Ling Nam was somewhat unusual, not only in being a restaurant with an explicit jazz policy, but also in presenting at different times a variety of jazz styles from mainstream to traditional, with a group led by Noel Gilmour, to the Port Jackson and Riverside JBs. The main venues for more modern forms of jazz in the early '60s continued to be El Rocco and the Sky Lounge, with short-lived ventures like the Jazz Workshop in Orwell St, Kings Cross, the Bird and Bottle in Paddington, and Basil Kirchin's big band at the Pigalle.

The national jazz slump from the mid-'60s led to widespread terminations of jazz policy incl. at institutions like Adams, the Criterion, and the Sky Lounge. A Jazz Appreciation Society established in Aug. 1967, and in which Doc Willis was a moving spirit, struggled for about a year before petering out. Live entertainment was now dominated by electric pop, and the professional musicians who were a significant repository of progressive jazz talents were increasingly earning a living in the commercial field of licensed clubs. It is important to make a parenthetical explanation of this term. The multitude of venues comprehended by the term 'club' in Australia is compounded in NSW by these entertainment venues. In this book I have referred to jazz clubs, which are centres of relatively well-informed and dedicated interest in jazz as opposed to all other forms of twentieth-century popular music, excluding rock, which in any case they generally predated. There have also been the 'casual' clubs, primarily a feature of Melbourne in the late '50s to early '60s, and discussed elsewhere. The phrase 'licensed club', particularly as used in NSW, generally refers to RSL or sporting clubs. In many instances these are virtually community entertainment centres, incorporating a fully equipped auditorium for floor shows which present top line local and imported performers, restaurants, bistros, bars, a cinema, and ancillary facilities like swimming pools and saunas. Primary sources of revenue are liquor sales and poker machines; it is the latter, licensed in 1956, and confined to NSW, which have enabled these clubs to become so economically significant in that state. From the '60s until the growth of home video and the advent of random breath testing in the '80s, these clubs (collectively known among musicians as 'clubland') have dominated live entertainment in Sydney, and have been probably the largest single

employer of musicians in the popular field, particularly the fully literate professional who is able to back acts at short notice and with minimum rehearsal. In terms of jazz, that means, most frequently, the musician with a command of mainstream conventions.

The rise of the clubs complemented the jazz slump of the late '60s in Sydney. On the one hand, they were a competing focus of musical entertainment. At the same time, they provided secure employment for many musicians who might otherwise have directed energy into support for jazz enterprises. Paradoxically, this slump created a situation which, although requiring further investigation, I believe encouraged a brief efflorescence of big band activity, embracing the Daly-Wilson band which was formed in 1969. One of the problems of holding a big band together is the competing offers from small groups which require no rehearsal, and which an accomplished freelancer can rely on in Sydney if there is a healthy jazz scene. If that scene withers, however, a big band with a committed jazz policy is ironically able to count on continuity of support from musicians with an emotional need to play in at least some approximation of a jazz context. In the late '60s, a number of big bands came into existence, and if they were unable to find regular employment, they nonetheless maintained a stable rehearsal existence which would probably not have been possible if the various members had easily been able to satisfy a wish to play jazz as a way of 'letting off steam' denied them in more commercial employment. In addition to the Daly-Wilson band, there were big groups being led by Dick Lowe (sax), Billy Weston and Edwin Duff, and Peter Lane.

The renaissance of jazz activity which spread across Australia in the '70s was being intimated in Sydney in a new expansion of the pub scene from the beginning of the decade. The Vanity Fair and Albury Hotels were two new jazz venues destined to enjoy long lives. They were followed by the Lord Roberts, the Forest Lodge, the White Horse in Newtown, and by 1974 the Grand National in Paddington and the durable Unity Hall in Balmain. The jazz pub has been a crucial element in Australia, especially since late night trading crept across the continent. It represents a distinct subculture which receives little public attention as compared with the cosmetically more acceptable jazz bistros and restaurants. The latter venues have the air of the slightly special occasion, the audiences are discrete in the sense that they vary from week to week, and on any given night, the various groups constituting that audience will be strangers to one another. The jazz pubs attract a different kind of clientele and fulfil a different social function: the audiences tend to constitute a coherent community which is cognate with the members of the band. The pub and its jazz sessions constitute a fixed point in the social life of that community, and not a novel or special occasion. The jazz pub is one of the unifying threads in the local culture, having a folk function far more akin to the earliest role played by the music than the rather artificial, but more publicized, concert situation, in which everything is done to emphasize the separateness of the musicians from the everyday life of the audience, who are atomized and static. The jazz pub is the primary sustaining force of the music as a component of everyday life, and although those who lead bands in these venues receive less recognition than the regular concert performers, they are playing a more important sustaining role. Sydney is fortunate to have a steadfast corps of musicians who maintain this vital underground foundation. In addition to others mentioned elsewhere, the bands assembled by musicians such as Dave Ridyard (clt), Rod Lawliss (clt), Don de Silva (bjo), Ian Barnes (tpt), Rex Gazey (tba), should be recorded, in both apposite senses.

The pubs generally favour traditional groups, and again I believe it is because in so many ways the earlier styles of the music are historically more compatible with the idea of an extended community. The '70s also saw the establishment of other key venues, primarily in the bistro style. Bruce Viles opened the Rocks Push (later a.k.a. Old Push) in Oct. 1971, and was later associated with The Basement (opened Aug. 1973). Other venues which began longstanding jazz policies in the seventies were Soup Plus (1974) and Red Ned's

(1975). All four of these maintained jazz for up to six nights a week. Jazz also received what might be called institutional support from various sources: the jazz studies programme at the NSW Conservatorium (1973), where the staff has included Howie Smith, Don Burrows, Bill Motzing, and Dick Montz; the inauguration of community radio (which, collectively and in some cases individually, presents more jazz than the established AM frequency stations), beginning with 2MBS–FM in 1974; the inauguration of Australia's first Jazz Action Society, in Sydney, 1974; the foundation of the Campbelltown Jazz Club in 1975, which had some effect in decentralizing the jazz movement by creating another focus well out of the city and inner suburbs; jazz representation in various government and privately sponsored festivals, including Shelley's Jazz Festival (1978), Sydney Festivals, and Sydney Jazz Festivals which grew out of the summer jazz clinics organized by Greg Quigley in the late '70s.

All of these developments have produced tangible results in the form of important musicians and bands, particularly in the mid-'70s. The Last Straw, Out to Lunch, Kindred Spirit, and Jazz Co-op which grew out of an informal alliance between The Basement and the conservatorium courses, became platforms for vigorous and creative innovation. The momentum thus developed was maintained subsequently in Sydney through the Young Northside Big Band, which provided an apprenticeship for some of the most important young musicians in the '80s, and regular festivals like the Manly Jazz Carnival, both of these being the work of pianist John Speight (b. 21/6/35). Although the spotlight has fallen mainly on Daly-Wilson, the Northside, and the Morrison Brothers Big Bad Band in succession, a number of others have maintained at the very least, and for varying periods of time, a rehearsal existence. In addition to the big bands set up in connection with the conservatorium, there have been groups led by Ian Boothey, Terry Rae, John Colborne-Veel, Craig Benjamin, the late Dick Lowe, George Brodbeck, and Adrian Ford, in styles ranging from that of the '30s to contemporary approaches.

The '80s have seen an explosion of Sydney's jazz

population on a scale not witnessed since the late '50s. Some of this has affected the traditional end of the spectrum, as for example, in the case of the revived Nat Oliver band. The inauguration of Sunday hotel trading in 1980 has provided many traditionally oriented bands with an extra opportunity, but bop and post-bop groups have also become visible in pubs on a new scale, as in the case of Joe Lane's groups in the Criterion. Indeed, the recent growth of jazz activity has been overwhelmingly in favour of post-traditional styles. The following representative list of new generation musicians who are making a mark includes an impressive number of graduates from the conservatorium, where the emphasis is on more 'modern' forms: Tony Buck (dms), Steve Elphick and Phil Scorgie (bs), Guy Strazullo, Steve McKenna, Tony Barnard, Ian Date (gtrs), Mark Isaacs, Michael Bartholomei, Kevin Hunt (ps.) —and of course others mentioned elsewhere. Barnard (son of Bob) and Date project a consciousness of earlier traditions, though the former is a more mainstream stylist.

Sydney's jazz population has also been augmented over the last decade by the arrival of newcomers from elsewhere and the return of musicians who have been working for long periods overseas, all this on a scale which invites a study of its own. The former category includes New Zealander Peter Cross, Americans John Hoffman (tpt), Vince Genova (p.), Steve Giordano (reeds), and Indonesians Bill Saraghi (p.), who first arrived in 1972 but only began to make his presence felt after returning from revisiting Indonesia in 1979, and young prodigy Indra Lesmana (p.), who has since settled in the US. Musicians arriving from elsewhere in Australia incl. John Callaghan (bs) who arrived from Melbourne in 1978 to join John Colborne-Veel. Returning musicians include Barry Canham (dms), Don Harper (vln), Eddie Bronson (reeds), Dave MacRae (p.), Andy Brown (bs.). The pool of local jazz talent has in the meantime continued to be fed by the usual influx of musicians from interstate (incl. Glenn Henrich and Paul Millard), wishing to take advantage of the more extensive musical opportunities in Sydney.

A special mention ought also to be made of the singers who are active in local jazz settings, partly because singers find themselves in a special situation for a number of reasons. A very large number of vocalists are essentially club musicians who are occasionally seen playing jazz concerts, and these cannot be said to be having a substantial effect in preserving a jazz tradition. There are others who confine themselves to more explicitly jazz work, and they face difficulties which instrumentalists of equivalent competence do not have to cope with. Above all, they tend to find that unless they lead their own groups, they will rarely be able to survive from week to week since there is less freelance employment for singers. The reasons are many and not always musical. They include matters of repertoire and sometimes unusual key signatures, but also a band's reluctance to share payment equally with someone who only performs, say, 60 per cent of the material presented. These are problems which singers everwhere face; they become prominent in Sydney simply because there are so many jazz-based singers. They tend to gain their exposure through 'guest' appearances or through that energetic activity known as 'sitting in'. Sandie White, Barbara Canham, Julie Amiet, Jan Adele, Trude Aspeling, Chris McNulty, Lorraine Silk, Joy Mulligan, Bobby Scott, Marie Wilson, all fit to some extent into this rarely publicized category, in spite of the fact that several of them have made highly praised recordings; Wilson's has enjoyed the rare compliment of being picked up for American distribution.

Some of the important new blood in Sydney since the late '70s has made its presence known more through important bands than on an individual basis. The Benders, John Rose's groups, James Morrison's groups, Roger Frampton's Intersection, John Hoffman's Big Band, and Supermarket, have all provided experience for some of the hugely talented young musicians emerging recently. Three groups making significant but in many ways distinctive contributions should be noted.

Crossfire is a fusion band, but with no ambiguity about its jazz commitment. Formed in the mid '70s, most of its material has been written by band members Jim Kelly (gtr) and Mick Kenny (tpt/keyboards). Other members are Ian Bloxsom, Tony Buchanan (reeds/fl), Greg Lyon (bs gtr), and Mark Riley (dms) (until he was killed in a motor cycle accident at the age of 23 in 1984). The band achieved considerable popularity yet without compromising its musical integrity; it has recorded extensively, including with Michael Franks, and has toured in SE Asia for Musica Viva, the US, and Europe, where it was acclaimed at the Montreux Jazz Festival in 1982. The band was in recess, 1983–84, and was revived in Nov. 1985 at The Basement, with David Jones (dms), Sunil de Silva (perc.), Victor Rounds (bs), Wayne Goodwin (vln), as well as Kenny, Kelly, Bloxsom, and Buchanan.

Women and Children First is a Sydney-based band which includes a number of ex-Melbourne musicians. Sandy Evans (reeds/fl./vcls) worked on the rock circuit with Great White Noise, and with an all-woman jazz group, the Midnight Toe Jammers, before she formed Women and Children in 1983. Although drawing some of its repertoire from the bop library, the band's importance lies more in its contemporary/experimental work, fertilized by the diversity of background of its members. This embraces rock, jazz, commercial pit band work, and multi-media experimentation. Steve Elphick (bs/tpt/vcl.) is, with Evans, the longest serving member, other originals having been Indra Lesmana (keyboards) and Tony Buck (dms). By 1985 the band had been augmented by Cleis Pearce (viola/perc./vcl), with Jamie Fielding and Jonathan Glass, both from Melbourne, replacing Lesmana and Buck respectively. Apart from the vitality of its music, one of the most encouraging aspects of the group is the considerable success it has achieved in presenting highly innovative music to audiences outside Sydney, in the course of tours throughout 1984 and 1985, including a national tour assisted by the touring and access fund of the Australia Council.

The Sydney Jazz Quintet brings together some of the young musicians who emerged from the conservatorium jazz course in Sydney. Brent Stanton (reeds), is the leader (and the band is sometimes presented under his name). He arrived from NZ in 1978, studied in New York in 1981, and has worked in other groups incl. Bruce Cale's. Warwick Alder from Newcastle is arguably the most interesting and consistently advancing of the young trumpet players to emerge in the late '70s, his work enriched by a grounding in the whole jazz tradition. He has led his own groups and works with the Morrison Brothers band. Likewise guitarist Steve Brien, who, with ex-Brisbane drummer Ron Lemke, has also worked with Keith Stirling. The band's repertoire and style extend historically from the bop period onward. Its particular importance lies in two things: one is the cohesion it has developed as a consequence of the stability of its personnel in a city where most groups in this style are pick-up bands; the other, and potentially more important, is that, with the backing of Musica Viva, the quintet regularly presents school lecture/recitals, thus introducing jazz to the only section of the population able to ensure the future of the music.

The 'institutional' support which Australian jazz has begun to enjoy has been fortified in Sydney through a number of non-government bodies, some of which, such as the Jazz Action Society, have been referred to elsewhere. In the early '80s, Keys Music Association fostered experimental and other forms of contemporary improvized work on a very considerable scale, and it is a puzzling circumstance that this organization has received so little notice. KMA grew out of a band called Keys, named in memory of young reed player Martin Keys who died in 1973. Personnel has fluctuated, with members including Mark Simmonds and Daniel Fine (reeds), Steve Elphick, Raoul Hawkins (tbn./bs), Robin Gador (bs), Azo Bell (gtr), Peter Fine (keyboards/reeds), Searle Indyk (vln/viola), Peter Dehlsen, and Greg Sheahan (dms). The band's commitment to contemporary jazz-based improvisation found further expression with the foundation of Keys Music Association in 1979. The aim of the association was to create performance possibilities for like-minded musicians, through the promotion of concerts and concert series, broadcasts on 2MBS–FM, and recordings. KMA introduced the Dale Barlow Quartet (later The Benders), and Mark Simmonds's Freeboppers.

Mark Bentley Simmonds (b. 21/7/59, Christchurch, NZ) is one of the most authoritative musicians to appear on Sydney's contemporary scene since the late '70s, with an eclectic background that includes the rock band Old '55, in which he replaced Wilbur Wild. Simmonds was a founder member of Keys, and while working with the Dynamic Hepnotics, a rock-based group, in 1986, was also preparing an album with the Freeboppers. KMA has provided support for virtually every major contemporary jazz musician to emerge in Sydney in the '80s, including Sandy Evans. By 1982 there were around 30 musicians involved in the association, and as recently as June 1985, KMA was still involved in concert promotion through The Benders' farewell concert at The Basement.

In 1983 Eric Myers began discussions which led to a public meeting on 25 July 1984 at which the Sydney Improvised Music Association was established with the objective of facilitating the 'performance and recording of contemporary improvised music'. The association was in part the expression of disappointment at what was felt to be the Jazz Action Society's retreat into more established names and styles. Although SIMA has only been in existence for two years at the time of writing, it has demonstrated great energy and administrative ability in mounting concert series with emphasis upon non-commercial music featuring both local and imported musicians, frequently with subsidies from government bodies. The work of SIMA has been particularly important in view of the difficulty even in Sydney of supporting regular jazz venues on a sustained diet of contemporary music. The Paradise Jazz Cellar wound down its first phase of jazz policy through 1983; Jenny's Wine Bar (later briefly reopened as Paco's) lasted for only 1982–84. Other attempts to sustain non-commercial venues, like the Jazzbah in Petersham (later used as the premises for Willie Qua's Quill Club) founder under the weight of local apathy to experiment, innovation, or even just forcefully uncompromising post-mainstream jazz.

Nonetheless, Sydney in 1986 is the most fertile centre for what is popularly known as 'modern' jazz. There are many reasons for this, and not all of them are a cause for satisfaction since they bode ill for the future of the full and rich range of jazz styles. One of these is the retirement of many musicians who are steadfast, even ideological, traditionalists. As the boom of the '60s receded, opportunities diminished and at the same time those who sustained the trad boom are now 20 years older. Jazz, especially traditional, is a high-energy music which, in the absence of extensive electronic/electric boosting, requires very considerable physical stamina, especially in the horns. Age wearies. Another important factor in the increasing predominance of later schools of jazz in Sydney is the stylistic development of many players. The flux, the comparatively high level of musical curiosity, the relatively low level of purist exclusivism, all encourage (or at the very least, do not discourage) stylistic advancement. To a greater extent than anywhere else in Australia, the musicians who formed the corps of traditional groups through the '60s and '70s are now working in mainstream to bop settings. The tendency in favour of later styles has also been accelerated by the bewildering proliferation of young musicians in contemporary styles, many of whom have come through apprenticeships—club workshops, conservatorium courses, clinics—which ignore pre-bop and even pre-Coltrane jazz history. The '80s have also seen the re-emergence into the jazz scene of many musicians who for 20 years or so have directed their energies into the non-jazz settings of the licensed clubs. In many instances this is simply a reflection of a new career phase: these musicians have seen their children grow up, have reached a stage of their lives where the regular income from club work is less important, and can now enjoy the luxury of more stimulating if less commercial work. In addition, the club scene has suffered from the effects of home video and random breath testing: at least the first of these (and, I intuit, the second) is more likely to affect the sensibility that had formerly found entertainment in the clubs, than that which is dedicated to jazz. All of these considerations have encouraged a movement of musicians from club work back to jazz, and the majority of these musicians, like Freddie Wilson (reeds) and Cliff Barnett, are mainstream in their jazz preferences.

In 1986 Sydney remains the major Australian centre for post-traditional jazz, and the only city in which such styles actually dominate the scene. The 'curve' which has contributed to this situation continues to rise: progressive jazz thinking has a clear future in Sydney, though inevitably the more innovative musicians continue to struggle for a public forum. As elsewhere, the biggest question mark in the Sydney jazz scene hangs over the long-term future of its traditional varieties.

Sydney Jazz Club (NSW)

The inaugural function of the Sydney Jazz Club took place at the Real Estate Institute, Martin Place, on 8 Aug. 1953 with music by the Paramount JB, for which the club had been established by Harry Harman to provide a venue. Harman was founding secretary and Fred Starkey the president. The club was an immediate success at a time when there were few other traditional jazz venues and the jazz concerts were entering a decline. It moved operations to the Ironworkers Building, George St, from 3 Sept. 1955 and in the same year began issuing a magazine, *Quarterly Rag*, with Bruce Hyland as editor, followed by Alan Burton (1956), Jim Conway (1959) and Kate Dunbar (1959). At the Ironworkers Hall the club's popularity increased, and it was able to present the Cootamundra JB (John Ansell p./ldr, Lloyd Jansson tpt, Greg Gibson clt, John Costelloe tbn., Kevin McArthur dms, Bob Cowle p.), in 1956. In 1958 the club extended its Ironworkers functions from fortnightly to weekly, with the Black Opal JB alternating with the Paramount. (Other house bands at various times incl. the Quayside JB, the Harbour City Jazzmakers, and from Nov. 1963 to April 1964, Graeme Bell, who became Patron of the club.) The club organized the AJCs in Sydney in 1958, 1962, and 1965. In Sept. 1961 it moved to the YWCA while fire regulation renovations were undertaken at the Ironworkers, where the club resumed in Jan. 1962. The jazz boom of the early '60s brought increasing crowds and the Ironworkers Hall became the focus for traditional jazz in Sydney: visitors sat in, and virtually every traditional musician based in Sydney since the '60s got her/his start there. By the mid-'60s the club was running two floors of the building simultaneously with the Paramount and the Black Opals playing at the same time. Crowds were numbering over 1000, and these included large sections from sporting clubs whose main interest was not jazz but simply the large-scale socializing. Violence became a problem, security guards had to be employed, and it became clear that the club had become too popular for its own good. Following serious threats of violence, a special meeting of the committee on 16 March 1967 cancelled all future functions, while new premises of manageable proportions were sought. During this period the club also altered its status to a co-operative, partly for purposes of indemnification, partly as a prelude to applying for its own liquor licence. A succession of venues followed as the club resumed operations, though it never obtained a liquor license, and it never again established the same fixed weekly base as the Ironworkers Building had been. Its continuity through the '70s was mainly ensured through the monthly late-night functions at the Abraham Mott Hall in the Rocks area. With Nick Boston leading the house band for much of this period, the Motts encouraged a sit-in policy, and a number of veterans recommenced playing here, and newcomers, including the nucleus of the later Keys Music Association, got their start. During the same decade the club adopted the Saturday afternoon sessions with the Eclipse Alley Five at Vanity Fair Hotel as its unofficial weekly headquarters, and by 1985 the pub displayed the club's bulletin board. During the '70s the club also revived jazz workshops which it had sponsored earlier, presented record evenings, and continued to hold monthly picnics at Berry Island Reserve. At the instigation of then president Graham Kellaway, *Quarterly Rag*, which had ceased publication in June 1967, was revived in 1976 first with an editorial committee of Kellaway, Kate Dunbar, Bill Haesler, Adrian Ford, Brian Hutchinson, Alison Johnson, Bruce Johnson, then with successive editors Bill Haesler (1976–77), Bruce Johnson

(1978–85), then Kate Dunbar. The changing character of the Sydney jazz scene led to changes in the club's function from the late '70s. The Motts finished in 1979, victim of competing late licensed pubs with bands, and a review of musicians' rates. So much jazz was now available in Sydney that many felt the club had outlived its usefulness. At the 1981 annual general meeting a motion to wind up the club's activities was carried, but since its incorporation as a co-operative, certain legal problems had accumulated. During the year in which president Neil Steeper was dealing with these, a renewal of support led to the recision of the terminating motion, and the club survived, gradually developing a new role. In addition to the monthly picnics, the Sydney Jazz Club has since embarked upon a programme of special promotions which provide alternative rather than competing functions in the Sydney jazz scene. These incl. imported musicians such as Dave Dallwitz, Ade Monsbourgh, and American Bill Dillard, and 'theme' functions such as a tribute to Sydney broadcasters. The Sydney Jazz Club has survived longer than any other such club in Australia, maintaining continuity for traditional jazz.

T

Tack, George William (Tacka, Fatso)   b. 14/5/19, Melbourne, Vic.   clt/ten./vcl
Began on brass in high school band, also some violin which he believes contributed to his unconventional intervals later on reeds. While at Melbourne University, 1939–43, the influence of Ade Monsbourgh led him to switch to clarinet and tenor, on which he played at dances. Also played sessions for the Musicians' Union Rhythm Club, using George Fong (a Jess Stacy-influenced pianist, later killed in the war), Don Banks, Charlie Blott. Replaced Pixie Roberts in the Bell band in 1947 while the former recuperated from motor cycle injuries. With Tony Newstead's South Side Gang throughout its whole existence; also a member of the Portsea Trio, with Bill Miller and Will McIntyre. Since the demise of the Southside Gang, mainly casual work with trios.

Taperell, Christopher Henry (Chris)
b. 19/4/43, Sydney, NSW   p./reeds
Born into a musical family, began piano at 10, clarinet in his teens. Began playing dances in the late '50s. First jazz work was as regular deputizer for Barry Bruce in the Riverside/King Fisher group, 1963–64. With Ray Price in the mid-'60s, then played reeds with Dick Hughes's trio at French's Tavern, 1966–68. In 1970, took his Diploma in Music Education. Freelancing in the early '70s, and toured with Wild Bill Davidson, Bobby Hackett, Clark Terry, during their visit, 1972. Founder member of the Bob Barnard JB, 1974, which has been his main commitment ever since. Has also freelanced and backed touring musicians, incl. Kenny Davern and Scott Hamilton in Brisbane, 1980. Resumed formal music study, 1985, at NSW Conservatorium, graduating with Bachelor of Music Education, 1985. In addition to work with Barnard, solo and freelancing as a sideman, in 1986 Taperell was engaged in some teaching.

Tasmania

The small population of Tasmania has been a major determinant in the development of its jazz. The brute statistical fact is that even assuming its cities and towns contained the same proportion of enthusiasts for different jazz styles as cities on the mainland, this would still leave scarcely enough people to support a broad range of the music. This circumstance has been exacerbated by the dispersal of the population among regional centres. Not only are there fewer people, but they are less concentrated in any potential centre for jazz development. The third major influence on local jazz is the state's isolation which has deprived audiences and musicians of fertile contact with other Australian and visiting musicians.

By an explicable, if elliptical, chain of consequences these circumstances have resulted in several distinctive characteristics in Tasmanian jazz. To a large extent, the difference between the history of the music in, say, Sydney and Brisbane, is one of degree: less of it and later, in Brisbane. But the magnitude of that difference of degree in Tasmania has created differences in kind; for example, on oral and written evidence, the initial jazz craze of the '20s barely touched the island. Similarly, the amount of bop activity in the '40s and '50s was so slight as to make that style negligible in the jazz activity of the period. In more general terms, there has been less stylistic development in Tasmanian jazz until comparatively recently when the state's musical isolation was overcome by institutionally imposed forces such as Jazz Action Societies subsidized by government funding. Accordingly, until about the late '70s, the preponderant style of Tasmanian jazz has fallen within the range from dixieland to late swing. There are many jazz musicians working with more progressive approaches, but they are unable to find sustained outlets, and work either in non-jazz settings or in bands based on earlier styles. So strong, indeed, is the dominance of the latter that even some musicians themselves have a conditioned inclination to regard later stylists like George Shearing as somehow not quite jazz.

A further circumstance arising in part from demography is the importance of bands rather than individuals. There is little freelance jazz work as such in Tasmania; to play jazz to jazz audiences on a regular basis, a musician has to be associated with a band, so that the history of Tasmanian jazz looks somewhat like iron filings clustered around a few magnets, unlike the fluid interchange in, say, Sydney where much of the jazz activity is produced by freelance musicians loosely tied to bands which are, to varying extents, pick-up units.

The most important of these bands were those which, since the '30s, have formed around Tom Pickering. In no other state in Australia has a single individual been so indispensable as the rallying point for jazz for so long. With the Pearce brothers, Cedric and Ian, Pickering was originally attracted to a broad range of what in the '30s was contemporary jazz as heard primarily on the radio. The importance of this medium for the dissemination of jazz has probably been greater in Tasmania, in the absence of the other kinds of access to the music enjoyed elsewhere. Ellis Blain and later Jack Smith with the ABC in Tasmania have had a significant influence on the dissemination of jazz in the face of commercial indifference. The first attempts to play jazz by the Pearce/Pickering group were not revivalist in spirit, but simply the recreation of a current minority music form which stood out from the general background of popular music. As they became more interested in jazz, they included among their models the Clarence Williams small groups. This was not, however, because they were perceived as more 'authentic' than Benny Good-man/Teddy Wilson sessions, but because the instrumentation of the Williams groups was similar to that of Pickering's Barrelhouse Four and, initially, the music was technically easier for these school-boys to reproduce. It was because of the dedication of Pickering and the Pearces that Tasmanians were exposed to the second wave of Australian jazz in the late '30s, earlier than most other parts of the country.

During the same period there was some 'hot' playing among the swing-based dance band musicians. The amount for public consumption was slight, and indeed only a handful of musicians enjoyed reputations as 'hot' players. Again, Tom Pickering was one, but others included Geoff (gtr) and Max Sweeney (dms), John Denholm (p.), Alf Stone (p.), Benny Cuebas (tbn.), Ron Roberts (bs), and most importantly Alan Brinkman (reeds). These were members of the local dance band profession with as much interest and competence in swing as could be sustained in comparative isolation. By any standard, however, Brinkman was an impressive improviser. The main venue for jam sessions through the late '30s was the Stage Door (known colloquially as the Snake Pit), owned by a colourful Canadian lumberjack and wrestler, Frank Fouché. Until he sold his interest in the place in 1946, Fouché's activities were the closest thing to live jazz promotion in Hobart. He encouraged jamming, and he also imported mainland entertainers which incl. musicians with varying credentials as hot players, such as Roy Sparks (p.) and Graeme Bell.

Although the Stage Door cabaret continued to function throughout the war, neither it nor Tasmania in general enjoyed the same level of stimulating interchange with new musicians that fertilized the music on the mainland. Many Tasmanian musicians however were posted north where they were exposed to more contemporary movements in popular music, and they returned home to inject new spirit into the local scene. Ironically however, the chief effect of this was to strengthen the revivalist consciousness rather than to bring the music closer to its contemporary developments. Ian Pearce, Tom Pickering, and Launceston musicians Ted Herron and Bill Sutcliffe all returned from the mainland with the traditional movement having made the strongest impression on them one way or another. Pickering in particular went on to emphasize the traditional characteristics in the music he was playing, and even Alan Brinkman who worked in swing groups in Melbourne, returned to Hobart with the idea more firmly fixed in his mind that jazz was to be equated with 'dixieland'. One of the main effects of the Second World War on Tasmanian jazz, then, was to lead it more in the direction of its traditional rather than its swing manifestations.

Even so, this never became a mouldy movement with the same narrowness of purpose that was often found in mainland capitals. In the post-war period Tasmanian jazz was always strongly tinged by commercial influences, for several reasons. One was that the prime movers such as Tom Pickering and Ted Herron were of necessity also dance band players. The local market did not allow them to indulge in more recondite jazz styles. The second reason was that, to put together a complete band, it was necessary to recruit musicians who were relatively poorly versed in 'righteous' jazz. This is not to be taken as a pejorative observation; the unique conditions in Tasmania have simply produced a jazz approach which has developed spontaneously out of what was available and possible, rather than being engineered doctrinally according to what it was felt jazz should be. The result has been a very supple music that is never at odds with its environment; that Pickering's group has been called the 'Good Time JB' is an indirect reflection of its sense of being cognate with rather than separate from its audiences.

From the late '40s and through the '50s, the Hobart jazz forum was the functions organized by Tom Pickering in the 7HT Theatrette, later moving to the Hobart TH. These were to Hobart what

A Pearce/Pickering group, probably late 1950s: from l. to r., Geoff Sweeney (gtr), Ced Pearce (dms), Ian Pearce, Kay Staveley, Col Wells, Tom Pickering, Benny Cuebas

the jazz concerts during the same period were on the mainland, with the differences that they were regular weekly and twice weekly events, that they were dances rather than just concerts, and that, inevitably, the range of music and number of musicians presented were smaller. Nonetheless, there was variety provided by the presence of other musicians like Alan Brinkman, Don Gurr, and Keith Breen (ten.) who would arrive to sit in after their other gigs had finished. There were other intermittent jazz activities during the period, including the formation of a West Coast-inspired group by Peter Webster in the late '50s. There was, however, no hard bop scene at all as it developed in Melbourne and Sydney in the late '40s.

The effect of rock 'n' roll in Tasmania was delayed and diluted, so that at the time mainland jazz was beginning to feel its effects, Pickering's TH dances were able to accommodate the initial low level demand by introducing new material into the band's repertoire, involving the leader doubling on guitar. Finally, at the beginning of the '60s, the combination of younger audiences and the erosion by new musical fashions led to the end of the longest reign of jazz in Tasmania. Some ripples of the trad boom spread across from the mainland, but they had lost their amplitude. A jazz festival was held in Hobart in 1960 and a jazz club was formed in 1961, with Tom Pickering as president and Peter Hicks as secretary. Hicks has been one of the most energetic and dedicated non-playing followers of jazz since the late '50s, serving regularly on club and AJC committees. A few other groups like the Sullivan's Cove JB were able to operate. In Launceston there was a flurry of jazz activity. Insofar as there had been jazz there, it had been almost invariably produced by bands assembled by Ted Herron under the name the Jazzmanians. But in 1964 a South Esk Jazz Club was set up by the Launceston Adult Education Board, and a house band, the South Esk JB, was formed under the leadership of Ian 'Toots' Totham (clt). The Gourlay brothers also formed a Modern Jazz Trio which was briefly able to present a more West Coast-oriented style.

Tasmania's equivalent of the jazz boom in the early '60s was, however, a brief episode, which in the long term failed to generate the significance that it had on the mainland. In Tasmania, the boom died leaving little trace, and the responsibility for the continued survival of jazz spontaneously reverted to the circle surrounding Pickering in Hobart and Herron in Launceston. Furthermore, the word 'survival' must be interpreted loosely. The late '60s slump was close to a coma in Tasmania, especially as compared with the balmy days of the 7HT Theatrette and the town hall dances. Intermittent activity was maintained, as always, around Pearce/Pickering aggregations assembled for occasional functions, brief residencies, and through the '60s, ABC broadcasts, with occasional bursts of interest following on from AJCs held in Hobart in 1967, 1971 and 1977 (the first had been in 1953).

From the late '70s Tasmania has enjoyed some increase in the level of jazz activity, largely in the wake of the Jazz Action Society which started under the presidency of Tom Pickering, with Peter Hicks as foundation secretary. The society has provided work for musicians who would otherwise have had no jazz exposure, such as Christine Lincoln (vcl), Simon Patterson (gtr). In Launceston the JAS was inaugurated as an offshoot of the JAS of Tasmania in 1978, largely under impetus provided by Peter Coleman, and the North West chapter was established by Viktor Zappner (p.). All these societies play an important role particularly through importing mainland groups with assistance from the Music Board of the Australia Council. Launceston had a jazz course at the Newnham Campus of the Tasmanian CAE, set up in 1984 under Jim Lade with the later addition of Michael Reynolds, but in 1986 it is still too early to assess its effects. On the east coast, Bruce Haley, formerly of Brisbane, has begun to emerge as an energetic organizer.

The centre of Tasmanian jazz continues to be Hobart. Civic support has been forthcoming in the form of jazz representation at the Salamanca Festival from 1983, and there is some through school orchestras, although with the exception of John Williams's (p.) group at the Matriculation College, the emphasis is more on reading swing arrangements. Mainland musicians have recently given workshops, and the Tasmanian jazz co-ordinator Alf Properjohn is attempting to address problems of funding and jazz education. The foregoing, and the establishment of the Contemporary Jazz Society in 1981 have given some impetus to the more progressive styles. Peter Webster's big band has occasional engagements, though it leavens the jazz content with pop material. Players like Tim Partridge (bs), Neil Heather (gtr), John Broadley (reeds), Alan Brown (bs) and others mentioned elsewhere are finding a marginal increase in jazz opportunities, though it still cannot be said that a post-traditional movement has any kind of continuous public support anywhere in Tasmania. The centre of jazz in Tasmania continues to be the Pearce/Pickering groups, the more so since the death of Ted Herron in 1986. In the latter circumstance there is a warning regarding the long-term future of the more traditional styles. The Pearce/Pickering group continues to play and record with not the slightest diminution of its vitality. Pickering continues to introduce new material into the repertoire, and the band manages to work regularly. Talking, listening to and playing with these durable pioneers of Tasmanian jazz, it is difficult to believe that they began playing together 50 years ago. Pearce and Pickering are approaching their seventies, and theirs is the only working jazz band in Hobart. Apart from the arrival of Noni Sadler (p.) and Mike Bellette (tba) from Melbourne, there has been virtually no infusion of new traditionalist blood since Bruce Dodgshun's move from Brisbane in 1964. The problem of the failure to attract young musicians to the traditional scene is apparent all over the country. In Tasmania, the reason that it *is* a problem will become most eloquently apparent.

Tattam, Richard Charles (Dick)   b. 27/7/35, Melbourne, Vic.   tpt
Began on cornet in 1944 in the East Kew School Band, joined the East Kew Citizens Band in 1945, the Kew City Band in 1949, and the Victorian Junior Symphony Orchestra in 1950. In 1951 be-

gan jam sessions and occasional gigs with Lachie Thomson (p.). His matriculation results incl. an honours grading in music, and while at Geelong Teachers' College he played in jazz groups for college dances, working with, *inter alia*, Owen Yateman and George Barby. Played concerts for the Geelong Modern Music Society in Melbourne at the Downbeat concerts, incl. with his own group, the Riverside Revellers, formed in 1955. On AJC committee, 1956; 1957–58 with Ade Monsbourgh groups and the Goulburn Valley JB. From 1959–60 he led the band at the 431 Club, and edited the *Melbourne Jazz News* from its beginning in 1958 until leaving for England in 1960. Tattam was active in England with a number of bands incl. the Georgian Jazzmen, Doug Richford's band (in which he replaced Nat Gonella), Mickey Ashman's, Monty Sunshine's, and Mac Duncan's. To Australia, 1965, joining Frank Traynor, 1966–69, then the Storyville Jazzmen, 1970–75. In 1986 he was leading his own Jazz Ensemble which he formed in 1976, and which was resident at Potter's Cottage restaurant, Warrandyte.

Tattersall, William Russell (Bill)   b. 21/9/27, Dartford, Kent, England   dms
Began drums 1950. Worked with groups in England incl. that of Johnny Scott and his own, before migrating to Perth. Tattersall was the most active drummer in the progressive movement of the '60s, with bands led by Bill Clowes, Keith Stirling, Bill Gumbleton, and Tony Ashford. With J.T. and the Jazzmen early '70s, and with the Will Upson Big Band during its residency at Pinocchio's. Currently freelancing.

Thomas, Bertram Henry Frederick (Fred, Freddy)   b. 28/8/20, Melbourne, Vic.
tpt/p./arr/ldr
Born into a musical family and began lessons at age nine from his father, the solo trombonist W.C.B. (Bert) Thomas. Later studied piano under Carl Bartling. Began touring Vic. with the Young Australia League orchestra (1934); became a champion cornettist and prize-winning pianist. Began casual dance band work in 1934 and joined Clarrie

Freddy Thomas, Palm Grove, c. 1949,
with Splinter Reeves in background

Gange at the Brunswick Palais, 1937, with Benny Featherstone in the same band. Played at Fawkner Park Kiosk, 1939, and at the 40 Club until March 1940. With Bob Gibson from the beginning of the Palm Grove engagement, remaining with the band while serving in an army entertainment unit from 1941. In 1948 became leader of Melbourne's first Kenton-style big band, which performed for the Modern Music Society on 25 July. On 30 Nov. 1948 the band played at the Jazz Parade concert organized by Graeme Bell and Charlie Blott at Collingwood TH. Led the Palm Grove orchestra from 2 Sept. 1949, featuring many progressive players incl. members of 'Splinter' Reeves's Splintette, with whom Thomas also played and recorded. Thomas also worked with groups led by Eddie Oxley and Bruce Clarke, and was Melbourne's leading bop trumpet player by the mid-'50s. He continued leading big bands through the jazz concert era, incl. for the Musicians' Union and the Downbeat series, and as support at the Gene Krupa concerts. Travelled overseas, 1954–55, then joined the ABC variety orchestra, leaving in 1958 and entering TV. MD for GTV-9 from 1961. Left TV, 1971, played session and symphony work and joined Smacka Fitzgibbon on piano, touring Hong Kong with the band in 1978. Retired from music, 1978, and now lives in Narooma, Vic. Thomas was one of the most-admired jazz soloists in the '50s, several times polling in *Music Maker* incl. Musician of the Year. His bands were also the main focal point of progressive jazz during Melbourne's jazz concert era.

Entry prepared by Thomas's
nephew Bradley, and John Whiteoak

Thompson, Lawrence Edward (Laurie)
b. 3/4/41, Melbourne, Vic.   dms
Began drums at 17. Led the Southern City JB in the late '50s and worked casually with Frank Johnson and Bob Barnard. Joined Derek Phillips's Port Phillip JB, early '60s. To Sydney to join Graeme Bell, Jan. 1964, until leaving for London, 1968. Back in Australia he worked in commercial areas incl. club work and government-sponsored tours of Vietnam from 1969. In the early '70s began jazz freelancing and worked with Col Nolan's quartet. Joined Don Burrows for 12 months, recording with him in 1974. Toured with Winifred Atwell in the mid-'70s, and spent periods as Bob Barnard's drummer. With Graeme Bell for the Salute to Satchmo tour, 1977. Joined Galapagos Duck, 1980, for about two years, then worked with Su Cruickshank at the Brasserie. Thompson has also been in great demand as a session drummer, in TV, and backing visiting musicians incl. Don Ewell, Rod McKuen, Billy Eckstine, Sammy Price. He has also worked with Judy Bailey, Tom Baker, James Morrison, and shares the drum chair with Alan Geddes in Graeme Bell's band.

Thomson, Jack Osman   b. 18/8/23, Toowoomba, Qld   reeds/p.
Started piano at 11 and played in school bands. First work was in dance bands. Took up clarinet during the war, and was in Sydney in 1943 playing with Al Hammett. In Brisbane from 1944, playing at Bellevue Hotel jam sessions, although most of his work was in dance bands, incl. Ernest Ritte's New Diggers and Eddie Colburn's Trocadero group, or cabaret such as the Ace of Clubs at Redcliffe (1946–47), the Capilana (1950), the Currum-

bin Playroom with Charlie Lees and Jim Somerville (1950) and the Cascades. He was one of the first choices when jazz work was available, playing with the Canecutters, and leading the local big band and smaller groups at the city hall concerts. Played the jazz clubs based in La Boheme, the Primitif, the Si Bon and the El Morocco in the late '50s and worked occasionally with the Varsity Five. For six years in the '60s he was MD for Channel 9 in a band that incl. his younger brother Vern, who earlier had led the Cloudland Ballroom orchestra. Early '70s, Jack had his own ABC radio series, *The Piano and Me*. Jack, and Vern, Thomson are representative of those versatile musicians who can play jazz more convincingly than many amateurs, but who, as professional musicians, need a steady income that frequently requires them to work in non-jazz settings in cities where public support for jazz, especially in its more progressive forms, is meagre.

Thomson, Lachlan Armstrong (Lachie or Lucky) b. 5/10/33, Finley, NSW   reeds/p.
Following a classical piano education, began clarinet while a cadet at Royal Military College, Duntroon. Formed the Federal City Seven, the only jazz band in the college's history. Performed at the 9th AJC, 1954. Following graduation, based in Vic., then Brisbane, and attended AJCs until posted to Bangkok in 1960, where he has spent many years. Founded Bangkok Jazz Club, 1962. Played with His Majesty the King of Thailand's Orchestra and formed his own band, the Thai Internationals, which have played in Australia (1979) and the US (1980). With Australian JB, which played Sacramento in 1982. During periods in Australia has played with the Varsity Five and groups led by Neville Stribling, Roger Bell, Allan Leake. His main impact has been through his New Whispering Gold Orchestra, a big band in the Harlem style of the '20s and '30s. This group, which has recorded, assembles only for AJCs and consists of musicians from most states in Australia. Thomson is currently residing in Melbourne, playing with Allan Leake's Swing Shift and pursuing his career as an army officer.

Thorncraft, Jack Edmund   b. 13/7/43, Sydney, NSW   bs
Began on violin as a child. Developed an interest in jazz and began bass in his teens, with lessons from Jerry Tranter and Chic Denny. Freelanced and jammed on the modern scene with, *inter alia*, Dave Levy, Tony Esterman, and at El Rocco with Len Young's quartet in 1966. In London, 1969–72, working with Jon Hendricks, Ray Warleigh, Don Rendell, Stan Tracey, Humphrey Lyttelton. Back in Sydney, became a member of Jazz Co-op, c. 1973–76, and the regular bass player with The Last Straw, including for reunion performances at Jenny's Wine Bar, 1983. From the late '70s was frequently associated with groups led by Laurie Bennett, John Hoffman (incl. his Big Band) and Julian Lee. Moved to Myocum, NSW, in 1985. Apart from his work with the important Jazz Co-op and The Last Straw, Thorncraft has been a member of that extended freelance musical population that characterizes the 'modern' scene in Sydney.

Toad's Krazy Kats (Melbourne, Vic.)
Founded in April 1976 by John 'The Toad' Sheldon (tbn./vcl), Mike Bellette (tba), Murray Kent (reeds), and Peter Milley (tpt), with Noni Sadler (p.) and Andy Stevens (bjo), shortly replaced by Clint Smith. In September Peter Arnold (dms) was added, and the band received enthusiastic national notice at the 1976 AJC in Brisbane. In Feb. 1977 Kevin Bolton replaced Arnold, and the band played at the Deniliquin Easter Jazz Weekend. This led to a two-year residency at the Golden Fleece Hotel, South Melbourne. Following a recording in 1978 some dissension led to changes of personnel and the band began to lose its original character, which had been modelled on the Tiny Parham band of the '20s. In 1980 Bellette and Sadler moved to Hobart and co-founded the De Luxe JB; Murray Kent subsequently moved to Perth and joined the Cornerhouse JB; the rest of the band dispersed into other jazz activities in Melbourne. Although it enjoyed only a brief life, the band deserves notice as an example, now increasingly rare, of the loving recreation of recondite early jazz, which at one time was a significant activity in Melbourne.

Tough, Bob   b. c. 1911, Hobart, Tas.;
d. 19/10/49, Melbourne, Vic.   reeds
To Melbourne 1924, and in the early '30s Tough and his brother Ern (bs/gtr/bjo) began working in dance bands. Bob was with Joe Aronson at Wattle Path, and in 1935 with Tony Hall's band at the Forty Club (formerly the Green Mill and later the Trocadero Palais). In Sept. 1935 he was leading a band consisting of Benny Featherstone (tbn./tpt), Mick Walker (dms/vcl), Don 'Pixie' McFarlane (bs), Bert Cooper (p./p. accord.) and Ern. Two at least, Bob Tough and Featherstone, and probably the others, played the first 3AW Swing Club cabaret at the Esplanade, St Kilda, on 9 Oct. 1936, and the band which Bob Tough was leading at Fawkner Park Kiosk for its Sunday afternoon sessions in 1937 incl. Featherstone, Cooper, Walker, and Ern Tough. Bob was a stalwart of the 3AW Swing Club jam sessions until moving to Sydney in 1938, when Ern took over leadership at the Fawkner Park Kiosk where jazz activity continued at least until 1940. In Sydney, Tough worked in night-club and theatre bands incl. the Carl Thomas Club, the Regent Theatre, and with Stan Bourne at the 400 Club. In Melbourne from 1942, with

Bob Tough

Bob Gibson at the Palm Grove. By 1947 he was seriously ill of tuberculosis, of which he died at the reported age of 37. With Featherstone, Tough was respected as one of the first fully fledged jazz improvisers in Melbourne, an assessment which tends to be confirmed by the very few surviving acetate recordings of his work.

Traynor, Frank   b. 8/8/27, d. 22/2/85,
Melbourne, Vic.   tbn./p./ldr
Received piano tuition as a child. In the mid-'40s he began playing jazz engagements, and had switched to trombone by the time he formed his first group in 1947. With Len Barnard's band in the early '50s, when he won a poll in the trombone category. In Oct. 1953 he left Barnard and joined Nevill Sherburn's Rhythm Kings, then in 1955 replaced the late Wocka Dyer in Frank Johnson's Dixielanders, at the same time working with Bob Barnard's band which revived the Mentone LSC functions. After Johnson's band, Traynor freelanced until forming his Jazz Preachers in 1958 as house band for the Melbourne Jazz Club. After opening night, on which he used Bob Barnard, Fred Parkes, Ron Williamson, Don Bentley (dms), the band stabilized with Roger Bell, Neville Stribling. Keith Cox (bs) and Wes Brown, with Graham Coyle, another first-nighter, though the personnel was often different for non-Jazz Club engagements. By 1960 the Preachers had residencies at the Society Club in Preston, and the Contemporary Club in South Yarra, with Roger Bell, John Tucker (clt), Coyle, Len Barnard, Don Standing (bjo), Williamson. Traynor ran his own folk and jazz club, 1961–76, presenting his and other jazz groups, and folk singers, as well as becoming the base for the Melbourne Jazz Club in its last phase. In 1961 Neil Macbeth and Ade Monsbourgh joined and Rex Green temporarily replaced Coyle. Other musicians who worked in the Preachers through the '60s incl. Peter Cleaver, Denis Ball, Don Boardman, Les Davis (bjo), Dick Tattam, Jim Loughnan, Jim Beale (dms). Traynor also used singers Helen Violaris and Judith Durham, the latter of whom subsequently enjoyed international celebrity as a member of The Seekers pop group

Frank Traynor's Jazz Preachers, c. 1978, with, l. to r., Mike Longhurst, Joe McConechy, Don Boardman, Peter Gaudion, Frank Traynor, Roger Hudson

before she finally returned to Australia and settled in Qld. In the early '70s Roger Hudson (p.) replaced guitarist Mike Edwards. Traynor used a succession of reed players from this time: Alex Hutchinson, Ian Harrowfield, Paul Martin, and Mike Longhurst. Heinz Bergmann (bs) was replaced by Joe McConechy in 1974, and when Don Boardman left in 1979, the drum chair was shared between John Turner and Ron Ellam. Peter Gaudion was replaced in 1978 by Keith Hounslow who left for Sydney in 1983. Roger Hudson left in 1980. The Jazz Preachers was one of the busiest bands in Australia, playing pub residencies, cabarets, balls, school concerts, festivals, and even appeared in an episode of the TV series *Cop Shop*, as well as recording prolifically. Traynor himself was an energetic jazz proselytizer, producing lecture/concert packages, writing as a jazz journalist, and teaching jazz theory on a variety of instruments.

Treloar, Phil   b. 7/12/46, Sydney, NSW
dms/perc./comp./arr/ldr
Began on side drum at seven, moving to full kit at 13. In 1968 won a scholarship in percussion at NSW Conservatorium, studying under Alard Maling. Studied harmony with Chuck Yates, 1972. Since early playing with Alan Lee, Treloar has worked as sideman to most major jazz musicians in Sydney incl. Errol Buddle, Bernie McGann, Bob Barnard, Bruce Cale, and with visiting musicians Barry Guy, Chico Freeman, David Friesen. He established a fruitful association with Roger Frampton, the two, with Peter Evans, spending long private sessions in playing and discussion. An important outcome was Jazz Co-op, of which Treloar was a member from its foundation in 1972 to 1976. Composition became a major activity, and, with David Tolley, Treloar had begun to work with electronics in the '70s. In 1981 formed Expansions (Frampton, Dale Barlow, Mike Bukovsky, Tony Hobbs, Lloyd Swanton, Steven Elphick, Carlinhos Goncalves, and James Easton, synthesizer, though personnel was fluid). In addition to playing standards, and originals by band members, this group became a vehicle for Treloar's compositions, including the extended work *Primal Communication*. Further experiments in multimedia presentation led to *Double Drummer* a piece for four channel tape, live percussion, and electronics and transparencies, combining Treloar's percussion with recordings of cicadas, and which won the 1982 Alfred Hill Award for electronic composition. Involved in jazz education since teaching in a programme directed by Howie Smith in 1975. In 1976 he participated in a series of master classes at the Victorian College of the Arts and in the '80s established his own teaching practice. Has continued formal study—in 1980, with assistance from the Australia Council, he studied with Billy Hart in New York; in 1982 he embarked upon the Bachelor of Music at NSW Conservatorium, majoring in composition; in 1984, in the wake of a tour of India with Frampton's Intersection, he studied both there and in Sri Lanka on Australia Council and Indian Council cultural relations grants. Since returning he has been involved in experimental performance with Sandy Evans and Joe Truman, and completed the music for a series commissioned by the ABC, *Music of the Sub-Continent*.

Treloar has always been prepared to articulate a philosophy underpinning his work, in a musical environment often suspicious of abstraction. In particular he has sought to base his music on two principles: the need to transcend received categories, and the importance of the Australian identity.

Turnbull, Alan Lawrence (Tom Terrific)
b. 23/11/43, Melbourne, Vic.   dms, with some
p./vibes/vcl
Took lessons from Graham Morgan and began playing professionally at 14 with musicians at Jazz Centre 44 incl. Graeme Lyall and Keith Stirling. In 1961 he worked at The Embers with Frank Smith, who exercised a most important influence on his thinking. To Sydney in the late '60s, working with Don Burrows and George Golla at El Rocco, before visiting Europe, c. 1968. Back in Sydney Turnbull was a ubiquitous freelancer through the '70s, recording prolifically, playing studio work, backing visiting musicians incl. Joe Henderson, Gary Burton, Phil Woods, Sonny Stitt, John Dankworth and Cleo Laine, Milt Jackson, and working with the Sydney Symphony and the Australian Pops Orchestras. Has worked in film, TV and radio. Has remained associated with Don Burrows for many concerts and tours, incl. Montreux and Newport Festivals and Carnegie Hall, in 1972. With Keith Stirling c. late 1979. Turnbull is the quintessential professional freelance musician who nonetheless remains visible in jazz performance; there is scarcely a post-traditional jazz musician in Sydney with whom he has not played.

Turville, Francis Maurice (Frank)   b. 13/4/35, d. 30/11/85, Melbourne, Vic.   tpt/ldr
Began trumpet at 17, learning from Norman D'Art. First band, 1953, was the Southern City Seven. To England, 1956, with Willie Watt, where he was impressed by Ken Colyer and the NO revival centred on musicians like George Lewis. Returned to Australia, 1957, and joined Llew Hird in the band which became the Melbourne NOJB. Except for a brief period in 1961, when he preceded the band to England, Turville remained with them until they disbanded, 1963. Returned to Melbourne and joined Graham Bennett's Hot Sands JB, which gradually devolved as the '60s jazz boom faded. Since 1968 Turville was in retirement and in later years suffered serious ill health. An important pioneer of the NO style in Australia, with a depth and sensitivity often missing from its more mannered exponents.

Alan Turnbull, 1978

Frank Turville, 1963

U

Ubelhor, Peter Claus (Umbrella) b. 2/10/31, Hamburg, Germany tpt/dms/vcl
Played drums with the St Vincent JB during its final period, then involved in the Tavern venues through the '60s. Has since been continuously active on drums and trumpet. With the house band of the Southern Jazz Club, the Pioneer JB, and the Adelaide Stompers, Ubelhor has been a mainstay of the week-to-week pub scene which keeps traditional jazz alive in Adelaide.

Upson, William Durrant (Will) b. 26/2/45, Middlesex, England p./ldr/arr/comp.
Began piano at seven in Hobart where his father led a band. To Perth via Brisbane in the mid-'50s. Took lessons from Ossie Sanderson, and in 1961 began to get together with Jim Cook, Murray Wilkins and others, leading to the band called the Traditionalists (later the Westlanders JB). Left in 1963, replaced by Jim Best on banjo, worked in the Embassy, and freelanced. Two trips overseas during the '60s increased enthusiasm and an interest in arranging. Early '70s, began rehearsing a band, trying out arrangements, when Harry Bluck invited Upson to submit arrangements for TV. Upson went on to direct the band from 1973, enlarged it for a charity telethon, and, as the Will Upson Big Band, it enjoyed considerable success, appearing on Jazz Jamborees, recording for the ABC, and starting a residency at Pinocchio's night-club in 1974. Apart from the shorter-lived band, Manteca, led by Uwe Stengel, Upson's group is the most visible jazz-based big band in recent years in Perth, and continues to perform in the mid '80s. Upson has also composed the *Indian Ocean Jazz Suite* which was performed by the WA Youth Jazz Orchestra under Pat Crichton.

V

Varney, Edward John (Jack) b. 5/1/18,
Melbourne, Vic. p./vibes/bjo/gtr/arr
In his teens played in dance bands and followed
local jazz players incl. Barney Marsh (p.) and the
Fawkner Park Kiosk musicians. Played with other
jazz enthusiasts incl. Hadyn Britton and Cy Watts,
and played piano in the Bell band at Heidelberg
TH. In 1947, with Stan Bourne at the Plaza and
Empress coffee lounges, and leading the band at
the Australia Hotel. Joined Bell for the first European tour, leaving the band in 1948 to establish a
chain of music schools in Gippsland, Vic. To
Melbourne, 1954, with Stan Bourne at the
Menzies, later under Isidore Goodman; also with
George Cadman at the Chevron, 1959–60, and
at the Powerhouse with Russ Jones in the mid-
'50s. Joined W & G record company, 1960–76.
Travelled overseas in 1977, then upon returning,
established an instrument retailing business, then a
music school, which he was still running in 1985.

*The Varsity Five, 1962, with, from l. to r., Mileham
Hayes, Len Little, Les Crosby, Peter Magee (rear), Doc
Willis, Ian Bloxsom, Rob McCulloch*

Varsity Five (Brisbane, Qld)
This band grew out of private sessions involving
Lachie Thomson, Ian Oliver, Mileham Hayes,
Peter Magee (p.), Ian Bloxsom (dms) from late 1957
through 1958. In 1959 they began playing at
parties among the undergraduate set and, with Sid
Bromley's encouragement, sitting in as a group at
his medical school gigs. The personnel incl. a solid
core of dedicated students of the classic jazz of the
'20s. They assumed the name Varsity Five and
played their first paid gig at a Rugby Club on 6
April 1960. Thomson, who had left Bromley to
join Varsity Five, was posted to Siam in 1960 and
was replaced by Lenny Little. Robin McCulloch
joined primarily as a bass player and for a few
months Allen Duff was on trombone. The band

based itself at the Pelican Tavern where it remained
throughout its lifetime except for a period at a
Waterside Workers' Hall from Sept. 1960 to Feb.
1961. The band's activities coincided with the
boom of the early '60s, and it achieved a degree of
success that aroused some resentment among some
who had been zealous jazz proselytizers for many
lean years. In addition to its regular functions at
the Pelican, it became something of a 'society
band', playing for exclusive private functions incl.
one at Government House when the King of Siam
sat in for a long session. Ian Oliver left for Melbourne in 1961, replaced by Les Crosby, and in
1962 Doc Willis joined the band, assuming leadership shortly after. The band broadcast frequently
and had its own TV programme in 1962. In 1963

some of its members were approaching graduation and beginning new careers, and they formally disbanded on Nov. 5, though bands using the name continued for a while to take occasional gigs. The Varsity Five was the flagship of Brisbane's trad boom, and as well as stimulating a general interest in the music, enjoyed the contributions of a number of the city's most important jazz musicians incl. (in addition to those already mentioned) Stan Walker, Bruce Dodgshun (tpt), Jack Thomson, Tich Bray.

Vining, Edward Norman (Ted)   b. 22/8/37, Melbourne, Vic. dms/ldr
Became interested in jazz after hearing Frank Johnson in the early '50s, then developed more modern tastes listening to after-hours sessions at the Blue Derby, St Kilda, and through contact with Brian Brown and Chuck Yates. Began playing in his teens and played at the Lido under the auspices of Jazz Centre 44 with John Foster, Freddie Wilson, John Doyle et al. MD for the TV show *Cool for Cats*, 1958, leading his own trio which incl. at various times John Doyle, Barry Buckley, Barry Edwards, John Adams. In 1959 he began two years with Frank Gow and a 10-year association with Alan Lee, and worked with Bob Gillette's trio. Guested with Bob Gebert, 1961, played for the Ray Taylor Show on ATV-0 in Melbourne from 1963, and in Sydney, 1965, with

Judy Bailey, John Sangster, Dave MacRae, Bernie McGann, David Martin, and recorded with Don Burrows. To Melbourne, late '65, he joined Brian Brown's quartet, for three years, in 1970 formed a trio with Sedergreen and Buckley, playing two years at Prospect Hill Hotel. During 1972–79 with Brian Brown and was involved with Alan Lee's Gallery concerts. In Sydney 1979–81, with Bob Bertles, McGann, Ray Martin et al., then to Brisbane, working energetically to establish a modern scene in particular with a group of young musicians recruited into his band Mussiikki Oy (titled from a sticker he picked up while touring Scandinavia with Brown). Vining was the first appointed Queensland jazz co-ordinator in 1983, but factionalism on the Brisbane jazz scene led to the position not being funded for 1986. His work is well represented on his trio recording from 1983, on Jazz Note.

Vintage Jazz and Blues Band (Brisbane, Qld)
Originally the Sun City Six, followed by variations around the word 'vintage', the band was formed by Andy Jenner, May 1973, with John Braben (tpt), Mike Hawthorne (tbn.), Jo Bloomfield (p.), Vic Sanderson (bjo), Ron Hawkins (bs), Duke McMaster (dms). Personnel changes have been as follows, the last named on each instrument being the current player: dms—Bob Brown 1974, Lorrie Webb 1974–75, Bob Watson; bs—Harry Cantle 1974, Horsley Dawson 1974–79, Ian Cocking 1979, Ron Hawkins 1979–80, Peter Freeman; bjo—Mileham Hayes 1975–76, John Cox. Andy Jenner was replaced by Tich Bray in 1981 and Paula Cox (vcl) joined in 1975. John (b. 12/1/48, Melbourne) and Paula Cox (b. 22/6/49, Holland) had played in Melbourne in the Junction JB in the '60s before moving in 1975 to Brisbane where they also worked with Mileham Hayes's Dr Jazz. John Braben (b. 27/9/44, Merseyside, England) had worked in traditional bands in England with Andy Jenner, before arriving in Brisbane in 1973. Bob Watson had formerly been active in TV and radio session work, incl. with the Happy Day Show Band which he founded in the early '60s, and with Ken Herron. The Vintage band has worked pri-

Ted Vining, 1986

marily in hotels, though one of its most successful residencies was at the Twelfth Night Theatre Club, from March 1974, with a two-month break, to early 1980. In August 1981 the band opened its own club, Jabbo's, which continued to the mid-'80s, presenting blues and rock groups as well as its own music, featuring various guests incl. Bob and Len Barnard, John Sangster, Graeme Bell, Paul Furniss, and John R.T. Davies from England. The band has played support for Stephane Grappelli, Kenny Ball, the Dutch Swing College, and worked a *Minghua* jazz cruise in March 1983. The most durable and one of the most successful bands working in Brisbane, the Vintage band was still playing a longstanding residency at St Paul's Town Inn in 1985.

Vogt, Dieter   b. 23/3/45, Basel, Switzerland bs/tpt.

Began trumpet at 12 in local brass band and in amateur jazz bands at 15. At 18, while apprenticed in a music shop, was offered a job on bass, which he took after practising for four nights on an instrument borrowed from stock. Performed with this band at International Amateur Jazz Festival in Zurich where one of the judges was Humphrey Lyttelton with whom Vogt would later tour Australia in 1978. 1964–66 with Oscar Klein quintet, Basel. To Australia, June 1966, and has worked with bands of Jeff St John, Ned Sutherland, John Costelloe, Bill Burton, and the Daly-Wilson Big Band. In Easter 1973 performed at Tauranga Jazz Festival, NZ, with Judy Bailey. 1973–74, playing in the US. 1974–75 toured Australia with Winifred Atwell. In 1975 joined Col Nolan with whom he toured throughout the Far East for Musica Viva. In 1984 with Graeme Bell, whom he joined in 1982. A highly adaptable stylist, extremely active in both the session and concert scene, and in pub/restaurant gigs.

The Vintage Jazz and Blues Band, Jabbo's Jazz Club, Brisbane, 1982, with, from l. to r., front row: *Tich Bray, John Braben, Mike Hawthorne*, back row: *Jo Bloomfield, John Cox, Paula Cox, Bob Watson*

Dieter Vogt in the Rocks Push, 1976, with Laurie Bennett in background

Walkear, Ian Lindsay (Growly) b. 13/6/38, Wangaratta, Vic. reeds
After two years on piano took up sax on moving to Melbourne at 16. In 1957 spent three months with National Service Training Band at Puckapunyal, where he developed an interest in jazz. From 1958–64, worked with a number of bands including the Melbourne City Ramblers, Liam Bradley's JB and the Apex NOJB, and at various of the clubs which proliferated during the jazz boom of that period. With the Swing Trio, 1964–70, followed by a two-year layoff. Joined the Storyville Jazzmen (later the Storyville All Stars) in 1972, with whom he has made several recordings. Currently still with this band.

Walker, Raymond John (Ray) b. 25/8/43, Belfast, N. Ireland gtr/vcl
Began violin at age eight, taking tuition under a series of scholarships. In 1958 family moved to England where Walker began guitar. To Perth, 1960, where Walker became interested in jazz. Active in the '60s progressive scene at venues such as the Hole in the Wall. Walker then moved into session work, joined Ossie Sanderson (p.) at the Parmelia Hotel for a decade. In the late '70s, Walker worked at Perth Jazz Club/Society functions, with Kaleidoscope and Collage, and backing visitors such as Maree Montgomery and Georgie Fame. In 1985, teaching in jazz studies courses, playing Romano's and presenting occasional concerts with the West Australian Youth Jazz Orchestra, including a recent performance of Rodrigo's *Concierto de Aranjuez*.

Walton, Glynn b. 19/9/29, Adelaide, SA bjo
Began in 1950, first band, the University Jazz Four

(1950–53). Foundation member of the Black Eagles, to its demise as the St Vincent JB in 1961. In Melbourne, 1962–67, mainly with Frank Turville and freelancing. Returned to Adelaide and worked with Vencatachellum Jazz Peppers 1968–69. Subsequently with Bill Munro, followed by four years with Captain Sturt's Old Colonial JB. Now retired from regular playing.

Webber, Ronald Charles (Ron) b. 4/1/28, Sydney, NSW dms
Began drums in his teens and playing publicly in 1946 with Jack Allan and Keith Silver at the California coffee lounge. Went on to play the Sydney night-club scene incl. Christy's and Golds' (1950–51) and for a period in the early '50s under Col Bergerson at the Trocadero. In 1954 he worked one of the earliest jazz pub venues, the Forest Inn, with Paula Langlands (vcl), Jack Craber, Rick Farbach. Was with the Australian All Stars with Don Burrows at the Sky Lounge residency beginning c. 1956, and during this period continued in other groups: with Julian Lee at the Ling Nam (1957), Terry Wilkinson at the Chevron, with the Port Jackson JB, incl. playing support for Kenny Ball's first tour in 1962, and during the early '60s with Dick Hughes's quartet. From the mid-'60s he worked primarily in the club and studio scene, incl. at South Sydney Leagues Club from 1968. In the US, 1976–78, during which time he played with Dick Saltzman in San Francisco. Since returning, Webber has worked on a freelance basis, in both jazz and commercial areas. In addition to his important tenure with the Australian All Stars (preserved on recordings), Webber has worked with most of the musicians active in the post-war era, incl. Graeme Bell, has played studio and

theatre work, cruise ships, and has backed numerous visitors incl. Al Hibbler and Jimmy Witherspoon.

Webster, Peter John (Ginger)   b. 14/8/37, Hobart, Tas.   reeds/fl./arr/ldr
Began in dance bands at 16, originally on violin, but switching to sax when the former was in decreasing demand. Late '50s, formed the first West Coast-styled group in Hobart with Neil Levis, Noel Addison (bs), Graham Clark (dms). There was little outlet for the group and in the early '60s, Webster moved to Melbourne. Replaced Heinz Mendelson in Barry McKimm's band at the Fat Black Pussycat until 1966, then at Club 21. Following a period of comparative inactivity he joined bands of Alan Eaton and Barry Veith. Webster returned to Hobart to play in Allan Deak's big band for its Wrest Point Casino residency, where fellow reed player Frank Smith was a source of inspiration. Webster returned to the mainland and in 1976, he embarked on a yoga-based programme which further developed his musicianship. Resettled in Hobart, 1981, joined Bebop Brothers with Bill Whitton (gtr), and formed a big band which provided opportunities for young musicians to gain experience. In 1985 he was still directing this band, Integrations, as well as playing back at Wrest Point Casino.

Wells, Colin Charles (Col)   b. 16/8/25, Hobart, Tas.   tpt/vcl
Lessons on cornet as a child then worked in brass and dance bands. During the Second World War, he served as a musician, and in 1946 was with Benny Cuebas in Fouché's Stage Door. Joined Ron Richards, 1947, and was recruited by Tom Pickering, with whom he has remained ever since when the band has used trumpet, except for a layoff for health reasons in the early '70s, when Bruce Dodgshun took his place. Over the same period Wells continued to work at commercial venues, incl. Wrest Point in the late '60s. Has played in most of the jazz groups in Hobart, incl. with Ian Pearce's more mainstream sextet (c. 1956–66), Sullivan's Cove (1975–77), the Ives Jazzmen (mid '70s), the Southern City (1981–82) and the Doghouse (1982–83) jazz bands. Wells has been the mainstay of jazz trumpet in Hobart since the war.

Wesley, John Jeremy (Jerry)   b. 7/6/40, Londonderry, Ireland   p./bs/vcl/comp./arr/ldr
Born Wesley-Smith, his family immigrated when he was about 18 months old. Received coaching from Roger Hudson whom he met through Adelaide Jazz Society functions. Formed a quartet for dances, 1957, and in 1958 began studying violin at the Elder Conservatorium. During this period he joined a band with John Lewis (tpt), Michael Price (bs), Tyrrel Talbot (dms) and Schmoe (clt), which became the University Jazz Group, with Danny Haines (tbn.) a later addition. During 1961 he deferred his conservatorium course and worked with the ABC Adelaide Singers. Towards the end of his course he formed the Bottom of the Garden Goblins. By 1964, with the University JB, incl. Bill Munro, Bob Lott (bs), Keith Conlon (dms) and, with Rod Porter (clt) and Joe Eltham (bjo), this became Jerry Wesley-Smith and the Campus Five, then the Campus Six with Ernie Alderslade (tbn.). This was one of the most durable traditional/mainstream groups in Adelaide, but playing mostly functions rather than in jazz venues, surviving until the mid-'70s, and incl. for long periods Norm Koch (bjo). Wesley remained with the band, with interruptions, throughout its lifetime, on p., bs, vcl, and arr. During the '60s, except for a year in 1965–66 playing orchestral horn in Brisbane, Wesley was increasingly involved with more contemporary styles and also wrote for film, TV, radio, and revues. In Bellingen, NSW, 1976, worked with Brett Iggulden, Ian Wallace (alt) and Graham Bennett at the Good Food Shop. A season in the show *Cold Comfort Cafe* at the Nimrod in Sydney marked the beginning of an association with Robyn Archer who, as Robyn Smith, had sung with jazz groups in Adelaide in the '60s. Wesley has played a number of shows with Archer, incl. in London. Currently residing in Adelaide, working casually with various groups incl. Phil Langford's Superband.

West Coast Dixielanders (Perth, WA)
The West Coast Dixielanders grew out of 6PR Teenagers' jazz sessions and coalesced in 1952 with Phil and George Batty (the latter replacing Rod Wyatt), Bob Anderson, Vince Holmes (gtr/bjo), Alan Blight (dms) succeeding Tom Bone, and alumni of the West Side Jazz Group Dick Hatton (tbn.) and Colin Goldsmith (bs.). Both groups appeared in a Battle of the Jazz Bands concert at the Capitol in May 1952. Other musicians were at various times associated with the Dixielanders incl. another ex-West Side member Bruce Wroth, and Brian Williams (tbn.). The group achieved considerable success performing in jazz concerts and winning a dance band competition in 1953. The departure of the Batty brothers in 1954 for overseas foreshadowed the ultimate collapse of the band although there is a reference to a group of this name making a concert appearance as late as 1959. The band was an important conduit for the stream of traditional jazz running through the post-war era, and many of those associated with it were to become members of the later Westport JB.

Weston, Billy   b. 13/3/22, Sydney, NSW
tbn./tpt/ldr/comp./arr
Moved to Perth as an infant. Began playing and touring in childhood with Youth Australia League band with which he won prizes. First dance band work was with the Collegiates, 1937. To Melbourne, 1939, at the 40 Club (later the Melbourne Trocadero Palais) with Clarrie Gange, Don Rankin, then Frank Coughlan. Joined Bob Gibson at Palm Grove, 1940; in 1941 led a band incl. Ralph Stock and Ivan Halsall at Plaza coffee lounge. During army service, 1942–46, led the Tasmaniacs entertainment unit, then served under Charlie Munro. Following discharge, briefly with Tom Davidson, then to Sydney 1947, with George Trevare at State Theatre and freelancing. Following a short period on the Gold Coast with Claude Carnell, Weston was back in Sydney, 1948, joined Ralph Mallen at the Gaiety and formed his own Stan Kenton-inspired big band which took over the Gaiety in 1949 and frequently appeared at jazz concerts until 1953. Also worked with Fred McIn-tosh for Jack Davey's 2GB shows and with Gaby Rogers at Romano's. Returned Melbourne 1954, then toured with the Ice Follies, 1955. While based in Melbourne, played session work for TV and briefly worked with Frank Smith at The Embers, 1959. Returned Sydney, early '60s, and mainly active in studios. He became active coaching young musicians, and by 1966 was primarily conducting rather than playing. Musical activity diminished through the '70s, and in 1984, still resident in Sydney, he retired.

Westport Jazz Band (Perth, WA)
Formed by Phil Batty in c. 1958, virtually out of the remnants of the West Coast Dixielanders, with Ross Nicholson, Barry Bruce, John Bartlett, Rick Bryant (dms), and John Archer (bjo/gtr). With Ivan Matthews (tbn.) the group attended the AJC at Melbourne in 1960 with the Riverside Jazz Group, and following their return there was some personnel movement between the two groups, leaving the Westport with Batty, Nicholson, Bob Anderson (p./tba), Bob Dixon, Archer, with Dave Ellis (bs). Ellis was later replaced by Murray Wilkins and Nicholson by Phil Batty's brother George. It continued to be a substantial presence throughout the '60s and signifies historically as one of the groups in the succession that maintained the continuity of traditional jazz in Perth through the post-war period.

Whittle, Keith Richard   b. 21/5/27, Perth, WA
dms
Began drums, 1935, taking lessons from Pete Sullivan, and after six weeks was working in George Evesson's orchestra, a youth band of 40–50 pieces. In 1941, on leaving school, he joined Chris Gosper (p.) in radio work. Whittle received valuable coaching from 'Skippy' Alexander when the two worked together at the United Services cabaret with Sylvia Caporn, and Dick Hatton, who worked with him at Nicholson's music store, introduced him to traditional/dixieland jazz. Whittle replaced Merv Rowston (but who remained MD) at the American Red Cross Centre, leaving in 1944 and joining the RAAF. Demobilized 1946, into the

Keith Whittle, Hyde Park Hotel, Perth, 1985

Embassy under Bill Kirkham, staying on when Sam Sharp took over, 1947. With Sharp's band he found the opportunity of playing the broadest range of popular music, incl. dixieland and Stan Kenton arrangements, at Jazz Jamborees and other Sharp concert enterprises. Also playing at the 6PR Teenagers' Jazz Club, and freelancing in cabarets. Left the Sharp orchestra in 1951. Throughout the '50s involved in organizing the Sunday jazz concerts at the Youth Australia League. Apart from another two-year period at the Embassy in the '60s, Whittle worked in hotel/cabaret until the foundation of the Perth Jazz Club/Society in 1973 generated a stronger jazz scene. Replaced John Regan in Barry Bruce's Chicago JB, and worked with most of the jazz groups in Perth. He has also been among 'first call' to back the increasing number of visitors to Perth, incl. Ruby Braff, Ralph Sutton, and George Masso. In 1986 he was the drummer with the Lazy River JB which he joined in 1979.

Wiard, Curtis Jack (Jack)   b. 12/5/43, Lodi, California, USA   reeds
Began in school bands under the influence of Benny Goodman. In San Francisco from 1965, working with the Original Inferior JB (1965–69), Pops Foster's Young NO Jazzmen (1966–67) and other local groups; also jammed with Turk Murphy and Joe Sullivan. Graduated from San Francisco State College with music majors, 1971, and arrived in Australia in May as a teacher. Freelanced in Sydney, and worked with Nick Boston, 1973–74, with whose band he played the opening of the Opera House. Left teaching, 1975, to play full time with Ray Price, at the same time playing frequently with Nancy Stuart at the Bondi Icebergers and the Abbey JB. Left Price, Sept. 1979 to join Graeme Bell on a full-time basis, having worked on occasions with him earlier. In 1986 the Bell band was still his main commitment. He also teaches, and has recorded with several groups in addition to Bell's, incl. the Abbey JB and Adrian Ford's Big Band.

Wickham, Walter Edward (Wally)   b. 10/2/24, Sydney, NSW   bs
Began on piano in 1935, then guitar, which he played in his group the Melody Boys, formed in 1940, with Don Andrews, Don Burrows, Ken Williams (dms) and Ken Flannery on banjo. Went on to study bass with Reg Robinson, then on a scholarship at the NSW Conservatorium. Busy with many groups, incl. Ron Falson's, through the concert era. In 1950 began an association with Gus Merzi which lasted into the '70s, incl. numerous concerts, and radio and TV work: the *Joan Wilton Show* (1953–54), *Here's Your Song* (1954), *Bonnington's Bunkhouse Show* (1954–63), *Ross Higgins Show* (1955), *Six o'Clock Rock* on Channel 2 (1961), the *Saturday Night Club* on 2UW (1963). Numerous tours of Vietnam with Merzi, 1964–71. Also very active in the night-club scene in the '50s until the advent of TV, at Sammy Lee's under Ron Gowans, Les Welch, Billy Antmann, and Wally Norman; Chequers, under Paul Lombard and Allan Woods; André's, under Ray Glover and Frank Smith. Joined Isidore Goodman's Channel 9 orchestra in 1956 and was later a member of the ABC Show Band. Wickham has been active in other forms of studio and session work, particularly with Bob Young for film music. Has worked with a wide range of visiting musicians incl. band leader James Last. In 1974 Wickham

became a founder member of the Bob Barnard JB, which became his main jazz commitment through to 1986, though he also freelances, incl. frequent gigs with Dick Hughes's Famous Five.

Wilkins, Murray Lloyd   b. 18/9/45, Perth, WA   bs

Took up bass to fill a gap in a schoolboy band. This band became The Traditionalists, and enjoyed exposure during the trad boom, following which Wilkins moved into more modern jazz with, *inter alia*, Jim Cook and Will Upson, and into pop-oriented music. Until the mid-'70s he was primarily occupied in commercial session work, but became active again in jazz following the foundation of the Perth Jazz Club/Society, joining the fusion style Kaleidoscope and subsequently Collage. He also works in more mainstream styles with Barry Bruce's Chicago JB, and freelances. Wilkins teaches and performs with the staff on the jazz studies courses at Mt Lawley College under Pat Crichton.

Wilkinson, Malcolm Henry (Mal)   b. 12/7/28, Adelaide, SA   tbn

As a child was a reluctant street busker with his father who, with others at the local Salvation Army, taught him brass instruments. In Len Perkins's JB, late '40s, then formed his own Gutbucket Boys, which brought him to the attention of Dave Dallwitz and the Southern Jazz Group, which he joined by 1951. President of 1951 and 1957 AJCs in Adelaide. To Melbourne, 1958, and temporarily withdrew from music to work in accountancy. Joined Len Barnard and recorded with his band on the celebrated *The Naked Dance* album. Worked and recorded with Frank Johnson and Roger Bell, but most frequently bands associated with the Victorian Fire Band, later enlarged to become the Datsun Dixielanders. Around 1970 formed his own rock-influenced Original NO Rock Band, and following its demise worked casually for some years. In 1983 with Peter Gaudion's Blues Express which he joined in 1978.

Wilkinson, Terence John (Terry)   b. 9/3/31, Sydney, NSW   p./arr

Began piano at six and developed interest in jazz at secondary school, associating with Bill Walker, Mark Bowden, Duke Farrell. His precocious talent was noted at 2KY Swing Club functions as early as 1945 and at Reg Lewis's Modern Music Club, 1946. Foundation member of Ralph Mallen's band, 1947 and with Ron Falson's bop groups at Wentworth Ballroom and for a recording (1948–49). Following a brief interlude with Wally Nash's group in Melbourne, and with the Boposophical Society jam sessions at the Galleone, Wilkinson played the Golden Key at Bondi and began working with Les Welch groups. In the early '50s, played the after midnight sessions at Chequers, with Wally Norman on the ABC, in the Orchid Room (later André's) with Welch, and in the Petersham Inn quartet. Backed Frank Sinatra, Sydney Stadium (1955), and many subsequent visiting performers. Continued broadcasting night-club, concert work through the '50s, and was a member of the Australian All Stars for

Terry Wilkinson in the 1950s, with Jim Shaw (dms) and Clare Bail (ten)

their long Sky Lounge residency and recordings. In 1960 became active as freelance arranger and began what became a six-night a week residency at the Chevron Hotel with various sidemen incl. Jack Iverson (tpt), John Blevins (dms), John Allen, Jan Gold (gtr), lasting to 1966. Also playing with different groups at El Rocco. At North Sydney Leagues Club, 1967–68, then visited the US. His career from the '70s on has been mostly in commercial areas, but he is occasionally seen in jazz settings on a freelance basis, and was a member of Richard Ochalski's Straight Ahead bop group in 1978.

Willis, Francis John (Doc)   b. 23/5/25, Melbourne, Vic.   tbn./comp./arr
Began on harmonica in the Bluebird Harmonica Band in 1934, switched to cornet at age 10 and to trombone in 1943. Studied with Frank Coughlan and Roger Smith. Started the Noone Brothers Band with Reuben Markovich (dms) then formed Doc's Syncopators (incl. Manny Papas tpt, Nick Polites, Alan Knight p., Stewart Speer) for the 1st AJC, at which he also played with Frank Johnson and Tony Newstead. Following a period in Adelaide and Perth with Keith Hounslow, joined Len Barnard (1949), the Steamboat Stompers (1952), then back with Barnard (1954) until that band's break-up during its tour of 1955. Settled in Sydney in August 1955 and joined the PJJB. Left to enter the army's Eastern Command Band, 1957, then moved to Brisbane following his discharge in Jan. 1962 where he took over leadership of the Varsity Five. Returned to Sydney in 1963, working extensively in clubs and theatre pit orchestras including for Billy Eckstine, Johnny Ray. Formed the Duke's Men, 1971–78, then joined Mike Hallam's Hot Six in 1978. Willis's independent and peripatetic lifestyle has put him at or near the centre of some of the most significant moments in Australian traditional jazz, including two important revivals, in the mid '50s and the early '60s, as well as making his name familiar in jazz circles throughout most of the country. In 1983 with Mike Hallam and working on a suite of compositions dedicated to various Australian musicians.

Wilson, Edward John (Ed, Milko)   b. 22/6/44, Sydney, NSW   tbn./comp./arr/ldr
Learned piano as a child, later took trumpet lessons. Played trombone with the Waratah Jazzmen (1959–c. 1964), which incl. brothers Bernie (tpt) and Dave (clt), and the Philpott brothers Ken (bs) and Ron (gtr), with Jim Macbeth (dms). During the late '60s, worked with Sydney Symphony Orchestra and Jim Gussey's ABC Dance Band with which he toured Vietnam. Also active at Chequers night-club and the Hilton Hotel. In 1968 joined up with Warren Daly to establish what became the Daly-Wilson Big Band. The band broke up in 1971 and Wilson freelanced as arr/comp. and also worked with a pop group until the re-formation of the Daly-Wilson orchestra in 1973. Wilson remained with the band until the partnership with Daly finished in Sept. 1983. He moved to Murwillumbah on Qld's Gold Coast and formed a big band and led small groups for local jazz clubs. Also arranged for Ricky May and on a freelance basis. In January 1986 Wilson became MD for Jupiter's, the new casino at Broadbeach on the Gold Coast.

Woods, Barry William   b. 22/7/39, Ngaruawahia, NZ   dms/comp./arr
Began drums, 1957, and toured NZ with a rock group, 1959. Associated with other musicians who later made the move to Australia, incl. Mike Nock, Rick Laird, Ned Sutherland, Laurie Lewis (sax), Judy Bailey. To Sydney, February 1960, with Laird, Sutherland, and Mike Walker (p.), preceded shortly by Dave MacRae, with whom Woods frequently worked. Woods played briefly at El Rocco, then on the Gold Coast, Qld (1960–61) with Laird and Brian Smith (sax). In a trio with Laird and Dave Levy, he replaced the 3-Out trio at El Rocco in May 1961, later joined by Bob Gillette and Bernie McGann. Played the Jazz Workshop, Orwell St Kings Cross with Levy, McGann, Smith, and Tony Curtis (vcl) and also worked in TV. To Melbourne, May 1962, to back the American Jazz Ballet. Returned to Melbourne in Jan. 1963 to open the Fat Black Pussycat with Barry McKimm's group. To Canada, 1964, where he

studied with Ed Thigpen, and while waiting for his union clearance, worked with a rock group in a Batman outfit to conceal his identity as an unauthorized musician. Left Canada, Jan. 1967, toured with a commercial group, and played session work in London, recording popular material with Tom Jones and Engelbert Humperdinck. Returned Australia, 1969, and formed a pop group. Returned to jazz on a freelance basis, 1972, studied at the NSW Conservatorium and played in the jazz studies big band, 1973. With the important contemporary group Out to Lunch, 1975–76, and through the late '70s freelanced with Serge Ermoll, Craig Benjamin, David Martin, and Peter Boothman. Returned to commercial work in 1980, with very occasional jazz work incl. with Eddie Bronson at Jenny's Wine Bar. In 1986 was with a commercial group, Jupiter. Although not continuously a part of the jazz scene, Woods was a participant in three important episodes in the modern jazz movement: El Rocco and the influx of a young generation of New Zealanders, the Fat Black Pussycat in Melbourne, and the early Basement scene.

Wright, Darcy James   b. 31/1/37, Terowie, SA
bs
Began bass in his early twenties with lessons from John Foster and early experience in dance bands, symphony work, and rock groups, and with Bruce Gray. Involved in the modern movement in Adelaide in the early '60s, playing for the Modern Jazz Society with Graham Schrader and Lee Sydenham (dms) and at Blinks (later the Cellar) with Billy Ross, Alan Slater (p.) and John Bayliss (vibes). In 1961 played The Embers with Frank Smith, who exercised great influence. This was followed by work in Sydney with Bill Burton (tpt) at Chequers, and touring with Winifred Atwell. Back in Adelaide by the mid-'60s, Wright worked at the Cellar with Billy Ross, and the Paprika nightclub with Ross, Ross Smith (tpt), and Bob Gebert, later replaced by Tony Gilbert. Played Adelaide's Freeway Hotel in the late '60s with Bob Davies (reeds). Wright returned to Sydney where he has remained based, played mostly session work since the '70s,

freelancing in jazz settings, and playing semi-commercial residencies incl. with Bill Burton at the Silver Spade Room in the Chevron Hotel. With Marilyn Mendez at the Rose Bay Hotel, 1983–84, and in 1985 primarily engaged in TV work, and resident at the Regent Hotel with Julian Lee and Chuck Yates.

Wright, Robert John (Bob)   b. 5/9/27, Adelaide, SA   tbn./dms/tba/vcl
After early experience in school bands, Wright entered the Adelaide College of Music on banjo-mandolin, shifting to tuba to play in the college's military band in his teens. In the meantime, in 1943, he had joined on drums with Bruce Gray, Bill Munro, and Colin Taylor (p.) to form a quartet for school concerts and which became the nucleus of Malcolm Bills's pioneering band. Entered the Southern Jazz Group after Dallwitz took over, and remained until the group disbanded, when he took up trombone to join Gray's All Stars. In 1954 dental problems led to his retirement from playing, but he returned on tuba as a founder member of the Black Eagle JB until its dissolution under the name St Vincent. During the early '60s, he was intermittently active with the Adelaide All Stars, and in 1969 with the Vencatachellum Jazz Peppers, then withdrew from music and concentrated on his career in metallurgical research and development. In the mid-'70s he joined Bruce Gray's Vintage JB, and, overlapping with this association, the Captain Sturt Colonial JB as well as various Dave Dallwitz groups. Shifted exclusively to trombone in 1981, remaining in the Captain Sturt where Brian Green replaced him on tuba, and has continued to be intermittently active to the present. A regular performer at AJCs, Wright also won its first original tune competition with 'Boot Hill'. His career spans the complete history of the revivalist movement in Adelaide, and his unsurpassed authority on tuba in the days of the Southern Jazz Group not only won him many polls but also established the local dominance of brass over string bass which remains scarcely challenged. He is one of the key figures in defining the Adelaide jazz tradition.

Terry Wynn (right), at the Southern Cross Club, Canberra, 1979, with Greg Gibson

Wynn, David Terence (Terry)   b. 5/8/35, Sydney, NSW   reeds/fl./vln/comp./arr
Moved to Canberra, 1940. Began violin at age seven. In 1954, joined Bill Wheatley (p.), Bob Brown (dms), Dave Tuffin (bs) at Eastlake Football Club. Wheatley's record collection stimulated Wynn's jazz interest, as did monthly concerts in Canberra by the Cootamundra JB, whose reed player Greg Gibson was an influence on Wynn, who began clarinet c. 1954. In 1955 attended his first AJC at Cootamundra, and joined Bruce Lansley's band. This was his main musical commitment into the '60s. Also played occasionally with the Cootamundra JB after the departure of Laurie Gooding, and with Sterling Primmer's quartet. During the '60s he worked casually in jazz settings with local modernists incl. Ross Clarke (p.), Jim Latta, John Stear, Derek Long, Alan Pennay, but mostly commercial groups. Joined the Fortified Few (1974), becoming their main arranger. Left in 1978, and rejoined in 1981, remaining until the band's dissolution. Wynn has also worked casually with other Canberra bands, incl. the Jerrabomberra, and has played theatre work.

Yarra Yarra Jazz Band (Melbourne, Vic.)
Richard Maurice (Maurie) Garbutt, b. 29/5/41, Melbourne, began trumpet and joined the Port Melbourne Brass Band in 1954. In 1955 he met Bob Brown who was then with the Richmond Brass Band; they began listening to the Melbourne NOJB at the Blue Heaven in St Kilda, and Roger Bell's band at the Mentone LSC. At the Pier Hotel in Frankston they listened to Ian Orr's band, with Dave Rankin, Eddie Robbins, Mary Carter (p.), Brian Carter (tba) and Kuzz Currie (dms). In 1958 Garbutt formed his NOJB with Brown, Robbins, Lee Treanor (bjo) and Sid Clayton followed by Ronnie Rae, then Don Hall, on drums. In 1959 he added Jeff Hawes (tbn.) when economics allowed. Garbutt's group began a year at a dance called

Dantes Inferno in 1959, and changed the band name to the Yarra Yarra JB, after the river that flows through Melbourne. At this stage Hawes had been replaced by Llew Hird, who was replaced, late 1959, by Les Fithall who stayed until 1965. In March 1966 the Yarras started at the Corroboree Club, Moorabbin, where Judy Jacques began singing with them. In 1961 they were playing regular engagements at dances in Glen Iris and Murrumbeena and Jazz Centre 44. When the Melbourne NOJB left for England in August, the Yarras took over much of their work incl. Thursdays at Jazz Centre 44 and the Gasworks in Kew. In October they inaugurated a formal jazz ball at the Palais Royale, running annually for six years. In 1962–63 the band consolidated its popularity with

The Yarra Yarra Jazz Band, Kew Town Hall, 1962; from l. to r., Denis Ball, Lee Treanor, Maurie Garbutt, Don Hall, Les Fithall, Bob Brown

a residency at the Boston Jazz Club (renamed the Yarra Yarra Centre) in Richmond, its first recordings, Jazz at City Hall concerts, radio shows. In 1962 Eddie Robbins left for England and was replaced by Denis Ball until 1964 when Nick Polites joined. Lee Treanor was replaced by John Brown in 1963 and Judy Jacques by Pat Purchase in November of the same year. In 1964 they moved from Jazz Centre 44 to the Downbeat Club, and Purchase was succeeded by Kerrie Male. Personnel changes through the '60s were as follows: Roger Janes replaced Fithall in 1965, but Fithall returned when Janes left for Sydney in 1967; Don Hall replaced by Graham Bennett in 1965; Kerrie Male replaced by Sue Jennings in 1965, followed by Lucille Newcombe in 1967; Bob Brown replaced by Ken Sluice in 1965, followed by Dave Myers in 1966. John Brown and Polites were replaced by Willie Watt and Paul Martin respectively in 1966 and Martin by Dave Bailey in 1968. By 1965 the peak of the boom had passed and the Yarra Yarra Centre closed. Work became more sporadic, although in 1967 the band began long residencies at the Post Office Hotel, Coburg, and the Caulfield Club Hotel.

In 1968 Garbutt planned an overseas tour and began consolidating an available line-up: Bailey, Myers, Andy Symes (bjo), Lynn Wallis (dms), with Adrian Ford (tbn./p.) to be picked up in Sydney. The band arrived in England on 13 March 1969 where Roger Janes rejoined them and Ford moved to piano. The Yarras played several tours of England and Europe with considerable success, in the course of which personnel fluctuated. In July 1971 they visited the US, played in Jimmy Ryans, New York City, and in NO. Returning to England, they sailed for Australia on the *Fairstar*, arriving Oct. 1971, the band now consisting of Garbutt, Janes, Ford, Wallis, and Ashley (bs) and Petra (bjo) Keating. Ford left in Sydney and the Keatings returned to their home state, Qld, so that when the Yarras played their first gig back in Melbourne at the Victorian Jazz Club, it was with Willie Watt, Frank Stewart (bs) and Kay Younger (vcl), the last two of whom had worked with the band in England.

The '70s brought a tour with visiting trumpet player from NO, Alvin Alcorn, 1973, a residency at the Abbey, St Kilda Road, 1975–78, the Chez Nous Restaurant, 1977, country tours and Mildura Jazz Jamborees. The main changes in personnel (with the year each joined after his name) were: bs—John Healy 1973, Don Heap 1974, Dave Myers c. 1977, Peter Grey 1980; clt—Nick Polites 1971, Dave Ridyard 1974, Paul Martin 1975, Karl Hird 1981, Pat Miller 1982, Col Elliott 1983; bjo—Paul Finnerty 1974, Andy Symes 1976; dms—Peter Clohesy 1975, Geoff Thomas 1981, Kevin Bolton 1984; tbn.—Hugh de Rosayro 1976, Les Fithall 1979, Hugh de Rosayro 1982. In 1984 the Yarras played a 25th anniversary concert attended by many alumni, and in 1985 were still playing at Bells Hotel where they had started in 1977.

The Yarra Yarra JB is the most durable rallying point for that NO style of jazz inspired by those veterans gathered together by George Lewis and Bunk Johnson. The Melbourne NOJB and the Hot Sands JB (from the latter of which the Yarras frequently drew personnel) and the Yarras have constituted the nucleus of the NO movement, with Geoff Bull providing its equivalent in Sydney—indeed, his bands in the '70s were almost wholly composed of Yarra's alumni. Maurie Garbutt himself remains the most important focus of NO jazz in Australia. His own playing has always maintained quality and authenticity in an idiom which produced many stultifyingly mannered musicians. With Frank Turville, Garbutt is the most convincing and lyrical NO trumpet player that Australia has produced.

The history of the Yarra Yarra JB has been documented by Eric Brown in a booklet published by The Australian Jazz Archives, and updated in *Jazzline*, vol. 17, no. 3 and vol. 18, no. 1, all of which has formed a basis for the foregoing entry.

Yates, Edward Martin (Chuck)   b. 4/2/36, Bendigo, Vic.   p./ldr
Began classical training as a child, but heard Jack Brokensha's group and moved to Melbourne to take lessons from its then piano player Ron Loughhead, in the early '50s. Turned professional

and worked in the traditional scene supported by casual dances and clubs in Melbourne through the late '50s. In 1963–66 led a modern trio in the Fat Black Pussycat with Barry Woods and Barry Edwards, augmented briefly by Bernie McGann. In 1966 moved to Sydney to work at the Mandarin Club with John Pochée, who in turn played with Yates's trio with Andy Brown at El Rocco. Yates then went into Chequers for nearly two years, and during the early '70s worked in various clubs incl. the Eastern Suburbs Leagues Club, with occasional jazz gigs. In the late '70s he spent a long period with Kindred Spirit. In 1980 studied in the US with Barry Harris, Norman Simmonds, Roland Hanna, Dave Liebman, and in recent years has found increasing jazz opportunities both as teacher and performer. Has played concerts for JAS, SIMA, recorded with Errol Buddle, Johnny Nicol, Bertles, and Marie Wilson, for whom he is regular pianist. In the '80s has been working with Darcy Wright at the Regent, as well as in larger groups at the Don Burrows Supper Club. Yates is an illustration of the unhappy truism that some of the least publicly recognized musicians are held in the highest respect within the profession. Apart from the admiration in which he is held as a performer, he is gratefully recalled by younger musicians as an inspirational teacher.

Sources of Illustrations

Where two individuals or institutions are listed for an illustration, the first is the source of the illustration, and the second is an acknowledgement of permission to reproduce the illustration.

Details for sources and acknowledgements of illustrations are as supplied. Every effort has been made to trace the original source of all material contained in this book. Where the attempt has been unsuccessful the author and publisher would be pleased to hear from the author/publisher concerned, to rectify any omission.

Further Reading

A comprehensive Australian jazz bibliography remains to be compiled. It would range from articles on specific aspects of the subject, to personal memoirs which make occasional or have only indirect reference to jazz in Australia, such as Audrey Blake's *A Proletarian Life*, Michael Pate's *An Entertaining War*, or J.C. Bendrodt's *A Man, A Dog, Two Horses*. The following is a selective list only.

Books and booklets

Bisset, Andrew, *Black Roots White Flowers. A History of Jazz in Australia*, Golden Press, Sydney, 1979.

Bonython, Kim, *Ladies' Legs and Lemonade*, Rigby, Adelaide, 1979.

Burke, Alexander James (Jim) *Wobbly Boots—Dance Halls and The Way We Were*, Bellbird, Artamon, 1983.

Davidson, Jim, *A Showman's Story—The Memoirs of Jim Davidson*, Rigby, Adelaide, 1983.

Hayes, Scribner, Magee, *The Encyclopedia of Australian Jazz* (no publishing information; appeared c. 1975).

Hughes, Richard, *Daddy's Practising Again—An Australian Jazzman Looks Back and Around*, Hutchinson, Richmond, 1977.

Jazz. Australian Compositions (produced by the Australia Music Centre, Sydney, 1978).

Linehan, Norman, *Norm Linehan's Australian Jazz Picture Book*, Child and Henry, Salisbury, 1980.

——(ed.), *Bob Barnard, Graeme Bell, Bill Haesler, John Sangster, on the Australian Jazz Convention*, AJC Trust Fund, 1981.

McCardell, Terry, *Jazz Speaks All Languages* (ed. by Norman Linehan), AJC Trust Fund, 1985.

Mitchell, Jack, *Australian Discography*, 2nd edn, pub. by Mitchell with the assistance of the 13th AJC Committee, 1960.

Williams, Mike, *The Australian Jazz Explosion*, Angus & Robertson, North Ryde, 1981.

Periodicals

Australian Band News. The earliest institutional holdings of this date from vol. 20, no. 5, 26 Jan. 1925, in the Mitchell Library. It went through several name changes.

From vol. 21, no. 1, 26 Sept. 1925: *Australian Band and Orchestra News*

From vol. 22, no. 1, 27 Sept. 1926: *Australasian Band and Orchestra News* (having absorbed *Australasian Bandsman*)

From vol. 32, no. 5, 26 Jan. 1937: *Australasian Band and Dance News*

From vol. 32, no. 7, 27 March 1937: *Australasian Dance and Brass Band News*

Following the last issue, vol. 34, no. 7, 26 March 1940, it was incorporated into *Music Maker*.

Australian Dance Band News, from vol. 1, no. 1, 1 June 1932. From 1 April 1933, changed name to *Australian Music Maker and Dance Band News*

From vol. 35, no. 9, 30 April 1940, *Music Maker, with which is incorporated The Australian Dance and Brass Band News*. Ceased, March 1952, with a new series running from June 1955. The last issue in magazine format was in Sept. 1972. From Sept. 1972, changed to a newspaper format, incorporating the title *Soundblast* on bottom margins from Jan./Feb. 1973 until cessation of publication in Aug. 1973.

Other consistently relevant periodicals have not exhibited such erratic histories, and can be identified more straightforwardly. I have only included in the following selective listing those which enjoyed a run of more than one issue.

Australian Jazz Quarterly
Beat
Jazz: The Australasian Contemporary Music Magazine
Jazz Down Under
Jazzline
Jazz Notes
Jazz Parade
Jazz Record Review (later, *Jazz Review*)
Melbourne Jazz News
Odd Note (journal of the Canberra Jazz Club)
Quarterly Rag
Southern Rag
Syncopation
Tempo
Western Australia's Music Maker

There are, in addition to the above, numerous other entertainment journals which intermittently provide useful information relevant to the history of Australian jazz, as well as various jazz club/society newsletters.

Index